St. Louis Community College

Forest Park
Florissant Valley
Meramec

Instructional Resources
St. Louis, Missouri

GAYLORD

A SHEARWATER BOOK

The Others

PAUL   SHEPARD

The Others

How Animals Made Us Human

ISLAND PRESS / Shearwater Books
Washington, D.C. / Covelo, California

Library of Congress Cataloging-in-Publication Data
Shepard, Paul, 1925–
    The others : how animals made us human / Paul Shepard.
      p.   cm.
    "A Shearwater book"—T.p. verso.
    Includes bibliographical references and index.
    ISBN 1-55963-433-2 (cloth)
    1. Human-animal relationships.   2. Human ecology.
   3. Philosophical anthropology.   I. Title.
    QL85.S49   1996
    304.2'7—dc20   95-32313
                                    CIP

Printed on recycled, acid-free paper

Manufactured in the United States of America

10 9 8 7 6 5 4 3 2 1

Contents

The Others

Introduction: The Encounter

*Forests are enchanted enough without elves or hobbits. Did you
ever see a ruby-throated hummingbird?*

SAM KEEN

BETWEEN RAINSHOWERS IN Kansas City in the spring of 1931
I was out on my scooter. Around the corner from our house on Kenwood,
in the middle of 60th Street I found a box turtle. My grandmother Grigsby,
sitting on the porch, said it was a "terrapin." It wouldn't do, she said, to put it
in the aquarium with the minnows, perch, little bass, and young pond turtles
which my father brought home from fishing. The turtle seemed terribly
strong, pushing with its front feet against my fingers, its taut, skinny neck
aiming the toothless head toward some destination as secret as the meaning
of the yellow lines and dots on its shell.

Three years later, by the time I was eight, we had moved to the Missouri
Ozarks. Great-uncle Jack made a hoop from a coat hanger and my mother
sewed cheesecloth on it to make a butterfly net. Collecting butterflies combined the pleasure of running with the joy of collecting. Each netted butterfly was killed with a squeeze and then pinned to cardboard. Like box turtles,
butterflies too had alphabet-like marks, and I think my feeling about them
was mixed up with learning to read. Every capture required a trip out and
could be brought inside; each had come from a chance encounter, was a tro-

phy and a mystery, a way into the world and a way of bringing the world into the household.

I next discovered the magic of bird eggs when collecting them was still possible and moral. Their colors gave me at ten the same palpable sensations that crayons had given when I was five. The exquisite markings on the shells, different in each species and varied in each egg, were messages. Like the eggs of the crested flycatcher with their purple scrawls, hidden in a hollow tree in a nest that always contained a cast snakeskin, every new egg was a summons and a wonder. Collecting was a grand Easter egg hunt that lasted all spring. These uncanny found objects could be pondered at odd moments throughout the year as they lay in their glass-lidded box, the blown shells on a sheet of cotton, each labeled in pencil on round, metal-rimmed tags. Each egg evoked a recollection of the moment of discovery in a particular nest or mere cup in gravel, a certain fencerow or tree hole. In one supreme moment, when I was eleven, Professor Rudolf Bennitt, of the University of Missouri, took a look at my egg collection along with dessert while dining with my parents—he was studying the bobwhite quail and my father was an avid hunter—and corrected some of my identifications. Some eggs were mislabeled; others I never did identify before the collection was lost in the years that followed. Even now as I write there seem in these letters to be echoes of those scrawls: tentative meanings that were possibly true or untrue—like this book—relics, traces of a search with its backtracking and comparing, re-collecting the mystery of the animal figure in human scrawls in a thousand books. Egg collecting is now illegal and unethical—too many collectors and not enough birds. I am still astonished at eggs and the birds who make them. The egg is something writ small, like bones and shells, little puzzles in a bigger mystery.

Many of my other early memories are of birds and other animals—of the aquarium lives, the brown rat which, cornered, bit me, the chipmunk someone gave me, a tiny owl on our porch, and all the frogs and pet crows and stalking with BB guns over the years. Another kindly professor at the University of Arkansas, W. J. Berg, became a pen pal by the time I was twelve on the subject of spiders, as had George Moore, chief naturalist of the Missouri State Parks, on the identity of cave salamanders.

Memories of boyhood collecting somehow inform my theory—a kind of miniature presentment—that the human species emerged enacting, dreaming, and thinking animals and cannot be fully itself without them. Looking back I can see that my work to this point has been a circling round this idea as though I had been imprinted by the movie westerns I saw as a boy with the

Indians circling the wagon train. My first book, *Man in the Landscape: A Historic View of the Esthetics of Nature*, was like the sweeping "pan" shot with which the scene opens. *The Tender Carnivore and the Sacred Game* was a somewhat rowdy assertion of the epitome of love in the heart of the hunter. That foragers were the first craftsmen of ideas, as animals became terms in a language we still use, was the subject of *Thinking Animals*. In *Nature and Madness* I was fumbling with the concept of ontogeny as the key to our relationships to nature. With *The Sacred Paw* I attempted to flesh out the human relationship to one species as if no area of study were irrelevant. In this book I return to the animals as Others in a world where otherness of all kinds is in danger, and in which otherness is essential to the discovery of the true self.

I am still haunted by ontogeny, a kind of necessary pattern of growth toward maturity in which we acquire respect for that which is unbridgeable between ourselves and the animals. It is an attitude of accepted separateness which I think characterizes both the great naturalists and primal peoples. Among such naturalists—chief influences on my own life—I would include Konrad Lorenz, Frank Darling, Ernest Thompson Seton, George Bird Grinnell, Adolf Portmann, Loren Eiseley, Edward O. Wilson, and Charles Darwin, and among the tribal peoples I would include my own Celtic ancestors. It is their humility to which I am attracted—not civilized "kindness" but rather curiosity, receptive courtesy, gratitude, and respect for the power of animals. The idea of "mercy" toward animals, with its detached overriding of nonhuman life and its assumptions about "lower" and "higher" life-forms, seems to me more dangerous and anemic than the robust, meat-eating, storytelling, primal peoples or the best of modern hunter-naturalists.

After World War II, I went to study wildlife conservation with Bennitt. In his class we used Aldo Leopold's 1935 text on game management. By the time I left Columbia, Missouri, in the summer of 1949, Leopold was dead and we had all seen his new book, *A Sand County Almanac*, published posthumously. Those three years and that book framed the question that has dogged me ever since.

The Cooperative Wildlife Unit at the university included not only faculty and students but representative biologists from the Missouri Conservation Commission, the U.S. Fish and Wildlife Service, and the Wildlife Management Institute. Everyone I knew there loved animals, and yet they were all in the "management" business of killing them. Leopold wrote about land ethics and in the same book spoke of the joy of seeing the kicking red legs of a shot duck dying in the morning sun. In the forty-five years since then, none of the people or programs for "saving" animals has seemed to me anywhere near as devoted and committed to the nonhuman world as those academics, mid-level bureaucrats, and their student candidates for jobs in state and federal game departments.

But I did not recognize the question for another twenty years, when I began reading the anthropological literature on hunting and gathering. Here I found the same paradox, only expressed in archaeological and religious terms. The only way I could resolve this contradiction of both loving and killing animals was, on the one hand, to try to understand "native" cosmologies (or their traces in the modern unconscious) in which killing and eating animals was a positive quality and, on the other hand, to seek the flaws in the "humane" movement in all its forms.

There were many surprises along the way—among them, how important animals are in the *intellectual* lives of "primitive" peoples. I began to wonder whether the human species had similar beginnings to my own, that is, whether human consciousness, intelligence, or ways of perceiving owed their existence to animals in some grand analogy to my personal experience. Admittedly, the analogy of the self and the species is a peculiar restatement of an old biology chestnut, "Ontogeny Repeats Phylogeny," the defunct theory that each child repeats in its development the evolutionary history of the human species. Even so, I suppose that fifty years of my life went more or less unwittingly to probing the idea that my personal relationship to nature was my best source of information on how we became human in a world of nonhuman others over millions of years.

As for the animal protection movement, with its high-sounding terms of "rights" and "ethics," how can I be ecologically related to the living community by means of such abstract connections, by a deliberate distancing and hands-off attitude? It is almost as though spectatorship is to be the philosophy of nature. In the perspective of the enormous history of life and the role of animals in human evolution for a million years, I feel only disconnected by the precept of untouchability, the peculiar sentiment that animals and I should be friends at a tidy remove rather than interacting in each other's physical and psychic domains, used and user. Valuing animals as though they were museum specimens reduces them to camera grist, intellectual ciphers, words, models for woolly toys, and monuments to esthetic detachment, as if wild animals were shrines or works of fine art.

Great naturalists and primal peoples were motivated not by the ideal of untouchability but by a cautious willingness to consume and be consumed, both literally and in a mythic sense. Everyone lives in a mythic world, however ignorant of it they may be. The most revealing source of information about how people conceive of themselves in relation to the nonhuman world is myth. In studying the perception of animals, I am led again and again back to storytelling and songlines, to narrative and music, which are basic to the mythic tale and its enactment as ceremony. All myths operate on three levels: one deeply personal, concerning an inner, unconscious life; another the social and ecological milieu; and third, the society of spiritual and eternal things in tales of creation. Typically a story we call mythic informs our individual lives with exemplary models, our relationship to others in standards of conduct, and a vision of the invisible and eternal powers that govern existence. Many of the student friends of my undergraduate years, who would spend their lives counting deer or planting habitat for wild turkeys, would not develop much of a personal, social, or religious philosophy, the way their primal counterparts would, and yet they had available a fabulous myth from which to work.

My emphasis on myth does not imply that I hate what is modern or Western. Indeed, I am passionately committed to a scientific form of an old tale of kinship to other life. It is the myth of biological evolution, a wonderful story about how things came to be, in what sense what is here now is still the original, what kinship to the other animals means in terms of sharing energy and form and genetic codes that change in spite of sameness, and how each species is a master of a particular way of being that foreshadows something about ourselves. Because they deal with actual descent, evolutionary tales—whether they are fossil, genetic, biogeographical, or anatomical—speak of kinfolk and ancestors in the larger sense, of the perspective of life in the universe as a continuation of its order-creating character. Evolutionary tales confirm difference in a way that relates us to animals but does not assume that we understand them. Our modern myths of Faustian Historical Man and monotheistic hubris are motivated by fear of death, a compelling avoidance of biological nature, and chosen exile into a fantasy world of man-the-conqueror.

At mid-century the intellectual ferment around the concept of evolution was greater than it had been for seventy-five years. The syncretic work of G. G. Simpson and Ernst Mayr, the ethological pioneering of Konrad Lorenz, the leadership of Julian Huxley, were dazzling. In graduate school at Yale I attended a seminar called "Evolution in the Light of Genetics, Biogeography, and Paleontology." It had the effect on me of a religious epiphany. Evolution does not answer the big questions as to where the world is going or why—myths don't have to explain everything. Evolutionary thinking gives me relatedness, continuity with the past, common ground with other life, a kind of celebration of diversity. It is much more humble than the eschatology of "world religions" or the arrogance of secular progress or literary humanism. Its signs are around me every day, not as the handiwork of a remote Great Craftsman or a celebrity artist, but in the weather and in eggshells and the lines on a turtle's back. In its broadest sense, evolution extends our kinship to the atoms and to the stars, confirms our continuity with the chemical elements and an extinct sun from which we come, although such things are too remote for much fellow-feeling. My relationship to plants and animals is more vivid than that.

In fact, the celestial big bang or the final illumination predicted from analysis of the subatomic entities, explaining everything, seem more like shields against the inescapable tenuousness of life as fragile as a wren's egg yet complex enough to resist our final control. Theories of a universe, presented by a priesthood of physics, converge with our old escapist religions of the sky, emphasizing the very big and the very small, from subatomic particles to the flux of galaxies, from devils to omniscient saviors, transcendent heavens, and final things, a metaphysics too grand for trees, crested flycatchers, or the middle ground where I live, in the words of W. H. Auden,

> where all visibles do have a definite
> outline they stick to and are undoubtedly
>   at rest or in motion, where lovers
>   recognize each other by their surface,

> where to all species except the talkative
> have been allotted the niche and diet that
>   become them. This, whatever micro-
>   biology may think, is the world we

really live in and that saves our sanity.[1]

What saves our sanity occupies landscapes. While animals are my subject, I must give plants their due. Plants play powerful roles in human life—as food, narcotics, fermenters, healers, sedatives, tools, shelters, mood-makers. Our evolutionary continuity with them is profound: we share bacterial ancestors. But our perception of them as presences is limited by their immobility and form, a patience bordering on indifference. They are true beings whose otherness is so profound that it tunes and tempers our instincts for cover and comfort and protected observation. Their "intentions" appear to be at once more general and more subtle than those of animals, who fill the world as intermediaries between us and plants, signifying not simply in their strangeness but in an uncanny likeness to us wrapped in a difference.

My sense that the marks on bird eggs were an unknown script is widely shared. People everywhere have long believed that animals bear secrets, that our kind once married them, and that they, being both familiar and extraordinary, are a means for charting our lives. Edward O. Wilson calls such feelings for other forms "biophilia," defined as "the innate tendency to affiliate with other living things." But he does not mean husbandry disguised as kindness or that the Others are dependent, lesser beings than ourselves. The modern marginalizing of wild animals is associated with their physical absence and our shifted attention, as though we had lost both the opportunity and the ability to see them. Deep in my heart is Ben, a boyhood friend who could hunt the cottontail rabbit with rocks because he could see it hunched in its grassy form. He would point toward the earth ten feet away, but I saw only the grass while Ben saw with an innocent and archaic vision.[2] In him the synapses connecting desire, perception, and significance had never been broken by glancing—by what Christopher Fry once called attaching "visual labels" to things, making them invisible.[3]

Ben's capacity to see was part of a rural childhood, and in a longer sense with foraging, a basic human ecology. He was living out the myth (which is what they are for) that nurtures the capacity to see. From color vision in ancestral primates to our forebears' terrestrial audacity in savannas, to the most abstruse ideas, the hunt made us human. That quest is mainly attention to slight cues, a roving look punctuated by focused intensity. My grandsons, Philip and Brandon, hunt elk with their father in Wyoming. When they visited Yellowstone National Park one fall the "wild" elk were easily visible, great supernormal signals, living black holes. Unlike the children of tourists, who were bored after ten minutes with animals who grazed quietly or lay in the sun, the boys were ecstatic, almost hypnotized, for nearly an hour, as though the elk were an epiphany. They were experiencing an ancient, vigilant vocation, graced by repose, of which only traces remain in the usual brief stare at zoo animals, our scrutiny of pets, and hunger for "nature" on our television screens.

The latent meaning of the itinerant box turtle, the hunched rabbit, and the charmed elk is somewhat like a piece of music to the listener, or a poem in its cadenced voice. Of the elk the watcher asks, as John Ciardi once asked, "How does a poem mean?" Another poet, Ted Hughes, answers in animal terms: "The special kind of excitement, the slightly mesmerized and quite involuntary concentration with which you make out the stirrings of a new poem in your mind, then the outline, the mass and color and clean final form of it, the unique living reality of it in the midst of the general lifelessness, all that is too familiar to mistake. This is hunting and the poem is a new species of creature, a new specimen of the life outside your own."[4]

Perhaps this is why poetry finds animals irresistible. The meaning of animals is implicit in what they do: eat, run, leap, crawl, display, call, fly, mate, fight, sing, swim, hide, slither, climb, and die. One or another animal does each of these things with more finesse and more expertise than people. Their keenness is reflected in the shapes of their bodies and the traces they leave. Animate signs and signatures move through our dreams and imagination, evoke our feelings, portray "us" in a kind of allegory. The signature of the animal is somehow more apt than the colossal hieroglyphic of the rocks, the silent autograph of the plants, or the calligram of the landscape itself. Like amusing, wise, terrible, curved mirrors, animals prefigure human society. The lion, for example, shares a primordial ecology with humankind, a long history of symbolic power linked to our own feelings "in the blood." Likewise, the bird is spirit and the snake is the earth of our most elemental self, our mundane world, and our imagination. Sometimes, as cultures change, these figures gain in complexity, as the lion melts into the sphinx, the bird becomes the angel, or the snake coalesces into the dragon. Or they can lose their immediacy, weakened by cultural decay, distanced from their origins by the loss of tradition, becoming shadows the way the sphinx dies into an architectural decoration on the library steps, the angel becomes the pageboy between a manlike god and chosen humans, and the dragon turns into a cartoon. But even as their images become obsolete, others emerge. More than monuments to human imagination, the whole panoply of their mythic, fantastic forms is based on a thousand millennia of watching and studying real, wild animals. This creative perception of animals is in us still, a perennial satisfaction and pleasure, one of the oldest human vocations, building on the complexity of natural history, limitless not only in gathered facts but the feeling that one is part of a gathering, a new understanding, a deepened participation. Children respond spontaneously to the details of nature and the names and movements of animals because animals were (and are) the path into categorical thought and, eventually, the terms of a philosophy or a cosmology.

In this book I have addressed these changes in consciousness. I have tried to write about childhood in this connection to nature as critical to human history. But I am leery of my own enthusiasm for writing. Is our relationship to animals essentially a branch of nature writing? Our bookstores and libraries are fat with accounts of the natural world, yet nature writing is flawed. The flattery of the printed word is as pushy as a German shepherd near a coffee table of sandwiches. It makes nature a subject matter and becomes a secret enemy of the natural world—from the death of trees for paper to its linear form, its misplaced concreteness, and the isolation of the writer. "Nature" is easily framed within the modern temper of alienation, a collusion with dispassionate apartness, the surreal estrangement of plants and animals as art. For example, the perfect nature writer is Henry Thoreau. When E. B. White called *Walden* "youth's best companion" he could only have meant an exceedingly perspicacious youth, perhaps as he himself was. For me Henry Thoreau is as tedious today as he was for many a century ago.[5] He is a maker of aphorisms, the favorite of professors in college English classes and the educated in a human-made world where nature, as art, is of value mainly as an embellishment. Nature writing nourishes the view of nature as esthetic abstraction—something like the sphinx on the library steps, the denizens of a bestiary whose charming irrelevance teases us out of the burdens of urban life and its stewpot of political and social drama where intellectuals have their true home. Nature writing breeds in the writer the greed for literary reputation and captivates its readers with a spurious substitute for experience in the natural world.

But that is an aside. Our species and our best observers emerged in watching the Others, participating in their world by eating and being eaten by them, suffering them as parasites, wearing their feathers and skins, making tools of their bones and antlers, and communicating their significance by dancing, sculpting, performing, imaging, narrating, and thinking them. Rachel Carson's book *Silent Spring* was not simply a warning against widespread pesticide use but against the deafened self, against emptiness. We must understand what to make of our encounter with the animals. Because as we ourselves prosper in unseemly numbers they vanish, and in the end our prosperity may amount to nothing without them. If art—writing—is to be a mere substitute, seeking to replace animals with an alternative reality, then let us seek instead an antiwriting against the seductive illusions of the "beauty" of nature. The account in this essay of a box turtle on a wet street in Kansas City in 1931 strives to transcend the event itself. Primal peoples know what we have forgotten: art can never replace, certainly not explain, that adventure among the Others which remains central to our lives, though it is the principal means of evoking it.

Cosmologically speaking, you pays your life and takes your choice. My experience tells me that neither the creationists nor the postmodern critics are right, one thinking that the world was made for us and the other that it is made by us. A better vision of the animals is that we are one among them. The only drama in town, to paraphrase G. E. Hutchinson, is the evolutionary play on the ecological stage; what holds the story together is the transformation of energy and substance. Somehow we must find a way into these exchanges in full awareness and discover how to cherish the world of life on its own terms. Thus do I return to the theme of love and death, the possibility that killing and eating could be the ultimate act of respect.

In keeping with the evolutionary myth as the core of my own cosmology, I have started this book with a section on assimilation in Part I, "The Animal Fare," in the sense that "taking the world into ourselves" shapes us, as it does all beings. The thought which emerges from that encounter is the subject of Part II, "Cognition." It follows that the words and concepts for the other animals makes possible the self-consciousness I discuss in Part III as "Identity." After incorporation, transformation becomes the essence of our being and my subject. "Change," Part IV, is the music of life, and animals represent and symbolize this subtle truth. Proceeding from the roles of animals in human biology and social life in the early sections, I have tried in Part V, "The Cosmos," to indicate by some selected examples from history how necessary the animal figure has been to our ideas of ultimate meaning. And in the last section, "Counterplayers," I have again chosen themes that illustrate the unlimited flow of nonhuman lives through the drama that being human means.

The nineteenth-century art of Jean Grandville, which I have selected to illuminate these chapters, is typically perceived as "bizarre" in modern eyes. Grandville does not fit easily into the conventions of our time. If our relationship to animals as true counterplayers is to break out of the banal stereotypes of "kindness" on the one hand and that of animals as mere automatons on the other, then radical ways of revisioning, at once unromantic and free of the old logic of hierarchy, are necessary. From time to time Grandville's cartoons are reproduced in literary reviews as satirical conversation pieces. But I see in them a persistent challenge to the old boundaries that have defined human and beast, the humor of our deflated pride, and insistence on an underlying continuity that demands redefinition of ourselves in the context of a larger animal world rather than as outsiders to it.

PART I

The Animal Fare

Being human has always meant perceiving ourselves in a circle of animals. The crucial event in this encounter has been ingestion. We have attended passionately to this consuming force until the idea of assimilation has permeated the nature of experience itself. To begin at the beginning is not rudimentary. It is the essence of all that follows.

1 | The Ecological Doorway to Symbolic Thought

*One must die into creaturehood, transcending the assurances of
"programmed cultural heroics." Being a self-conscious animal
means "to know that one is food for worms." This is the terror:
to have emerged from nothing, to have a name, consciousness of
self, deep inner feelings, and excruciating inner yearning for life
and self-expression—and with all this yet to die.*

ERNEST BECKER

THE HUMAN MIND is the result of a long series of interactions
with other animals. The mind is inseparable from the brain, which evolved
among our primate ancestors as part of an ecological heritage. That heritage
began with life on the ground, continued in the trees, and millions of years
later came back to the ground. This upstairs-downstairs legacy, arboreal and
terrestrial, is not unique to our descent but is widespread among monkeys
and apes, so that to understand our kind of consciousness—higher-level
thinking, artistic expression, and abstraction—requires some further explanation and is linked to our perception of animals in a roundabout way.

What follows is a possible scenario, spanning some sixty million years,
beginning with the earliest primate, a ground-squirrel-like form living

among the roots and leafy debris beneath the great ceiling of tropical forest. As this creature was nocturnal in habit, its sensory life was dominated by smell and sound rather than sight, by sniffing scent marks and the trails of beetles or listening to the scurry of feet, the faint popping of worms, or the calls and rustle that meant danger. Like redolent tracks making a course through the underbrush, the sequence of sounds was a trail of exclamations marking the position of some shifting body, a series informing the listener—the little, furry, pre-primate grubbers in the dark—of the movement of food, danger, others. Repeated sounds, remembered, revealed changes in position and movement, signaling whether something was approaching or receding or passing by. It required a good brain to hold that pattern in mind—marking the creation of an auditory world in which rhythms and musical phrases such as the successive notes of birdsong are heard as a melody. The writer and paleontologist Loren Eiseley referred to this as "time-binding."

Some descendents of those first tree-shrew ancestors went from the ground into the trees, perhaps as the flowering trees replaced coniferous forests over much of the tropics. In time, some of their descendents would be monkeys. In their brains were the neural networks making the recognition of patterns of sound possible. Remembering from one moment to the next—time-binding—by the listening mind was extended to vision, providing the primates with a mental basis for eliding sight into visual fields and visual fields into a visual world.[1] The first step in our thinking was in that arboreal reorganization of an obsolete sound/smell memory by transferring its function to the eyes. Color vision returned—that extraordinary reptilian adjunct to keen sight, surrendered by the early mammals in the dark, lost still to most families of mammals. The early monkeys, in those tropical forest canopies, also altered the central nervous system by linking emotion to awareness of novelty and change, creating monkey excitability and sentient behavior.[2] A great leap in brain size, unmatched among most other mammals, succeeded our ancestral primates' ascent into the trees, associated with pioneering a hazardous life in three dimensions, coping with storms, leopards, snakes, eagles, and gravity.

Speed in the trees nourished intelligence. Other animals live in trees without becoming large-brained, but for the primates life in three dimensions seems to have entailed the fourth, a subjectivity of time itself, implicated in their life strategy. That strategy included an intense and unrelenting social association. If any generalization can be made of simians that illuminates our own nature and our relations to other animals it is an "attention structure" which turns all information into social significance and is keyed to nuance and change in the visual world.

The drama which began when some small, nose-twitching mammals left the night shift in the basement of the tropics and scurried up into the branching trails of the bright upper canopy climaxed when their descendents took their shrill and panicky ways, their color vision, excellent brains, and devotion to the social context of all information, back down from the trees, out of the deep forest, and into the savannas. There the venatic (hunting/eluding) game of the predator and prey was played. Mind would be the child of the hunt. The prehumans joined a long-standing cybernetic system in which two great groups of savanna mammals—large carnivores and the herbivores upon which they preyed—linked destinies in a game of escalating wit.

In order to appreciate that game we must interrupt this account of our prehistory to examine the Game as it had been played for forty million years. Mind was evolving among the large predatory and prey mammals in open country. The fossil skulls of these hoofed and pawed forms reveal their progressively increasing brain size and the physical evidence of continuous improvement in the ability of each group to think through its situation and anticipate the moves of those to whom it was bound by the eternal ties of the hunt. As the complexity of their brains increased and their cognition became more subtle, they began to know the world by constructing mental representations of reality rather than simply responding to signals. How rare this event was can be seen in brain evolution in general. We brainy primates think of cognitive ability as being so valuable that evolution should always favor larger brain size; but beyond the size needed for basic bodily functions, brains are biologically expensive and risky. Flexibility in behavior and a capacity for imagining reality do not usually outweigh the disadvantages of making bad decisions and the time learning takes. Flexible behavior puts a huge array of choices before the individual with the proportional likelihood for terminal error. Brains are especially vulnerable to developmental impairment due to malnutrition, disease, traumatic experience, and injury, and they require large amounts of energy and investment of parental care and protection for a long time. The range of psychopathology increases with brain complexity. Life is improved for most animals by the other route: genetically modulated, refined niche adaptations plus behavioral and breeding strategies more tightly keyed to signals. If culture is a characteristic of complex intelligence, then its rareness among animal species is perhaps a sign of its long-term fallibility.

The Game accompanied the proliferation of flowering plants in open country, in associations dominated by the grasses, with deep soils and a diverse invertebrate fauna, making possible the energy demands of the drama of cerebral counterpoint among the game animals. The savanna and prairie's potential would ultimately be evident in the great herds. In the beginning, hunters and hunted worked through chance encounters: the stupid predator's random search and the stupid prey's contingent vulnerability. One became food for the other as a statistical probability of crossed paths—like the photons of light falling among the chlorophyll bodies in a leaf or the random taking of grass stems by grazing animals. Consistency was the major virtue of this fortuitous style. Then, fueled by the nutritional and energetic resources of the grasslands, there slowly emerged a progressively more focused pursuit and better-timed escape, through tuned sensibilities.

Nothing in nature had previously matched the latent, mental possibilities of energy stored in the seeds of grasses and the nutritional quality of their associated plants. Herds of large grazing animals and packs of social carnivores following them became possible. These social predators enhanced the likelihood of the paths of the two groups intersecting by their ability to remember and anticipate, while their prey improved the means to avoid and escape. Chance encounters would diminish in importance for both groups. The sheer abundance of life—the fertility of the grasslands and the ecological stability it provided—made this pursuit of the risky brain feasible. As time passed, contact gradually included sensing the Other at a distance; proximity was no longer necessary to force or avoid an encounter. In order to recognize the Others by their signs—odors, droppings, footprints, sounds—and read those signs as an ongoing event, remembering and communicating individual experience became increasingly important. This required more brain, more storage and association tracts for integrating information.

Up to a point, increase in overall size enabled animals to carry bigger heads and brains and to muster the strength and speed to elude and escape, chase and subdue. The fossil record confirms this gradual enlargement in certain groups of mammals, such as horses and lions. Larger brains are psychologically more complex, making possible what we experience as expectancy, anticipation, and problem solving. Once the thought of the Other could be kept in mind, trail following by the hunter, as well as vigilance and more guarded movements by the hunted, became possible. The sequence of signs was itself a path, the way evidence gives direction to a mystery.

Each began to know the place and time of expected appearance of the other. Better minds made scanning the field, ambushing, and stalking by the use of defilade possible. Awareness of routes was the key to intercepting the prey; the prey answered by improved capacity to recognize and avoid dangerous terrain features. The predator developed a sense of potential cover; the prey answered by varying its passage among a variety of routes, hiding its young, and decoying the hunter away from the newborn. Such initiatives by predator and prey were a new coevolutionary stream. As more thought entered the process, the predator masqueraded, imitated, or attracted its quarry by odd behavior and the prey learned in turn to recognize deception, to distinguish stalking manner and body postures from leisurely onlooking, and to know its own safe distance from the lurking carnivores.

Both hunter and hunted needed better mental maps of the terrain, whether for driving prey or for choosing escape routes. The gameboard expanded, from acres to square miles, along with body size and mobility. Because of the strength of the great herbivores, the carnivores began to teach their young how to stalk, attack, subdue, and retrieve, and the young learned these skills over a period of years. The comparable skills of the hunted were the adroit use of horns, antlers, and hooves and movement in concert, forcing the predator in turn to estimate the prey's potential resistance in order to select the weakest, disabled, callow, or panicked individuals. Hunters became concerned with recognition of vulnerable individuals: the pregnant, young, old, ill, or isolated. The hunted needed to discern the predator's intention, to read the future in the rhythms of the distant lion's moves. The potential prey's response to being stalked could intimidate inexperienced predators or lead to joint action in fending them off.

So much learned behavior required that the young of both predator and prey develop slowly in a series of age-grade activities, including play that prepared adult agility and sharpened their bluffing, catching, or avoiding skills. It also required longer care by the parents and their intentional demonstrations of attack and protection. Both sides earned the capacity to "keep in mind" cue fragments—to perceive and respond to the cries of other animals such as the alarm calls of birds and rodents. Cooperation between individuals improved, especially if the prey were large or the predators numerous. The social hunters and the hunted eventually developed roles, often by sex or age, for coordinated attack or mutual protection. Predators developed techniques for relayed pursuit; the prey developed techniques for inducing the hunter away from the more vulnerable individuals or, perhaps, even for selecting the individual among their own to be lost or "volunteer" for self-sacrifice.

Such "progressive strategies," linked in tandem, were a minuet whose steps became more baroque with time, beginning with the bumbling mobility of minimal forethought and proceeding to the nimble elegance of master performers. These tactics evolved over thousands of centuries from simple responses to remembering and thinking ahead—the before-and-after-thinker—in veldt or savanna. Such interlocking brain evolution is not generally characteristic of animals; brains, as noted earlier, are no bigger than necessary. Most animals have relatively little cognition other than essential physical, social, and ecological processes keyed to the larger rhythms of the body or the season. Learning is important, but it is limited.

This Cenozoic mutuality of mammalian hunter and hunted is one of the few long-standing and conspicuous episodes of reciprocal mental evolution. It is based on the measurement of fossil crania spanning more than forty million years. It took place in the energy-rich, open spaces with dispersed cover, spaces in which life depended on a radical attention to other animal species. Strangely, the Game was played in no other environments than the savannas and perhaps the sea (or at least it is less discernible by modern scholars in other habitats). The brains of whales and dolphins evolved in an intensely social, three-dimensional world, where long-distance communication is possible, paralleling that of the monkeys. The dolphins became group hunters whose food quests pitted them against fish who, if not as smart as antelopes, had a 200-million-year head start in perfecting the musculature and alert reflexes of escape. This perfection of the fish was like a whetstone to the minds and bodies of the cetaceans. Large size, the protection and education of slow-growing young, movement over large distances—all are part of the evolution of the larger savanna mammals and the marine dolphins. Among these special mammals of the savanna and the sea—and the human newcomers who joined them—there is an almost clubby quality, as though some story were being told among themselves about an elaborate quest.

Now it is possible to return to our own story, in which our predecessors, already unusually cerebral, entered the drama in progress. They were like Americans arriving, decades late, on the world's soccer fields. The newcomers skittered about the fringes of play at first, perhaps more prey than predator. But the outcome in time was a surge in human brains, a saga in which a multitude of large, keen, dangerous animals—lions, antelopes, rhinos, elephants, horses, aurochs—loom in protohuman thought, and early humans went into open country afoot and occupied an ecological niche. At first, at the edge of the forest, on an expanded new diet of roots, stems, and seeds, they had entered a niche somewhat like raccoons: munching turtles and frogs, gleaning the residue of lions' meals, opportunist scavengers and part-time small-game hunters. Their primate sociality and big brains served them well and, almost suddenly, they were transformed into our species, beginning to speak, sing, dance, and above all cooperate as full members of the great savanna show.

The circumstances in which a series of large carnivores and herbivores became more thoughtful, by watching, pursuing, evading, stalking, hiding, mimicking, and otherwise seeking to comprehend and anticipate each other, set the stage and the terms of our presence, as though we had won a role in a play that had been running for years or married into an ancient lineage. When our ancestors moved away from the forest and began to forage, scavenge, and hunt we joined an adventure with ancient rules. To this union of large, fleet herbivores and powerful mammalian carnivores we brought primate social relations. It was already a wily band of frog and cicada munchers, would-be meat-eaters who would challenge their fanged competitors and chase the sly prey in the ongoing venatic Game, and who, with their chimpanzee-sized brains, would parlay cognition into new realms.

The humans brought features not common among the hoofed and clawed members of the community: diminished smell and hearing and a society germinated in the treetops. The cat-, dog-, and hyena-like players already there had great olfactory bulbs in their heads, a sense of smell so keen that a world could be built on it, along with ears and sometimes eyes to match. Our ancestors had abandoned the nose and neglected the ear, having become mainly visual like birds. Their primate legacy included a well-developed vocal system with its potential for speech and, eventually, the assignment of words to categories of the visible world, thereby coding the world as objects which could be evoked in the mind's eye or between individuals by speech.

Crucial to any coding is the nature of an "object." Every cognitive system depends on assumptions about the nature of entities. A "thing" to the mind of a frightened lizard may not be a protective rock but the rock's lower edge together with the shadow of an overhang and space in which to crawl. To the wolf pursuing a caribou calf the "object" may be composed of a small caribou together with the smell of its urine, its awkward movements, and its bleat of fear. Among the prespeech humans there were probably already vocalizations, as among many monkeys, triggered by the profile of a soaring eagle or the sinuous movement of a snake. Lacking the fine discriminations of nose and ear, these prehuman prairie novices would finesse what they had, play the hand dealt by their biology, and match the seen (and felt) world with their own sounds. For example, their new ecology carried them into a larger geographical range than any primates had previously experienced. They would navigate under different conditions and seasons and learn the boundaries and places in detail. They may have marked places with words as the wolves marked them with urine. This would enable these bipeds, who could carry bits of the environment in their hands, to carry terms or references to location in their heads as well. Animals themselves could also be mimicked and inferred, or referred to, by human sounds. But the subtleties of sorting a great variety of animals would require more explicit reference. Their new hunter's meditation on signs—not only as objects in their own right but as indications of an unseen animal—unlocked speech and symbol from the call and sign system. The animals would be evoked again and again, so that the human imagination would in fact become more densely populated by recollected, imagined, represented, and dreamed forms than by tangible presences. A leap in mind was occurring in which meanings could have echoes in other realms, perhaps based initially on analogies between themselves and the other species, as when they danced the fighting. Humans tracked into a new world of double meaning, based on an amplified relationship to plants and animals.

Although they joined the exclusive club of the open-country carnivores and prey, they did not lose their floral affiliations. Indeed, they acquired the universal attention of omnivory, the soul of which was the prospect of an infinite world of latent meanings. As primates, the early humans had an old taste for tender greens, fruits, nuts, berries, and a new interest in roots and seeds. They relished herbs as food and as signs of large animals—the bitten twig, the seed in a dropping—indicating a class of ideas that could become "objects" of thought. Phenology—the seasonal timing of life processes: sprouting, leafing, blooming, fruiting, quiescence—must surely have been one of the first sciences, a plant key to animal emergence, birthing, migration, and hibernation. Plants reveal the hidden qualities of particular soils and the subtle properties of place and site. In airborne pollen, leaves, and winged seeds they give air visibility and demonstrate the mysteries of season and sequence. They chronicle the year, keepers not only of their own periodicities but those of animals who depend on them. By their presence or absence, plants seem to regulate the passages of animal life, becoming cues or metasignals.

Our species' capacity to key into such signs was a crucial step. The animal world provided models for the very idea of thought. The bear, for example, was traditionally the forager whose movements were the gnosis of a wise eater, whose inner clock ticks away in resonance with an evanescent larder. Bears follow the paths of a seasonal geography based on a food quest. For bears as for humans, the moving object or thinking prey is more interesting if not more tasty. And so the bear itself became, for many Paleolithic peoples, a model of ourselves. Where there were no bears there were elephants or lions or whales. Indeed, no single species was necessary, for the gift of mind enabled the hunter/gatherers to identify aspects of themselves in many species.

The complementarity of plants and animals in the diets of omnivores is also a set of signs pointing to each other. The totem pole with its nested animal figures is in the form of a tree, for example, the plant that connects the animals one to another. The biblical tree of the knowledge of good and evil, for example, with its fruit, its snake, and its human players, implies connections beyond the words of Genesis. The life of bears, like that of coyotes, foxes, jackals, and pigs, is a gift of thought to humans who see in them a reflection of their own omnivorous versatility and who perceive such omnivores as mentor "cousins" or spiritual guides on other branches on the tree of life, who know when to join the chase or when and where to find berries and fruit.

The human world of men and women is intimately bound up with this plant/animal dialogue. The hunt was not a man's club, excluding women and children, but a mode of perception, a form of the quest which included all members of the group and all food. For open-country primates among whom the mothers would at times be alone with children, a sly sense of defense and knowledge of the style of dangerous predators must surely have characterized the wisdom and sociality of the gatherer. Mental maps and the capacity to read vegetational signs, the memorial embrace of the seasonal cycle, the teaching of small children, the mentoring of play, the significance of the Game as a story, the recollection of custom and experience—these did not belong only to one sex or one age group. Even in its maternal twist these aspects of emerging humanity were part of the hunt because it was in relationship to animals that mind beyond that of monkeys and apes prospered. Sexual dimorphism is sharply defined among terrestrial primates, including humans. This "division of labor," by physical and psychological differences, was not the means of inclusion or exclusion so much as a way toward success by partnership in the Game. High degrees of sexual difference could also be found among other species in the club—the great hoofed aurochs, horses, elephants, rhinos, buffalos, wolves, lions, canids, and a panoply of now-extinct players on both sides that joined and then left the Game, giving way to more sagacious replacements.

We were latecomers in a well-established process that had gone on for fifty million years. The four-legged carnivores and their prey had long since learned that an animal, watched long enough, gradually dissolved into signs. It left the marks that came to represent it: footprints, urine, secretions, feces, molted antlers, scratchings and rubbings, gnawed stems, bones, feathers, beds, diggings, nests, tracks, and bits of fur as well as an immense range of sounds and smells unavailable to us. Language and art meant that those signs, as "objects," could be transported and therefore shared with those who, for example, had never come upon the bleached leg bone of an antelope near a water hole, broken by hyenas, pithed by vultures, and chiseled by mice, but who could hear the story and see its representations for what it was in the wilderness and what it might signify as an analogy of a seamless social form. Our human specialty was to dislodge such signs from a momentary stuckness in place and time and build a mental world of them that could be played over at times of our choosing.

As signatures these animal signs were the beginning of abstraction and symbolizing in a drama we primates already "knew" was social at heart: gestures, expressions, innuendos. In less than three million years, all these categories of the self and society were shaped by the traits of animals observed, the dangerous, competitive, beautiful, tasty, scrounging Others. The human hunter or hunted shared recollections through stories, song, and performance, nouning and verbing the Others and, by indirection and insight, themselves. Tools for capture and for defense were created, although, beyond a cutting edge, their flinty remains give an exaggerated impression of their importance, for it was the questing mind that was refined. Hunting and escaping were rehearsed and planned. Butchering of large prey led to autopsy, comparisons, and reconstructed lives. Food sharing expanded from nutritional to social and sacred symbols. The nuances of animal body language and the subtleties of the food chains were represented in art and ritual—and words.

There are odd connotations in this cognitive ecology. Large prey, in the predator's grasp and in eye contact with it, cease to struggle, as though surrendering to ecological necessity. A human observer may see in this quiescence an augury of self-sacrifice. The larger carnivores display a prudence about killing, in contrast to our fantasy of the rampaging beast, living instead frugally without excess or leftovers. The older among them lose avidity for the chase, as though foreshadowing their counterparts among aging sportsmen, tribal elders, or southern gentlemen who become keepers or members of conservation groups, observers reflecting on the hunter/ hunted life in an ethical sense that may be impossible for those who have never hunted.

There is something about human omnivory, with its roots in strategic carnivory and referential speech as categorical tags designating "reality," that plays like a muted echo in the impassioned social life of human primates. Flurries of emotion, subtleties of posture, glance, and intention, pulse through the group. Even in repose, typical ground monkeys are bathed in the to-and-fro of provocations, tempered fellowship, and contested priorities. Human society emerged from this tumult of shifting fidelity and status as individuals move through life. Proximity, grooming, sharing food, and sexual mounting have signaling functions that go beyond their primary functions. Full consciousness of these moods required speech; and speech, in turn, needed concrete reference. Like chimpanzees, our ancestors did not lack nuance in their lives but a verbal code for externalizing it as an object of discourse and contemplation. Judging from savanna monkeys, the prehuman troop, more self-consciousness, mercurial, fiercely social, and anxious because of its extended ecology, needed only categorical references by voice to become sensitive to life as a double stream with its metaphoric parallels of social and ecological reality.

Venatic life, the Game, places animals at the heart of human symbolism. As metaphor, venatic thought is the reading of tracks—as when Pueblo Indians speculate on the footprint of the roadrunner, a bird whose foot has two toes pointing forward and two backward, unlike most other birds with three forward and one backward. The result is a puzzling track: which way was the bird actually going? It could suggest deception, which is characteristic of advanced hunters and their prey, the bird being both a hunter of lizards and the prey of coyotes. Pointing both ways at once, it signifies a contradance, a movement in both directions that closes a circle. Ritualized, incorporated in healing clans, the roadrunner's sign has power for the Indians in the recovery of health, as a hex in war fostering a conclusion, and as a transitional fetish in funereal rites that bring the individual back to a new beginning.[3]

The human mind came into existence tracking, which for us creates a land of named places and fosters narration, the tale of adventure. Perhaps the quest began as a food search. But in scrutinizing the details of the potential prey, competitors, and predators upon ourselves, and all the signs they leave, it seems more abstract, like scientific curiosity, communicated in art and narrated as myth. All kinds of data are said to be better remembered when "spatially encoded," mentally mapped. Orientation in space, the sound coding of places as an analogy to anatomy—locations as parts of a body—and mathematical computations are intimately interrelated. Among recent tribal peoples, the story of a mythic hunt or the latest pursuit is an account and a rehearsal of the landmarks in a known world. The tale tracks events and seems always to tease logic with a strangeness. We, seeing lethargic animals in the zoo or fragmented in film, have forgotten how surprising they can be, but there must have been a long time when the novelty that so animates the monkeys could still flush us with excitement. The unexpected turns in the epic quest have their origin in traces of the wonder of things, itself the object of our tracking mentality.

Eventually, as humans, the primates created visual metaphors of smelling and listening to the unknown. Objects represent something other than themselves, or "the primal experience of 'seeing-as' " in three ways.[4] These three ways are tracks, shadows, and simulacra. Tracks distanced the predator from the prey in such a way that signs became the grist of thought. Shadows are resemblances, not so much cues as a kind of facsimile. Seven hundred thousand years of cunning shadow use by the hunter and hunted and of watching shadows play around campfires or move across the land beneath clouds are behind our foreheads. E. H. Gombrich has suggested that shadows are such forms that, if traced by hand, involve an artist in making simulacra which stand for—or stand in for—that which they represent. This representation may be the primordial basis for image making in art. In the old caves of France the sculpted and drawn figures of animals are implicit in the contours of the natural features of the rock itself, seemingly in motion as the flicker of torches animated the stone in light and shadow, as when the head of a bear or the profile of a horse are given painted eyes and other living details by humans. In some caves these animal figures were juxtaposed to painted outlines of human hands, to the puzzlement of modern scholars who suppose them to be individual signatures. Hands make gestures which could evoke images, draw lines, etch the forms of the beasts. Long before they drew the animals, people were perhaps pantomiming their lineaments, "drawing" in air. It was as though the long heritage of primate facial recognition—touching and reading the features, bony surfaces, lines, shadows, the movement of musculature under the skin—was transferred to "reading" the whole animal. The first makers of animal art, like the sign-watchers who preceded them, sought not only to identify but to discern circumstances, to tell a story. Twenty thousand years ago, the "tracks" made by animals or men—by erosion in rocks, the shadows cast by campfires, or carved in stone—were cues to different parts of a living reality whose connection might require some speculation: "A horse associated with a serpent, a vulva, a fish, a human pregnancy, or a horse that is symbolically killed is defined not only by its recognizable shape but also by the contexts within which it functions."[5]

The whole sequence of brain and mind evolution by attention to animals constitutes a unique twist in the primate obsession with the self and society. The reciprocity of two vital areas that comprise the setting of human intelligence—social and ecological—is implicit in that Pleistocene cave art in which the great game mammals are shown as paired species. The game of mateship, like the ecological intercourse of predator and prey, has reciprocal customs and stratagems. Mateship and the cultural facilitation of courting are both *play* in its larger sense. The Game in both senses is central to our genesis. Our ancestors and recent hunting/gathering peoples looked upon marriage and the hunt as formalities subject to protocols of assimilation. Sex and the kill are gifts. The love between man and woman and between the humans and the other animals is addressed through an allusive semantics. A poetry of consummation as an act of unity with the Other was the inspired legacy of omnivory, for the plants play their role in thought as their fragrance, beauty, and pharmaceutics generated the vision quest. The flower on the body of the dead is an affirmation: not compliance with a raw reality, but a loving closure.

Venery, says David Guss, "symbolically transforms the child's meat into spirit, and the hunt transforms the animal's spirit into meat." Thus killing is "a moment of creation for it brings potential into actuality."[6] The paradoxical relationship between flowered, reproductive, social sex and the animalized, thoughtful hunt is an idea conceived by primate omnivores who are themselves male and female, prey and predator, thinker on plants and on animals. The convergence of love and death produces, not a love of killing, but their shared identity in the cycle of transformation.

2 | The Swallow

Once every year, the Deer catch human beings. They do various things which irresistibly draw men near them; each one selects a certain man. The Deer shoots the man, who is then compelled to skin it and carry its meat home and eat it. Then the Deer is inside the man. He waits and hides in there, but the man doesn't know it. When enough Deer have occupied enough men, they will strike all at once. The men who don't have Deer in them will also be taken by surprise, and everything will change some. This is called "takeover from inside."

GARY SNYDER

PURISTS ARGUE THAT we do not "think ideas" in the usual sense of verb and object but, rather, that ideas occur within us. Yet without a sensed world there could be no ideas of it. Perhaps there are steps to mental life, an ingestion or perception followed by an unconscious translation, generating an idea which may seem freshly created. A primal source, the sensed world lends itself to conversion or metathesis, such as the maw of a great beast as the image of the abyss of hell.

For two thousand years the sacred symbol of pain and torment as body-rending carnivory has identified evil and demonology with wild predators. Hell's jaws are a picket fence of terrible teeth like those of snakes, wolves, and lions, not the dull dentures of cows. Our fear of monsters in the night probably has its origins far back in the evolution of our primate ancestors, whose tribes were pruned by horrors whose shadows continue to elicit our monkey screams in dark theaters. Yet we are endlessly attracted to the symbolic role of slavering monsters as evil. In literature the pleasure of fear was associated with the late Renaissance discovery of the joys of sublime terror, but the hunger for frightening spectacles of fearful onslaught must surely be a more elemental part of our being than can be explained by our awareness that we are actually safe in our cinema seats or at home with our books.

Images of the great bear remind us that the ecological contract, celebrated in ceremonial feasts, includes the possibility of being eaten. Our primate and prehuman ancestors were hunted by large cats and probably by hyenas, wolves, and sabertooths long before we ourselves ate horses. Humanity's first venture into North America from Asia may have been delayed, for instance, by a huge carnivorous bear, *Arctodus simus*, stronger and more dangerous than the brown bear of Alaska today.[1] But even if true, that experience by those early Americans may have had no physical connection to chronic fear and terrible pain. As surely as we hear the blood in our ears, the echoes of a million midnight shrieks of monkeys, whose last sight of the world was the eyes of a panther, have their traces in our nervous systems. Modern fiction is rich in allusions to the terror of the "victims" in the jaws of raging brutes. But the teeth of a predator may be painless for the prey because of brain-made endorphins, so that such a death may be euphoric, even a kind of epiphany. To understand this, we must stand outside the stereotypes of raw gore, of good and evil among eaters, of innocent victims and bloody demons.

To do so is to contemplate the alimentary tract itself. When I was a college student studying biology, certain images stuck in my mind and grounded my thinking. One was a series of different animals with shorter and shorter intestines, beginning with long, complicated alimentary systems like those of rabbits and ending in the simple tube running the length of an earthworm. In elaborate systems teeth, tongues, stomachs, appendixes, and other anatomical details, even the specialized food habits, all become accessories united in an elaborated assimilation of the world. But the worm is like a hose sliding through the soil, its ends open as mouth and anus, through which part of the environment passes. Such forward-feeding puts one's brain near where one's mouth is, and, for us speakers, vice versa, extending the principle of nutrition to a kenning, or knowing.

Substances taken in become food for thought, meaty thoughts and kernels of ideas in our mind-bodies. Whomsoever we eat we contemplate. I would worm my way further into your thought. Swallows on the wing, focused on tiny flying insects, do not simply scoop with mouth open but snap up chosen bits of the world, the birds verbing themselves, swallow-ing. Compared to swallows, the passengers eating lunch on a modern airliner are like accelerated worms, taking what comes, thrust across the sky by engines gulping air. The birds and insects are engaged together in an intricate dance of life and death. There is more minding about it by both partners than the peristaltic feeding flow of the worm. The birds anticipate, estimate, remember, and adjust to dispersed and moving quarry, which become subjects of thought, the insects themselves mustering such mental alacrity to escape as they can. The bird and its tidy counterplayers are bits of the real world so distant, so esthetic, that we can watch them with no qualms at all.

As for ourselves, the history of human eating is a primate story. At the zoo, the "lower primates," the prosimians, have little family resemblance to us and are merely curiosities, like raccoons. Apes are formidable, yet pacific, immobilizing us watchers as though we were trying to remember something. Our story starts with an ancestor more like a monkey. It is difficult for us to accept a monkeylike ancestor. And our feelings of repugnance are not entirely Charles Darwin's fault. Monkeys are funny perhaps because humor is the anodyne for our repressed anxiety about their distorted reflections of ourselves. Even when their babies draw our affection they are difficult to love. Monkeys' habits are disgusting because they act on the physical impulses that we restrain and make private. Among the ground-dwelling species their capricious meanness is especially harsh, their bite full of infection, hatreds unrelenting, personalities paranoid, and neuroses frighteningly familiar. Rounds of contention and intimidation are typical of ground-dwelling forms such as baboons, who seem more like humans in their agitated ways than do the apes, especially in captivity, where commotion degenerates into depravity: freaky panic and squabbling alternate with desperate affection, curiosity, play, and periods of lassitude.

Tracking backwards through our past reveals how multilayered we are. At every stage, food has been a basic generator of consciousness and qualities of mind. Quietly watching, plant-eaters combine a roving eye with a pedestrian drudging and grubbing. Concentrating enough vitamins and amino acids from vegetation takes hours and hours, making an eater vulnerable in the open. But the tranquility of herbivores is a human fantasy, a kind of bucolic dream. Baboons are a hard lesson for those who think that a vegetarian diet produces a benign personality. Our prehuman forebears, combining vigilance against being eaten with venturing into the open grasslands, became food for social predators as well as foragers and were very likely as truculent, irascible, and passionate as modern baboons.

In the evolution of prehuman feeding, insects, frogs, flowers, leaves, and buds came first in the trees of the deep forest, then large fruits, followed by stems and roots, then seeds and the flesh of large animals last. Grass seeds opened the play of savanna ecology. Concentrating huge amounts of energy, seeds made the fabulous masses of large-bodied, hoofed animals possible, the coinhabitants of the savannas with our ancestors. Much has been claimed for our granivorousness—erect posture, bipedality, increased size, and the associated modifications of the skeleton, especially hands, pelvis, ankles, even the neuromusculature of the lips and tongue, which made speech and human society possible.[2] But our story does not end there. The eating of grass seeds may have been a crucial step in human evolution, but it brought us only to the threshold of the great semicarnivorous venture. Recent advocacy of total granivorousness, in an attempt to make cereals the center of our health and gustatory lives, falls short of the whole story.

Tender leaves and buds have an archaic appeal, answering an old monkey appetite. We still like greens and fruit, which need no preparation, keep well in their own skins, and come in profligate quantities. Tender corms or basal parts of grasses sought by terrestrial monkeys are another matter, as they require extended gathering and exposure in open country. Root eating extended that adventure, as tubers must be dug and would eventually be pounded, peeled, soaked, leached, or cooked, eaten as relish, sweets, nectar, wine, narcotics, and intoxicants. Insects, buds, rootlets, berries, and leaves are eaten whole or simply pulled apart. While they have insides—soft tissues, pits, seeds, and cores—there is about them none of that palpitation which suggests independent activity.

We are each, so to speak, a society of organs, and we can also imagine ourselves as parts of a larger, living whole. Imagining one as a "group" seems unlikely to have occurred to primate eaters until they—we—began to open the bodies of other mammals and birds and discovered organs, bones, things recently eaten, and embryos like those seen in the torn or dissected bodies of our own dead. Only then would it occur to us that the world is:

the monstrous mystery play of life,
which is called, "Now you eat me."[3]

Such an idea may be a final affirmation or a final horror. The biblical paradise, where there was no carnivory, marked our attempt to avoid the choice.
Perhaps the loss of innocence by Adam and Eve was the discovery that in
eating animals we kill and eat other sentient beings. We lost our unself-
consciousness, the end of our monkey titbit omnivory, and discovered the
ravishing taste of baby rabbits and gained the eye-rolling vision of a new primate personality, an inside-looker's view: gut-wrenching awareness of the
subterranean similarity of ourselves and our victim.

From seeds to meat in ever-larger packages, from beetles to bison, never
abandoning salads, in the rare company of other extraordinary mammals—
coyotes, pigs, and bears—our progenitors discovered omnivory and the inner life. As the early human butchers of large herbivores explored the car-
casses of their prey, the organs they found were like a fauna. They saw the
hot, convulsive livingness of intestines, hearts, lungs, muscles, glands,
bones, and other tissues. They developed gastronomic preferences of a kind
already exercised by other predators. Intellectually, it went further—the discovery of the correspondence of similar organs in different animals, a revelation in "hidden" kinship, a paradox in which essence contradicted the external diversity of plumage and pelt. The parts themselves were like different
species, transcendent categories. The differences between squirrels and cats
to an anatomist are less important than the common ground of stomachs,
lungs, diaphragms, and hearts. Old, widespread stories tell of the organs' independent lives at the beginning of the world. In Siberian myths the visceral
parts of the elk once lived free in the forest and tundra. Later, the liver,
stomach, lungs, heart, and kidneys then came to live together in the bodies
of mammals, birds, and humans. Proper butchering in traditional societies
is a matter not only of skill but the recollection of such metaphysics. Dismemberment is proscribed "to assist the reconstruction of the animals from
the pieces into which they have been broken for the purposes of consumption, thus ensuring the regeneration of that on which human life depends."[4]
Further speculation from thinking on the body cavity: the reverse idea, that
all the different species began as a unitary being, which explains why differences among the animals now are limited, as though they were modeled on
a larger, cosmic "body," or part of the body of the earth, perhaps. The digging stick may have been the first human tool, opening the earth to reveal an
inside like that of an animal or human body.

Large animals are cut up to be carried and cooked and to make nutritional and social distinctions in the feast. Butchering transcends mere pulling apart or mauling. It requires a graphic concept or image of the whole, tools to cut and remove the skin, to disembowel, to separate major parts at the joints, to detach organs by severing connective tissues and slicing muscle masses. It can be done with a sharpened stone only if the skeletal structure and the body's own assembly patterns are followed. Pleistocene peoples and today's hunter/gatherers, who eat as many as two dozen kinds of vertebrate animals, probe these inner landscapes, from tiny birds to elephants, with the finesse of surgeons.

We cannot know when humans first spoke the names and classified bones, organs, and muscles. Hunters of large game everywhere, with their technical skill, professional detachment, and protomedical savoir faire, typically ruminate on variations in the details of the anatomy. Drawings of animals made by tribal and archaic peoples indicate that the interior of the body, like the cosmic underworld, has a topography of its own and that the two are related. Therefore entry into the other world is assisted by a ceremonial butchering. Among the Buriats, for example, a people of the Russian-Mongolian border, a new shaman is consecrated at a sacred feast and a trance in which his soul is borne away by a spirit horse (in other tribes by a bird) to the world of sacred powers. An animal is dismembered; the head is set aside as witness; the parts are then "fed" to the fire. The animal's soul is made up of parts corresponding to body parts. Since these match those of the shaman, he is thereby dismembered by proxy. In the fire the organs and muscles are returned to the respective spirits who control their health, and each regains its autonomy and strength. Awakening after his ceremonial dismemberment, the shaman is reborn whole, his organs renewed. Through these rites he gains the power to heal as the knowledge of the organs, bones, and muscles is united in him.

Different plants and animals have their places in nature, to which the viscera show a parallel order. Anatomy is studied by our carnivorous side and then, as omnivores, we make analogies like "the heart of the forest" and "a tongue of flame." To the human eater and anatomist, hunter and healer, predator and prey, these correspondences of organ and animal, inside and outside, link what is otherwise made separate in our naming and butchery. Like the community of animal species, we are each a community of life whose organs are constrained by their mutual integration much as wild animals are themselves subordinate to the natural community in which they live. Each organ, like each species of animal, is nutritionally or behaviorally unique.

In the course of thousands of generations, the gastronomy of the new omnivore primate led to the belief that each animal has a spirit guardian who receives the soul of the dead animal and reincarnates it. It may later return as new food to the hunters, or not, depending on their state of grace. Hence the people in small-scale societies take care to neither waste nor casually destroy other beings or their body parts. Not even blood or bones or pelts are carelessly treated. With the discovery of our insides we became consciously embodied, organically incarnate. Organs are to the organism what organisms are to the social body. In our modern teeth and bones and digestive enzymes are the continuation of those ancient regimes, representing not only the food chains or niche to which we belong, but the whole forest-edge landscape of which we are still part. We measure its distances with anatomic units such as "feet," its durations by our pulse as "seconds," and so embody time and space.

Seeds, fruits, or small animals may be devoured directly without losing their identity, but the meat from large animals is no longer clearly the antelope or bison. We may eat a rabbit or grouse, but instead of a deer we eat "venison." Butchery makes new categories by abstracting "meat" from the whole animal, creating a perceptual gap between the food and the thing eaten. Abstractions imply contrary possibilities: one is disjunction, by which we disguise from ourselves the ugly death of the slaughterhouse steer to get "beef"; or, alternatively, in eating venison instead of a deer, we assimilate spiritual qualities that transcend the individual animal, belonging to all deer living and dead, the sacramental meal.

Before we were eaters of animals we (monkeylike) were their prey, subject to going inside them. Over geological epochs we have been engaged with animals in multiple passages and encounters, entering those who eat us, assimilating those whom we eat, and watching the rest who endlessly exemplify variations on that double flow.[5] Horrendous "memories" of being eaten and eater are within us, scripted in our genes, which are themselves already "inside." Our genetic program anticipates eating, fleeing, and perhaps even the notion of mythic levels—the underworld of the dark inside and the upper world of visible things. Inherited from ancient animals, our genes encode feeding behavior as if food were a kind of mnemonic, as in the words of the Christian Holy Communion: "Take and eat this in remembrance of me." So may shamans and other intermediaries have spoken, in a thousand ceremonies of sacred blood and flesh.

In a sense our own lives became available to consciousness by means of the universal use of animal signs for charting the experience of the world. That experience is essentially cyclic. Feasts of communion precede a resurrection. Among primal peoples the wild prey is said to offer itself, divinity embodied, a self-willed gift which will return, perennially renewed, so long as gratitude is formally expressed by the recipients. Among boreal tribes the slain bear participates in the festival at which it is both honored guest and the main course. The celebrants dine on it in an exculpation and union with the deity, followed by its spiritual departure and later incarnation. People "swallow the god" as a vital principle of the round of life, the eater and eaten echoing ecological reality in spiritual reciprocity. Meat is a consummate danger and blessing to which humans—not monkeys—are heir, a terrible and beautiful mystery and an infinitely repeatable Last Supper.

Animals dying in the name of a deity represent an incisive affirmation of nature, a theophagy of sacred energy flow, the primal religious act. In sacramental feasting, humans not only participate but acknowledge their place in a system in which death is no less essential than life, a system not subject to arbitrary human disposition or judgment. In the ceremony of the slain bear, the animal is understood to be a gift: supreme but, in the same sense as all other foods, self-given.

When, in historical times, humanized gods subdued the pantheon of earth, plant, and animal spirits, sacrifice replaced thanksgiving with a tender of negotiation. Such an offering, usually associated with agricultural and military fortunes, was intended to deflect evil and sacred wrath as a gift to greedy gods or as tithes in the maw of death, which demanded blood in order to yield new life in the land or to spare the living in battle. Animal figures on the oldest coins suggest a commerce in which their bodies became part of an exchange rather than sacramental participants.

WE HAVE FORGOTTEN that the coin was a substitute for death. We hide death, whose most dramatic expression is the interspecies drama of the eater and eaten. And we prefer not to acknowledge the eater as eaten except as physical necessity. I remember the revulsion among the parasitology students at the University of Neuchâtel when Professor Jean von Baer revealed that he grew tapeworms in his own body for research. Parasites repell us, as does the thought of being nibbled by beetles, shredded by vultures, gnawed by wolves, or ripped by sharks. Mortuary practices which preserve the body reflect our aversion to the organismic agents of decay, cyclicity, and transformation, and at the same time they appeal to the dream of resurrection as literal restitution of the body. Cremation avoids the conflict between preserving the body with poisons and its dispersal into a host of hungry creatures, although the flames are simply a little hotter and faster than those in the belly of worms. Being eaten by fire invokes the cosmology of solar deism typical of otherworldly religions, which repudiate the sacredness of organic cycles and physical life on earth.

Canadian Arctic Eskimos speak of hunting and eating not animals but souls. Eating and being eaten, from doughnuts and animal crackers to the ecological Eucharist, from sucking mother's milk and the sacrament of the body and blood to sacrifice, show that the ambiguity of "eating souls" can never be entirely resolved because we are both swallowers and swallowed. The Fall from the infantile dream of Paradise, where all animals lived on plants (whose sentience was not allowed), was a descent into a world of predator and prey.

Eating the fruit of the tree of knowledge of good and evil led not only to the awareness of sex but reality. The snake in the garden was the old temple guardian and consort of Asian goddesses, anathema to the Hebrews. Apples, like all plant products, tend to disguise the death that goes into their making. Yet that apple was radical meat for thought (perhaps because of the worm it contained). Our human appetite for nibbling at the world has become huge, and our conscience is panicked with doubt about how to participate in the celebrations and mythic narratives that join life and death and preserve the world for others as well as ourselves. Our primal beginnings in personal infancy and species history endow us with enthralling intimations of ingestion and engulfment, drawing us to the terrifying story or film of frightful monsters. It is no wonder that our ethical concern about animals should focus on how to kill them.

Death is a tender subject, with its imagined pain and terror, vistas of roaring carnivores killing beautiful deer and lions raging among themselves over bloody bones. Images of predation as the power of the strongest confuse our monkey politics and its endless skirmishing for power with food chains in ecology, making the false analogy of nature to violence and war. Not surprisingly, "civilized" nature lovers decline to participate in carnivorous food chains, much as women may decline to breast-feed their infants. Yet the madonna giving her milk is not so "devoured" by the infant as some psychiatrists have suggested, as she is an ardent bearer of an urgent flow. Similarly, the grass eaten by the buffalo and the flesh of the buffalo eaten by the wolf we imagine as taken by force. But the milk, grass, and buffalo, even the madonna and wolf, transmit something more important than themselves. In the ethos of the ancient conjunction of "to prey on" and "to pray to," the hunt is not a seizure but a voluntary immolation. Hunters preserve a lore of wild things who oversee the ethics of their own transformation into food, observe atonements, and return again and again.

Food chains are fundamental connectors. Howard Norman envisions the energy from the sun as a great push, not "trapped" by the plants but pouring into them, not taken or caught in a succession of life-forms but enveloping them. Swollen to capacity, like the milk-filled breast, with this stream from the sun, plants fill the earth. Their turgidity finds joyous release as food for herbivores. The solar tide carries this onrush of energy beyond them into the bodies of the carnivores and through them, overflowing into bigger predators, scavengers, parasites, and microbes. The dream of perfection of the soft-eyed deer and the singing grasshopper is to burst like a small sun into the blood of the wolf and bluebird, who, surrounded by prey, accept each bite as a gift from the sun-parent. The whole drama is enacted in a thousand different scenes of transfer which include ourselves. Before its force finally dissipates through a myriad of beings, from beetles to bacteria, and so radiates back into the receptive universe as heat, a little of the energy roils into pure meat-eaters like the leopard, lying in wait on a tree limb, a kind of deliverer for its prey:

> And those that are hunted
> Know this as their life,
> Their reward: to walk
>
> Under such trees in full knowledge
> Of what is in glory above them,
> And to feel no fear,
> But acceptance, compliance,
> Fulfilling themselves without pain.[6]

Carnivores, such as leopards, are specialists. Their attention is an intense, narrow beam. Herbivores are specialists narrowly absorbed with large quantities of a few things. Omnivores, like foxes, are not so single-minded as either. As the Greek poet Archilochus said, "The fox knows many things, but the hedgehog knows one big thing." We omnivores diffuse the carnivore's laser mentality but keep its leisure to reflect on "big things." Human omnivores combine a snacker's relish tray with a gourmand's casserole and the connoisseur's brisket. In addition to the inherent primate expertise in matters of fruit, the acquired skill in the selection of seeds and cereals, we have learned the savoriness of fat and the satisfactions of the hot meal. The world has become a delectable signifier, a banquet of signs for investigative foraging, the gustatory quest. Omnivory leads from food choices and habitat options to speculation on the traffic of food and the lives which incarnate it. Plants and animals emerge from edibility to thinkability, and so meaning is shaped and tempered by omnivory.

By this route our attention is drawn to all plants and animals. Some are strangely powerful in their attraction. From sign to insight, the butterfly's color and pattern seem to assume the existence of a delighted watcher: they seem to have been "designed to be seen."[7] The walkingstick, whose body "looks" like a twig, is shaped to be seen and misinterpreted in a dangerous world of predatory watchers. Bees "presume" a hunter to whom its yellow marks call attention as if to be seen and recognized; yellow-banded flies who are stingless and moths whose wings are marked with imitation owl eyes misinform the stalkers by mimicking real warnings. Among all such prey and their predators such marks and colors may be only signals equivalent to "yes" or "no," but to the reflecting mind of the omnivore they constitute conspicuous cases in a world of signs which attract, repel, or deceive but always contain information about fundamental relationships that may be grasped as abstract categories. A vast trail system opens up, like the brief streak of a bee in flight, pointing from a flower toward a distant, hidden, honey-filled hive, or like the sixteen-hour-old hoofprint of a solitary, pregnant wildebeast, headed toward a river crossing, heavy, hesitant, lagging, hungry.

In two or three million years of human evolution such tracking of nature replicated itself in the neural elaboration of the human brain. The perception of animals ceased to be only a recognition. First swallowed in substance, then swallowed in thought, they were finally incorporated in psychic structures. In an experiment exposing human subjects to the brief—tachistoscopic—image of a tree in which one branch outlines the figure of a duck, the participants were unconscious of what they had "seen." Yet when asked to tell a story they made an improbable number of references to feathers, eggs, nests, and so on. The concealed duck was "unperceived but registered," a "perceptually recessive" figure.[8] The duck image experiment does not prove that we have latent ducks in mind, or that any other particular creature is straining to leap by association into narrative allusions, but something less explicit. Why should the investigators have chosen a duck as the primal sign rather than a chair? The anticipation that bushes contain concealed animals is part of our world, a world of food habits and the risk of becoming food as the primal means of widening intellectual horizons.[9] Thought grew from metonym to metaphor. Energy became information, imagination, and thought as we evolved, encompassing herbivore grounding with carnivore leisure, taking the high middle ground with omnivorous attention to everything. The arc of the swallow's flight is a gesture in a world of counterplayers, marking a career in which assimilation leads from bodily metabolism to spiritual parabolism.

When the world became our apple and sin was the worm inside, our species made the break out of herbivory and its garden. The defect of all gardens is the insulating, stultifying effect of their walls.[10] The outside to which we were banished according to Genesis was a wilderness. The wild savanna was entered in thought as well as diet with all its possibilities of ingestion. "Eden," geographically and linguistically, is an area in the valley of the Tigris River. It was indeed a great grassland with islands of trees. There the humans, like the carnivores, included animal insides in their diet. Having passed through a series of stages, first as runners following tree limb trails, then upright leapers, larger-bodied swingers, prowlers on the ground, scavengers of meat, extractors of nuts and seeds, and stalkers and butchers of big game—having escaped from the tropical garden—having done all this, only then did cosmology and philosophy become possible.

Early cosmology took its quest for finality from the forager's search, dissection, the semantics of naming anatomy and naming animals, the poetic recognition of parallel inner and outer worlds, and the artistic representations of them. The first tools of horn, bone, wood, and stone were not constructions like the beaver's dam or the bird's nest but parts of animals. Digging sticks, blades, picks, mauls, saws—means for opening things to get at roots, kernels, through skin, into joints and marrow—were not so much for killing as extracting. Tools revealed not death but clues to the life of animals, a society within the body, a mirror of our own. Once opened, insides become new outsides. What is around us in daily life can also be imagined as some great "insides." For the Kunwinjku in Australia, the ancestors' body parts became the visible features of the terrain around them. Anatomy was perceived as landscape, produced from mythic origins and experienced by every generation in the planned itinerary of clan walkabouts.[11]

The insides were not only parts but meat. "Meat" is from Indo-European *mad*, from which comes Old German *gemate*, or "mate," one with whom food is shared. The Sanskrit is *madati*, meaning "rejoice." Meat is therefore "celebrated sharing." After the milky infancy of all mammals, among them only the meat-eaters regularly carry food to the young and mark social status by their food sharing. Meat distribution is the language of social obligation and kinship in primal societies as well as in ancient cultures and mythology. It involved eating their god in Thracian-Dionysiac rites of the sacrificial animal. The sacred meal, or *agape*, was part of the worship of the Great Mother. The Mithraic banquet, the Zoroastrian *Haoma* ceremony, and the Christian Eucharist share common roots in the *tauroctonus*: the Vedic and Iranian rites of the sacred bull and its capture and death, its blood liberating all life and securing human salvation. Hunting for meat established later modes and eventually the basic metaphor of humankind's assimilative relations with the world.

Our roots are in the tissue of life itself, fundamentally realized in microorganisms and plants. Plants mediate the edaphic world, translate geology and climate, connect hot and cold, dry and moist, light and dark, distribute nutrient elements. Plants make animals possible and, in the end, they eat animals by decomposing their corpses. We are subject finally to a kind of silent philosophy of vegetation. Plants incorporate while animals eat. To embody is both active and passive, absorption and feeding, assimilation and being assimilated. Ultimate ecology, the flow system of nature, is ambidextrous.

Consummation and communion in all their nuances are ritualized in food and drink, and whole philosophies of life and death are expressed as quests for food and its transformation. The hunt sharpened our intelligence. And the crucial turn was the inwardness which began with the scrutiny of the opened body.

PART II

Cognition

Mind emerges from encounters that matter and, more subtly, from intimations of encounters. Voice becomes speech with naming the creatures, and the imagination dances thereafter with images that arise in "thinking" animals. Their presence and absence give both time and space reality in our forethought and afterthought of them.

3 | The Skills of Cognition: Pigeonholes, Dinosaurs, and Hobbyhorses

The diversity of species provides the most available intuitive picture and direct manifestation of ultimate discontinuity of reality: the sensed expression of objective coding.

CLAUDE LÉVI-STRAUSS

THERE ARE SAID to be twenty million stamp collectors in America alone. Their hobby has something in common with that of all collectors, including dinosaur enthusiasts, for at heart it is an exercise in nomenclature. For some, collecting no doubt includes many other pleasures—esthetic, informative, curatorial, competitive. But the naming and positioning of a stamp on the right page, series, date, and country, as well as its individual variation from its class, is the common calling of all such extravagant assemblages, from fossils to wineglasses to horse tack. However diverse the objects, this organizing and naming of a family of miscellany has two origins, both deep in the life of the mind. The first is personal and begins with an innate agenda in early childhood which takes each of us through body parts, then animals, then on through our individual life experience and whatever hobbies we choose to pleasure our taxonomic bone. The second origin is that of speech and conceptual thought in our species, which may help explain why, as children, we are so determined to name the animals.

All infants are fascinated with their own bodies and the corresponding anatomy of adults. Adults in turn supervise early talk in babies by tickling and talking to babies about their eyes, noses, lips, cheeks, toes, fingers, and arms. Like the infants, the adults are playing their part spontaneously in a scenario dealt by their biology. This play talk, often at bath or dressing, is joyously shared by the amused parent and the attentive infant, who gradually builds a repertoire of terms and becomes conscious of the similarity of its own discrete anatomy to that of others. At about ten months the child begins to repeat the terms aloud.

Isolating bodily parts in this way is a garish parody of butchering, the body dissected, as though the child could only construct a whole image of itself by first disarticulating itself perceptually along the natural demarcations and joints. Infants also learn to name the different people in their lives by a similar rapt attention to isolated facial details. This is important to them, not because their survival depends on it, or did so in the past, but because classifying constitutes a basic skill, like walking. The capacity for thought, for cognition, for organizing the sensate world, begins in this way. The role of animals in this activity is rather surprising, since it has little to do with our direct relationships to animals. Other animals besides ourselves distinguish different species around them, but the distinctive human expression of this experience is naming them, a diagnostic procedure based on the recognition of body parts and their mental imagery.

So, in their second year, children become ecstatically absorbed with animals, not only as beings but as types with names. Naming is based on visible anatomical clues—less often sound, feel, or smell—an attention to parts already practiced and now ritualized. Confronted with a strange animal, the ritual unfolds as follows: there is first a staring, silent look; then a turn of the head to the caregiver who responds by saying the animal's name; then the child's repetition aloud of the name; and then further staring. It begins with isolating and naming parts of the body, pigeonholing, making "collections" of parts—noses, tails, legs, claws, ears. These exist not only as bits of creatures but as categories, a nose being the concrete example of the abstraction "noses." This procedure is little noticed by parents because they are part of it, having experienced it themselves as children, and are predisposed to fall into the pattern as adults.

As the child learns to visually isolate and name parts of the body, a second, less conspicuous, process of perceptual discrimination is also emerging. This is the establishment of a repertoire of basic animal shapes. Generalized forms, such as "bird," "dog" (for all mammals), and "fish," are established. These general forms represent the clearest cases of collective membership and share many traits within the group and few with other groups. Recognition of such inclusive groups is the first step in cognition. In a body of medium or large size, the cluster of four legs, round head, and whiskers is the tangible representation of "cat."

Body shape, size, and the form of appendages provide the first general division; then body parts furnish the labels for category spaces which we seem "born to discover." As infants we are predisposed to label the world by practicing a kind of imaginary dissection on classes of things as though they were born with dotted lines marked "cut here." And behold: an already-categorized world is found, first in our own bodily anatomy and then in the compelling realm of animals.

As a result of these two steps, children identify animals first as basic prototypes and then specifically by key parts. Cats in general are recognized as a basic form, but the kind of cat—lion, lynx, tiger—depends on a few specific anatomical details or cues. The two steps are "the result of psychological principles" of imaginary disjoining and lumping, combined with speech or naming, set in the calendar of child development.[1] Because this agenda is universal, the names of animals among peoples in many different cultures tend to be binomial, such as the "golden cat." The more inclusive "cat" also corresponds closely to the scientific idea of the biological family, the most abstract collective that can be easily pictured in the mind's eye. "Cats" and "dogs" (the families Felidae and Canidae) comprise many species, and both are members in turn of a still more inclusive order, Carnivora, which includes weasels, bears, and seals. One can picture a cat without denoting its species, but it is difficult to envision a "carnivore" that is not a bear, dog, weasel, seal, or cat. This is because the large categories like carnivores, mammals, and vertebrates are more abstract and less visible to the inner eye. The all-encompassing group, the "animal," is curiously hard to imagine. Almost all humans recognize the same sets of species, though the logic of their grouping may vary. For example, their ability to fly may mean that bats are classified as "birds." But the concept of the named, prototypical, or family group and the distinctions within it by species is widespread.

A shift from one-word naming to grammar occurs abruptly at about the age of eighteen months.[2] Anxiety caused by the mother's occasional absence may help to account for this early phrasing by which the child calms itself, evoking her image by such sentences as "Mommy gone; mommy come back."[3] Naming may "predicate on a previous presence" or on the "subsequent presence in a current absence." When the little human primates, mother-bound like no others, contain their fears by representing the absent mother by speaking her name, the first step is taken in bringing all the creatures into an imagined presence.

Named animals, in the evolution of thought and speech, became synonymous with special sounds or movements—making the name work like a verb, calling forth the characteristic behavior of a particular species, and opening the door to imitating and, later in life, acting and dancing the animal. In the thousands of centuries of human evolution this predication of the animal upon the self—in which one declares, "See, I am now the bear!"—may have been done at first not by children but by adults who were creating the earliest sense of self and human identity, becoming in time a way into thought, as the means of mentally handling categories. As the centuries of human emergence passed, it was initiated earlier and earlier in individual life, rolled back to infancy, where it linked the "idea" of identity with taxonomy and speech.

But why should the naming and predication of animals start early in individual development? Why is speech there at all in a two-year-old? Communication by words is not at stake in the first years any more than it is at stake in the life of chimpanzees, who, without speech, can communicate their needs, fears, and discomforts. The answer is that the early acquisition of speech does not have to do with communication but with cognition. The mosaic of animal kinds is the supreme concrete model upon which this skill is achieved, and, as an added benefit, being alive, they keep before us an organic figure of reality, a world of kindred beings as the basis of a purposeful, living cosmos. The identity, names, and behavior of animals give us some of the first satisfactions of the mind. Animals are unusual parts of our environment: they are radically different from people and from each other and have movement as well as anatomical properties as though the cosmos had dreamed up a "functional core concept" for us. In "feature abstraction theory" it is said that "children learn an impressively large number of 'natural concepts' during their early years, yet they have great difficulty in learning the artificial concepts used in laboratory tasks" until later.[4] Categorization as described here develops out of the sensory perception of tangible objects; principles such as momentum and volume are scientific abstractions.

Breaking up the world in thought, attending to its diversity and discontinuity, discriminating differences in order to think—all this clearly threatens its continuity and wholeness. Learning the morphology of bodies has been likened to a kind of dissection. The butchering analogy extends as well to the naming of the internal parts of the body. Oddly enough, it is the insides of animals that work against the tendency of the world to fragment. As a result of drawing the internal organs, modern urban fifth graders get the larger picture: "Only around age ten is the inside of a person seen as composed of many different body parts, just as the outside of the body is."[5]

Such direct knowledge of internal anatomy is delayed in modern society where children, unlike their tribal counterparts, lack the opportunity to examine the insides of bodies, often influenced by their parents who themselves are repelled by the sight of viscera. All carnivores recognize the different organs, just as they do species, and have preferences among brains, organs of the viscera, bone marrow, fat, and muscle. But referential thought, beginning with naming the innards, speaks not only to nutritional value and taste but to some further use of this nomenclature. The children of our hunting/gathering ancestors not only engaged in the naming of animals; they watched and helped with the butchering. According to Claude Lévi-Strauss, the "ultimate discontinuity of reality" implied by parts and species is balanced by the obvious kinship of organs. For no matter how diverse the exteriors, recognizable hearts beat within the body, enclosed by ribs, flanked by lungs. Organic membership, of difference and grouping, has this internal aspect that confers kinship among the different animals (including humankind), affirmed by the correspondence of their internal organs.

This connection to animals is characteristic of human experience. The parts of animals each do something—ears hear, mouths chew, hands and paws grasp. The verbing of parts seems inescapable. Writing of the relationship of his poetry to his boyhood animal watching and hunting, the poet Ted Hughes says: "I think of poems as a sort of animal. . . . They know something special, something perhaps which we are very curious to learn. Maybe my concern has been to capture not animals particularly and not poems but simply things which have a vivid life of their own. . . . How can a poem, about a walk in the rain, for instance, be like an animal? It cannot look like a giraffe or an emu or an octopus or anything you might find in a menagerie. It is better to call it an assembly of living parts moved by a single spirit. The living parts are the words, the images, the rhythms. The spirit is the life which inhabits them when they all work together. . . . As a poet you have to make sure that all these parts over which you have control are alive."[6]

Nothing is so easily classifiable as animals, because they are perceptually distinct among themselves as a result of the needs of their own visual recognition.[7] In his field guides to the birds, Roger Tory Peterson collects all the heads of the warblers together on a single page so that the reader can compare the code of lines and colors by which the birds themselves discriminate at a glance. In addition to their own need for swift identity, their differences, especially of bills and feet, conform to shapes that adapt them for specific ways of life. The many species of sandpipers, for example, are clearly differentiated by the length and form of legs and bills according to the depths of water in which they feed and the depths of the food organisms for which they probe. Ecologists have calculated the exact degree of difference and the disadvantage of overlap between two species. Because such characteristics are visible to people, birds and other animals are the perfect embodiments of diversity in unity and unity in diversity. Anciently connected to category making as the archetypes for the cognitive act itself, they constitute the child's practice ground in making categories that balance the poles of likeness and difference.

AS THE PROTOTYPES of categories, animals serve as a scaffold for all kinds of classifying. Yoga positions, for example, are defined as *vjrasana*, the butterfly; *matsyasana*, the fish; *bsujhangasana*, the cobra; *salagasana*, the locust; *simhasana*, the lion; *vidalasana*, the cat; *gomukasana*, the cow; *ustraasana*, the camel; *garudasana*, the eagle; *swanasana*, the swan; *hanumanasana*, the monkey; *virischikasana*, the scorpion. There are many kinds of pain, but how are we to define them? As Virginia Woolf once said, "Let a sufferer try to describe a pain in his head to a doctor and the language at once runs dry."[8]

Evidently she was not familiar with the Ainu on the Japanese island of Hokkaido, who identify headaches and boils by placing them, perceptually, on the same screen of animal categories. Their headache, or *sapa*, types include *sapa araka*, "bear" headache—like the heavy footfalls of the bear's steps; *opokay sapa araka*, "musk deer" headache—a light galloping; *nina sapa araka*, "woodpecker" headache—like drilling; *seta sapa arfaka*, "dog" headache—gnawing; *akoype sapa araka*, "octopus" headache— sucking, chilled; *takahka sapa araka*, "crab" headache—prickling, chilled; *ikurupe sapa araka*, "lamprey" headache—digging and cold. Their boils are the *otah caru araka*, shark mouth boil—on the tongue; *keputenka huhpe*, bat boil—painful like a bat's cry; *sumari huhpe*, fox boil—reddish color; *tomakaci huhpe*, bee boil— looks like a beehive; *uriri huhpe*, cormorant boil—with a black ring; *kurupe huhpe*, lamprey boil—digging into the flesh; *takahka huhpe*, crab boil—crawling and bright red; *ahkahka huhpe*, octopus boil—dark color; *hewnay huhpe*, sea anemone boil—reddish.[9] The Ilongots of the Philippines identify the different wild orchids according to parts of the (animal) body: thigh, knee, fingernail, braid, thumb, finger, elbow, calf, tail, whiskers. This is the "lexical set" through which the body is extended to plants.[10]

Animal taxonomy provides terms in regional folklore, signs of seasonal chores, levels of bureaucracy, burlesques of political types, and indicators of weather and chance. There are motif clusters of disarticulated parts signifying autocrats by their coats of arms, beasts in the signs above shops and inns, and human surnames when William decreed that the conquered English must have a second or sur-name and they took not only place and occupations but—after a Continental practice—Fox, Raven, Bird, Rook, Marlin, Swan, Wolf, Wren, Martin, Lamb, and so on, giving physical shape to chosen associations, the eagle as the spirit of Roman emperors, the centaur as Achaean cavalry, all tools of thought in the construction of the self as a category. Even modern urban people, without many animals in their lives, cling to animal typology for athletic team names and automobile models. Since there is really little difference between these teams or cars, the different names seem to heighten their relative distinction.

The zoological taxonomic system is also projected to create parts and an orderly structure which is not evident, as in the sky. Apart from the movement of sun, moon, and planets, the night sky is one of the most nebulous parts of our environment. Its dense complexity is made comprehensible by *organ*-izing it with animals as the zodiac, a word derived from an Indo-European root meaning "life" and Greek words for "animal" and "a circle of figures." The ancient Greek mystery religion following Orpheus conceived such a sidereal bestiary. One of the most important ideas of astrology was the celestial cycle of time, the snake with its tail in its mouth as the constellation Draco. The Greeks and Romans had twelve zodiacal divinities in the constellations. Hebrew cosmology enlivened "the heavens" with animal spirits of seven kinds. A heretical Christian sect, the Gnostics, filled the sky in layers with archons, guardian spirits represented as animal-headed, which were held to be superior to the angels. At first gatekeepers, the archons became punishers and demonic rulers. Human souls, after death, failing to pass one or another of the gates, were returned to earth and reanimated as the earthly animal representing that power—lion, bull, eagle, snake, bear, dog, and ass. The Gnostics were influenced in this by the Egyptians, many of whose sacred animals were the same. In late medieval times, making sense of the chaos of the sky carried long traditions of partitioning the clusters of stars as distinct entities, of which forty-three constellations were species of animals. (Most of the others were tools such as "sextant," "workshop," "telescope," and "octant," obviously added late in the history of the naming of constellations.)

The daytime sky, too, is refashioned as a habitat with its animals. As we gaze upon the clouds they become a parade of whales, horses, giraffes, fishes, and birds. Their forms change in passing, and the whale becomes a snake, the horse a camel. Why are there not baskets and chairs, plants or eating utensils? Other things do "appear," of course, but the prevalence of animals suggests that a common principle is at work. Animals move by their own volition and thus seem lifelike and purposeful: primal, archetypal, the most perfect prototypes. Moving across the vault in a common direction, the different forms seem to be chasing and fleeing, as though fulfilling their lives in heaven as on earth in the great round of predator and prey. We "see" them as though we arrive prepared to do so. Like living beings, the clouds melt and disintegrate into fragments, assuming different states, as though contradicting all fixed definitions. Like the living stream of life on earth, the heavenly denizens are transient, altered by the energy which they represent as it passes from one to another, disappearing and reappearing as they decay into mere puffs of detritus and reform. It is as though, in their life as animals, clouds enact a profound lesson of the cosmic passage—unity in the flow of changing forms. The gigantic cloud animals reflect a terrestrial reality: the world is coherent, alive, sentient, responsive, beautiful, powerful, a comity of powers like the organs of a body or the creatures of a vast savanna.

The joy of naming clouds is an example of the larger pleasure of using the animals as an order-making game. The exercise of the skills of grouping animals by structure and name is not only intellectually but esthetically satisfying as well. It depends, of course, on an existing system of animal identities with which the child is familiar. For cultures which have not lost touch with the sky, whose members spend much of their lives in the open, the need to perceive order in the immense, omnipresent ceiling must be a substantial part of the desire for coherence. For adults the sky also demonstrates meteorological patterns. But perhaps children enter upon this game of discovering meaning in the most direct way, by projecting upon it the most basic cognitive categories. It may also happen that the animal clouds are a kind of foundation structure for a philosophy of meaning and purpose in the great overworld, which is less obviously organic than the middle world of living things.

Time too has its mealy grain which can be broken up as "species." The Chinese name each year after an animal; hence "the year of the dragon," "the year of the dog," "the year of the pig," "the year of the horse," and so on. The "notational" scratches on hundreds of etched figures of animals on bone and horn from the Old World Mesolithic (34,000 to 11,000 years ago) are the earliest calendars. Decorated tools, pendants, amulets, batons, and stones from a single deposit at La Marche, France, are a "library" of temporal events. Its script is the tiny etched lines on carvings of horses, incised in the form of sets and subsets of tiny marks, "microscopic, sequential, and cognitive," unquestionably made by people. These are not tallies or scores but lunar tables associated with periodic rituals, "an almost perfect observational notational phrasing" of the moon's phases. Everywhere there were records for scheduling tribal activities.[11] Our own calendars are illustrated by seasonal events—flowering, migrations, births, emergence from dens, signs by which humans set the tempo of their lives. Compared to the regal shift of the stars north and south or the lunar procession, typified in the flocking of ducks and geese, the occasional storm seems as impulsive as a thundering herd of caribou or horses, free within the larger framework of the seasonal march.

THE BLISSFUL ACT of categorizing may help explain why children are fascinated with dinosaurs. Given all the possible living and fossil animals available to the minds of eight-year-olds, why is there so much enthusiasm for a group of extinct reptiles? One psychologist argues that it is the yearning of a small person for power and play in imaginary fear. Bigness is indeed exhilarating, as any who have encountered a moose or an elephant in the wild can attest, but moose and elephants do not receive the same degree of devotion. Dinosaurs, being extinct, are not threatening. Yet the issue is not their physical presence, in any case, but their representations in pictures and models.

The answer is perhaps in the genesis of speech as nomenclature, the tasks of order making, names as conjuring images. One attraction is in the many forms, which represent the world in small, presented as a perfect schema. The great array of dinosaurs is somewhat like a garden of great riches, a kindergarten of the intellect. Minding them is a sugarplum pie in the naming process, which is ordinarily meat and potatoes, too easy among farm animals, too hard among beetles or birds, too limited among mammals. In general very large animals have few species and very small animals have minute details beyond the skills and patience of children. Ideally, there is an optimum number of basic forms whose many subtypes are flagged by specific traits or cues. To identify an animal one first recognizes the size and basic shape, given by the contours of bone and muscle. But nature does not always cooperate so that both prototype and details stand out clearly. Feathers and fur can mislead as well as inform.

The dinosaur is a perfect balance between a few prototypical models, easily recognizable, and a diversity of species ranging from little birdlike forms to giants. The species' characteristic details have "high cue validity"—a small combination of distinct traits separating them while indicating kinship within a few basic types. Dinosaurs are tailor-made for practicing the important perceptual tasks basic to thought itself, to which the child is inherently committed and zealously engaged. Bones make the basic body shape, the first step in identification, with the result that an observer recognizes the general group to which large animals belong more easily. An engineering principle dictates that large size among land animals is accompanied by a disproportionate amount of bone. As bodies increase on a linear scale, bone increases exponentially, so that the skeleton makes up a greater part of an elephant's body than it does of a mouse's. Although many large animals are available (whales, bears, gorillas, elephants, horned and antlered forms), none matches the dinosaurs in its combination of distinctive basic form or definitive detail and few have such diversity within the group.

The perceptual process underlying our delight in dinosaurs can be illustrated by the arts of effigy in which the objective is to create an unambiguous likeness of a particular person, to be seen at a distance, identifiable even though the image on the retina of the eye is relatively small. The sculptor or painter first makes a model or sketch, which requires that spurious detail be omitted.[12] Such is the statue of a mounted hero that can be recognized on the top of a building or in the far corner of a park. In making the miniature, the artist keeps only the significant details of clothing and emphasizes the main features of the hero or horse, defined by skeletal structure. From this model the sculptor makes the final, enlarged figure. The likeness goes from being shrunk and losing its excessive detail to being stretched with its essential lineaments beyond life size, so that the completed statue is simplified. Most dinosaurs look as though they had been conceived as toys and then made larger with careful attention given to the bony understructure, so that its prominence is magnified. Like dinosaurs, great statues seem simplified and stripped of confusing detail, ready for the public whose recognition of the basic form is assured and who will then proceed to specific recognition by perceptually butchering the hero.

Proceeding from the general "hero" or "dinosaur" to specific recognition and naming, the eye searches for cues. Again a physical principle is at work favoring the large forms, which results from their having proportionately less body surface than small forms. The surface-to-volume ratio of a cube four inches on a side is only 3:2, for example, while that of a cube two inches on a side is 3:1. Since an animal's heat exchange with the air is dependent on this ratio, large animals in hot environments need additional surface to shed excess heat. They often have enlarged ears or other extended appendages, knobs, crests, folds, fringes, spines, and other protrusions that increase surface area. Such spectacular projections are characteristic of dinosaurs, too, serving as specific cues to identity. Even if their function were social rather than physiological—meant to be seen among themselves rather than for temperature control—these knobs and fins supply exactly the keys that inform the human observer. Indeed, the naked skin of reptiles exposes these diagnostic cues by the clarity of the junctures of parts. There is no feathery blending of wing to body or legs hidden in flight, no mouth shape obscured by whiskers, no furry blur. The diversity of appendages enhances rather than confuses their identities because the joints are so clear.

The dinosaur, like the heroic sculpture of the general on his horse, can be duplicated, sold, and handled by its owner only in small. Dinosaurs reduce to toys and pictures without losing their characteristics, unlike, say, African antelopes or American birds. This resilient scale has much to do with their charm. The real play with dinosaurs is with small models. Thought goes back and forth between miniatures in the shop and the reconstructed skeletons and full-body recreations in museums and parks, to even larger-than-life models in roadside amusements. Some of the satisfaction may be that they are extinct—in the same sense that a story which begins "Once upon a time" protects the listener from the ogres and dragons in the story. Looming over his little plastic dinosaurs, the child is like an observer at a distance or a giant creator, making up pretend lives for his creatures. The possibility of its physical reality shifts the appeal of the Loch Ness monster—possibly a kind of dinosaur—to adults, as it contributes little to the cognitive pleasure of naming and lacks the "as if" shield of play. Just as giant reptiles can be reduced to miniatures and then enlarged, we can imagine yet another upward octave in which the full-sized statue and the mighty dinosaur are themselves miniatures of something. Does the full-sized dinosaur not also suggest shaping by a yet larger "hand"?

SUCH IDENTIFYING AND naming, or pigeonholing, filing away for future reference, undoubtedly comes from the care of the common pigeon or rock dove which fanciers keep in a dovecot. The dovecot is divided into numerous chambers so that each bird, returning from flight, enters its own compartment or, after being handled, is put into its own "hole." Keeping and breeding pigeons is typically a hobby rather than a true vocation, thus invoking a connection to horses and hawks. The hobby is a small falcon, also kept as a pastime, while a hobbyhorse was originally a pony characterized by an ambling or rocking gait. Such were the mock horses "ridden" in medieval mummers' performances as well as the rocking and broomstick horses in children's play. Webster's Dictionary refers to a hobby as "a topic unduly occupying one's attention." Perhaps one may surmise from this pastiche of definitions that the themes represented unconsciously in the morris dances, burlesques and pantomimes by mummers, expressed preoccupations whose enactment on festival days disguised their true content in stylized actions, paralleling the child's hidden motives, furiously riding the rockinghorse, energized by repressed emotion. Anyone who has experienced the passion of the child on his rockinghorse may likewise have suspected that the pounding thump was the effect of a driven theme.

In modern speech a hobby usually means a highly routinized pastime, like the solace of collecting and pigeonholing, with that same driven quality. Like many childhood pleasures this game of nomenclature does not disappear with age but may be transformed into a new pastime. The early satisfaction in naming and building an animal repertoire is never fully forgotten. Identifying, naming, and classifying—often accompanied by collecting—continue throughout life, just as walking and running extend the physical attunement and vestibular pleasure of the first steps.

Category making based on animals, linked to speech, was at the center of the evolution of the human mind and the beginning of language itself. Subsistence peoples today continue to extend and enlarge their repertoire of taxonomic groups avidly—indeed, we might speak of them as hobbyists or naturalists.[13] Tribal peoples around the world know hundreds of plants and animals by name and natural history. The Nuba of Africa identify more than forty species of locusts (biologists recognize only ten) and twenty-seven varieties of sorghum which are botanically but three. "In the two preliterate societies in which I have carried out field research," says one anthropologist, "knowledge of the biological world constitutes—I would claim—a greater chunk than all other types of knowledge combined."[14] He calculates that primitive tribes have an average of 1,000 to 1,200 kinds of plants and animals in their vocabulary, and he goes on to point out that familiarity with this great diversity of organisms is not primarily because of their economic usefulness. The alternative explanation—that tribal peoples are simply keen naturalists—requires that we ask what makes such knowledge "good for its own sake" and how the knowing is exercised. The single universal characteristic is that the organisms are named. This naming makes it possible to "handle" them cognitively, that is, making them objects of discourse, and such activity carries the inexplicable pleasure we get from healthy function of an organ or organ systems.

Taxonomy persists, not only as an essential process in child development, but as an esthetic activity in adults, a luxury of thought by which order is continually brought forth from disorder, even to becoming a light madness. The bird-watchers or fern collectors or gem hounds rampage mildly in the sweet bliss of infinite typology. Some bring the objects back as a collection, arranging them, adding another schematic geometry to the geography in which they were found. Compared to practical knowledge, this activity is a kind of taxonomic overrun, a collecting hobby. Historically, modern art or natural history collection began with the ancient Greeks and flowered in the Renaissance, but its true roots, in light of the widespread numbers of nonliterate "collectors," can be traced to a kind of intellectual pack-rat phenomenon that extends the individual satisfactions of cognitive utility, preceding modern science, innate and evolutionary.

Collecting stamps, coins, antiques, Indian artifacts, fossils, cut glass, skeletons, beetles, or gems extends an activity central to the first five years of life. It combines physical and esthetic pleasure with the satisfactions of accumulation, ownership, and social fellowship in clubs and publications. Merrill Moore, a lifelong shell collector, explores this taxonomic frenzy in his own life, asking himself, "Why do I like shells?" He recalls that his childhood was enhanced by his collecting during outings and the gifts of shells from family members—a "mother's shell" and a "father's shell," some sent from distant places or brought home by travelers as a gift. As an adolescent, showing the collection to family and friends was the focus of his social participation. Handling and examining the shells gave him visual and tactile delight, reminding him of places and outings, and opened up the complex lives of the mollusks themselves who seemed to him separate personalities, "like people." Moore's collection was put away or forgotten for long periods, brought out, added to, and examined from time to time. Collecting took him on excursions, put him in touch with others for swapping, made beautiful gifts, involved cleaning and arranging, showing, receiving, a search in books for identities and life histories, exercise in the logic of their relations and natural history, an emotional outlet, sublimations and catharsis in times of stress. Thinking about the shells, Moore made periodic new "associations." They served, he says, as a training for "sleeping on" problems. Shells, Moore comments, are "a world within a world. . . . If I wish to see richness in Nature, variety in pattern, and seemingly endless alterations in color and line, all I need to do is take them out and look at them. Among the riches of the earth I am amazed by Molluscan wealth, the treasures of the Sea. Neptune rules over a mighty kingdom. More of us earth-bound mortals would do well to pay him court. Sea shells are his emissaries."[15]

Few of us can articulate the connection of taxonomy and collecting with a view of the world as well as Moore, but we can remember the occasions on which items in our collections—the coins, bones, dolls, and marbles—were obtained, the tactile pleasure of their physical qualities, and the challenge of their classification and name. Shells, inorganic objects of calcium and other minerals, are made by living things, manifesting a relationship between the mineral and organic realms just as Moore's experience connects nonhuman objects to events in his family life. His avocation requires classification but has no limit, a mnemonic device for reconstructing a world. As the collector becomes more selective, the taxonomy becomes esoteric and the links of the objects to the world more complicated. Yet the process is basically an extension of the means by which we first perceive order in a diverse world as children and, perhaps like play itself, it continues to fulfill our adult lives.

Objets trouvés, found objects, are a special case of classifying and collecting. These are the odd bits of the world that turn up on beaches or in refuse piles, disconnected from their origins. Their isolation is both physical and categorical and is itself part of their attraction. As detritus they have a special taxonomic ambiguity, the cognitive gaps of unrelated objects.[16] As human pack rats, we pick up parts of old machines, driftwood, bird feathers, and bottles as though, contrary to the categorical impulse that complies with place, function, and ecology, we are also mesmerized by the order of disorder, the small chaos of things beyond ordinary associations.[17] If collecting a set of clearly related objects is music to the mind seeking coherence, perhaps objets trouvés are a mental counterpoint, the antiphony that laughs at the vanity of finality.

This trickster-like aspect of the found object reminds us that collecting disengages things from their places and their context, making them objects of value in themselves. From the perspective of systematic collecting, a bird egg is potentially a lavish display of connections to a clutch, a unique nest and brooding mother, a certain limb in an individual tree in a particular season, and so on. But distanced from such associations by the battering of time, it tends to be valued in its own right, to be scrutinized esthetically, judged by the formal relationships of its parts. Moore's ruminations on shells, which attempt to bind not only natural history but social, cosmological, and esthetic qualities, are a clue to the odd flotsam on the mantel or in the curio cabinet, which test our sense of connections with outrageous evidence, challenging us to bind the least-ordered bits and pieces.

The dissociated enjoyment of such shards for their intrinsic qualities is what is sometimes meant by surrealism. In this sense, the most difficult of all categorical challenges is also the most risky, for there are serious faults in surrealism: the moral risk of enjoying an object shorn of its functional connections, geometrized, allowed to stun us with its schizoid effects. In reality, nothing is without connections, however hard to discern they may be. Should one collect the skulls of executed rapists, the debris of an air catastrophe, the tattered remains of a household half-melted in an explosion? On these grounds Susan Sontag has criticized the admiration of "great photographs" of suffering humans.[18] She abhors the commitment of the human mind and heart to such images as simply light and shadow, romantic figures or geometric shapes, textures revealed by the reflection of light on surface or skin, compositional principles, planes and curves, and other retinal attractions of wounded life that displace humane concern.

Perhaps this final distancing induced by esthetic abstraction is the dark side of our best sense, our passion for order gone haywire when context is ignored. Most stamp collectors would probably be revolted by an issue on "great murderers" or "major diseases," however beautiful their design and colors. Merrill Moore avoids a disintegrating analysis, however steeped it may be in sophisticated connoisseurship, by "collecting" the lives of the mollusks in his collection—their links to each other, to the places from which they came, and to life in the ocean. He puzzles over the scars marking the conditions of their individual experience and all the occasions in his own life of which the shells are mementos. He loves the life of the sea of which they are the concretions.

By contrast, the surrealistic venture raises our doubts about the extension of the taxonomic impulse when it denies such connections, turns away from the composite structure of the world. Theory in the name of Art invites the taxonomic mind to abjure subject matter, content, or relations.[19] Animals are a natural antidote to such intellectual dandyism. Like plants and microbes, they are subordinate to ever larger systems, and the study of them is a doorway to engagement, enlarging meaning and consciousness. They importune us ethical beings to minding with a commitment of the heart.

4 | Savanna Dreaming: The Fox at the Fringe of the Field

Every viable organic entity must include an ordered base and an element of chaotic instability.

PHILIP SLATER

HUNDREDS OF CENTURIES of human experience have generated our drive to master the skills of category making as we learn to speak, first in nouns and verbs. Animals are a primordial ground for this endeavor because they are the most nearly perfect set of distinct but related entities, and perhaps because they are alive like us. Naming has its calendar. At first the criteria are anatomical. Using anatomy, we instinctively employ a two-step process in which we identify animals according to general body form, such as "duck," and then tighten the identity by specific cues, such as "green-winged teal." As our knowledge of the animals increases we incorporate secondary characteristics that become part of the identity of each. But the system has a catch: some animals do not fit the criteria. This defect—their marginality—is turned into a virtue. Just as ordinary animals are perfect models of assortment, the exceptions are the prototypes of marginality itself. Much of the universe is ambiguous, and we learn not only to make the best of it but to acknowledge ambiguity's place in the world.

Defining and naming types has its limits because many things in the world overlap or blend with others. Animals are the best concrete model of types because the species of the larger forms are usually distinct. But only if we keep our eye on chosen cues, for in many other details the differences disappear. Biologically, two closely related species in a line of descent are like the positive poles of two adjacent magnets, repelling each other, developing characteristics that reduce confusion among themselves. Ecologically, however, various major groups of animals are represented by species in similar niches, with the result that physical and behavioral convergence occurs, as in flight by mammals, birds, and even reptiles, or as in swimming, when other mammals go to the sea with fish and birds.

Thus does evolution play a game of difference and likeness. Categories defined by human observers inevitably collide with animals at the edges of categories. Such confusing forms elicit strong responses, but even the ambiguous forms may be classified. Such "misfit" animals may be seen as anomalies, superior or diabolical, more interesting than the rest, for they challenge the very grounding of our thought in category making. Classification systems do not easily accommodate change, so these species at the edge can be perceived as on the way to becoming one thing or the other. In this they exemplify change and attitudes toward them reflect feelings about ambiguity and transformation in a larger sense, as characteristic of the world. In this way, animals provide us with a code and an imagery not only for naming things but for coping with contradictory experience as well.

Whatever variable features we use to separate species, there are animals that combine them. As we age and our inner encyclopedia of zoology grows, we discover these borderline exceptions to that satisfying classification upon which our psychic comfort depends. The bat is a major example because it contradicts the break between mammal and bird (and between creatures of day and night, the underground and the air). Such exceptions either transcend the logical order or add confusion and stress. Bats create spontaneous arousal that is the basis of both the repugnance that some societies, such as Europeans, feel toward it and the reverence that other peoples in Central America see in it as superior to the ordinary order of things. Marginal animals are therefore psychologically provocative, and most societies find ways of exploiting the agitation we experience in seeing them for purposes of indoctrination in cosmology or cultural values.

Another source of fringe animals is those not native to "this" place. For millions of years human life was centered in a home range of a few hundred square miles, and lifetimes were spent without encountering many unfamiliar animals. Because of that heritage, by about twelve years of age we have learned to identify and name the local fauna. Along with speech itself, an in-built timer gradually closes the door on this semantic set of names—a psychogenic closure. After this point all new animals will be seen as variants or combinations of the familiar types, and naming systems must be stretched to include them, reasons invented for their existence, cautionary rules employed for dealing with them, and customs established for using them as synonyms for irregularity.

Both types of marginal forms—those which are familiar, but do not fit our definitions, and those which we may encounter as travelers—disturb the primal model of an orderly world, producing taxonomic crisis and cognitive dissonance. Being paradoxical or "out of place" means discord, a wrenching of the cosmos, as when a wolf pack invades the village or a bear wanders into the city. This perplexity seems to call for explanation, for meaning that can be symbolically interpreted and spiritually significant. We must account for centers that do not hold, boundaries crossed, and rules broken. Anthropologists have shown that what is culturally unclear is perceived as unclean.[1] If we repress these organisms, like all repressed material they can erupt into our dreams or stir our unconscious life in powerful ways. Official society tends to prohibit or regulate their expression. Yet they are precious; their very ambiguity seems to sharpen other definitions.

The animals of "the edges" have a curious parallel in visual physiology. Margins make sight possible. The human eye vibrates left and right in excess of 180 times a second. The edges of images on the retina trigger an on/off effect on the nervous impulses in visual cells. If the object is made to vibrate in synchrony with the eye so that its image on the retina remains in the same spot and its edges are not shifted, the object disappears. Similarly, forms which are themselves at the edges of groups become the focus of accentuated attention and deliberation. Just as edgeless entities threaten visual chaos, types without borders, ambiguous in their relationships, subvert cognition. Animals and people who are not clearly classifiable may become the object of anathema as signs of corruption or disarray, or they may be seen as sacred mediators, but in either case the dubious forms create excitement, thoughtful deliberation, and a rich mine of metaphorical ore.

ODDLY ENOUGH, TWO canids, the fox and the dog, are among the most anomalous animals for different reasons. The fox seems to challenge the taxonomic boundary between its own family, the canids, and the family of the cats or felids. Typically canids and felids are sharply divergent. Between wolf and lynx, wild dog and lion, coyote and cougar, domestic dog and cat, the differences are conspicuous in head and body shape, day and night activity, calls, food habits, endurance hunting, and surprise hunting—all reinforced by the seeming hostility between them. As captives and pets, their personalities seem to emphasize these differences.

Although biologically a canid, the fox's doglike features are melded with felinity, as though the old antagonists, dog and cat, seem magically blended. The pupil of the fox's eye is slit like that of the cat, indicating the capacity for both day and night activity, and its hunting behavior is catlike in stalking and pouncing rather than persistent pursuit. Its body is light and delicate. Like the small cats, the fox is both the prey of larger carnivores and eagles and a predator on smaller forms. Although a canid, customarily genderized as "masculine," it fills a catlike, "feminine" niche. Its feline adaptations—lithe body, whiskers, movements, mouse chasing, tree climbing—contrast to its doggish bushy tail, hole living, pointed muzzle, barking, and trotting.

The fox traverses copse, woodland, fields, roads, and barnyards. Wild, it raids domestic coops and prowls the gardens, moving back and forth across fencerows, forest margins, and suburban yards. It is hunted yet not eaten. It crosses the lines separating town and country, an exception and offense to our typology, a wild denial of the ordering of the world in which Adam, Orpheus, and Linnaeus, each in the major myth of their time, named the animals.

On the theme of purity of type, the Bible prohibits crossing horses with asses or creating other hybrids. In a great many societies purity is equated with unqualified devotion as opposed to pollution of thought or act. Among peoples with traditions that abominate racial or class mixing, the fox is a skulking embodiment of pollution, a living amphiboly, confusing our grammar with its biology, creating semantic stress because we are not sure who the fox really is. In the folktales and literature of both Europe and the Orient it is a schemer and deceiver, divided in loyalties. For Machiavelli the fox is the perfect complement to the lion, both combined in the personality of the ultimate ruler.

Samson, the strong man of the Bible, caught three hundred foxes, tied their tails together two by two, set them afire, and released them in the wheat fields of the Philistines. The terrified animals scattered flames through vineyards and olive orchards.[2] There is a lot of careless killing in the Bible, but the fox's peculiar vulnerability, its association with grain and killing of fowls and their eggs, omnivorous garden raiding, poaching in the storehouse, and stealing of fruit were already well established. The Song of Solomon reads:

"Take us the foxes, the little foxes that spoil the vines: for our vines have tender grapes."[3] While actually referring to grape eating, this poem may also allude to the corruption of the young by cynical, sensual, selfish adults, perhaps the female fox or "Vixen," a snappish and malicious woman according to the Solomonic attitude toward both nature and women.

Hostility toward the fox as a boundary form is typical not only of Judeo-Christian cultures. In an eighth-century Chinese story, "The Fox Fancy," a dog forces the revelation of the fox maiden's hidden identity. The fox as deceiver is found in both East and West, although ambiguous animals in the Orient seem not to become symbols of evil as often, perhaps because of a wider acceptance of ambiguity as a normal part of life. Several Oriental stories of transformation pay homage to the vixen. Shifting from animal to woman, she is not like the unfortunate werewolf, but is superior and beneficent, just as apes and monkeys are also seen in Southeast Asia as marginal humans in a good sense.

IN CONTRAST TO the fox, the dog brings its ambivalent qualities much closer to home and is therefore the object of an immense history of anathema. It is a borderline animal in so many ways that its marginality has mythic proportions, especially in connection with the geography of chaos. To understand this in an era of kennel clubs and beloved pets, we must realize that throughout most of its history of at least ten thousand years, dogs have seldom fared as well as they do now. Although they have been cherished for their good qualities—hunter, guard, herder, friend, worker—the inverse dog is the spoiler of human graves and eater of corpses, the keeper of hell's gates, the carrier of rabies, the mad dog of August heat, the black death as "hellhound," and the half-wild howlers like the winter wind.

"Dog" is from the Indo-European root word *kuon*, from which we also get "cynicism," with its snarling doubt of people's motives, and "cynosure," which refers to the dog star, or guide star, probably a mariner's adaptation of the tracking hound. The antitype of the dependable servant at the doorstep is the untamed, bastardized outsider, all those hangdogs who have circled human settlements for millennia, wolfing scraps, harassing livestock, and scavenging from the battlefields, prototypes of antigods at the fringes of the known world. In the sky the celestial dogs guard paths and orbits. On earth they are guides to the underworld, travelers between the dead and living, wild and tame, the messengers who became the Egyptian jackal-headed Anubis (who, combined with Hermes, the Greek dog-headed guide to the underworld, became Hermanubis). Atrum, Egypt's dog-headed demon, was said to devour the souls of the dead. The legend of the Nestorian Saint Christopher—a giant third-century dog's head—tells of his conversion and martyrdom in the context of early missionaries to Persia and refers to the Far East, where Asian myths depicted people with dog's heads who had descended from human-wolf and human-dog matings. In India the outcast "dog-eaters" and "dog-milkers" were also associated with stories of miscegenation. A "divine bitch" accompanied the Meruts, Hindu gods of the winds, whose principal lord was Rudra, a great hunter and "howler." One of the six Indian seasons, lasting sixty-one days, the period of the dog's gestation, was celebrated by rites of dog sacrifice and scapegoating. For Indian Brahmans, dog saliva was an extreme form of pollution. Although the dog myths of India are not unequivocal, the predominant anathema toward the dog is, in balance, undeniable. Its mixture is an entail of history, with traces of old, Vedic, pastoral, masculinist hound fancying, later diluted almost to invisibility by the indigenous Hindu contempt for the loathsome village mongrel.

The natural basis for dog-heads, or cynocephali, is baboons, especially the hamadryas, which appear to combine a humanlike body with the dog's head. Reports and perhaps skins of these animals, carried out of Africa, undoubtedly lent credibility to the idea of such people at the frontiers of the world. The Greeks celebrated a festival, the "massacre of dogs," as a sacrifice to Apollo's son Linos who was eaten by dogs. It corresponded to a similar legend of Saint Mercurius, the great saint of the Coptic Church, who was said to have been accompanied by two cynocephali who ate his grandfather and then relented and were converted, and to a Roman celebration in which russet dogs were sacrificed to the goddess Furina, the bitch who hounded animals' souls. The same day, July 25, became Saint Christopher's day in Roman Christianity.

In the Classical world the underworld rulers, Erinyes, Ceres, Hades, and Hecate, were either dogs or dog-headed—derived not only from Africa but from Central Asia, where the dead were said to be herded by pairs of demon dogs in myths that spread across Europe from Indo-European roots, generating appeasement rites in burial ceremonies as far west as Ireland. As in India the roaring winter solstice winds were "child-eaters," hounds of the gods of the dead, Saturn and Woden.

Their travel and death connections, their feral terror in winter packs, their arrival with "dog-headed cannibals" invading from Central Asia, mark the nadir of the dog in the Western World, everywhere the most liminal of animals because of the tension between its civilized associations and its degraded state in the wild. The dog was "the archetypal social pariah whose wildness is as much a result of his exile as his exile of his wildness."[4] On the one hand the dog is "man's best friend," valorized as the companion of wandering ascetics, redeemable, welcomer of the dawn, mediator to the other world, a Neolithic deity, just as the wilderness itself was a place of refuge and contemplation. On the other hand the dog is the alien monster and hypocrite, fallen and hateful, the most corrupt of animals. This negative side undoubtedly has its own natural history. Wild animals, after all, are not "pretenders" to civilized status. Consider how the dog may offend human standards: their howling heralds strangers, winter, and death at the town's edge; they run as a dangerous pack, like berserk warriors, randomly biting when diseased, snapping as ill-natured bitches or the tangled bluster of males around the female in heat; they scavenge and dig up human dead and carry skulls and other bones about, congregating on battlefields and places of contagion and epidemic, even eating each other's bodies; they lick their genitals and anus, urinate, defecate, and mate without shame, and attempt to mount people; they kill sheep, cats, and other domestic animals; and they display a tyrannical social system, becoming wild and deadly in the dog days of August.

Dogs seems to go over from ambiguity to duality, their gross bestiality representing all that is opposed to humanity and civilization.[5] In modern affluent societies, dogs fare well; but in times and places that far outnumber prosperity, they have been the most hated of animals. It is far worse to be called a "dog" in this world than a "pig" or a "skunk." As the first of the domesticated animals, perhaps the dog initially represented one side of a divided universe among the early farming peoples, whose hunter/gatherer ancestors had seen the world less in terms of opposition than complementarity. But with sedentism the known world became smaller, bringing the outside closer. This "presence of an absence" of chaos became an obsession to be confronted by a victory.[6] There is evidence that duality and the problem of evil became more important in the cosmologies of agricultural and village life, as opposed to that of nomadic peoples.

WE CAN SUMMARIZE the forms of ambiguity which give marginal animals status in the ordered world of named things. First are the simple Boundary Forms, such as the fox, living at the edges of two habitats. This category helps explain our anathema toward frogs and toads, who are not only aquatic and terrestrial but shift in shape from swimming larvae to hopping tetrapods, and toward all the auroral and crepuscular creatures living at the dusky edge of dawn and evening: bats, cats, nightjars, and owls.

Then there are the Amorphous Forms, such as earthworms, leeches, and the larvae of many insects, which lack anatomical traits of the kind that define membership among larger animals. Our special horror of the shapeless state is evident in tales and films of terror in which evil oozes, constantly changing, taking arbitrary forms as it attacks its victims.

The Shape-Shifters, like caterpillars or frog tadpoles, are plural or complex beings, latent and metamorphic. In European tradition the werewolf is perhaps the most familiar legendary nether beast between the wolf and human. In South America there are were-jaguars, and other countries have their were-forms. Characteristic features cannot be relied upon, as the individual may turn into something else. Since similarities in structure, as in the skulls of all mammals, seem constantly to threaten the distinctions made in classification, perhaps were-forms are a mythic way of explaining the overlapping characteristics of different but related species.

The Dislocated Forms are, like the unclean forms in the Jewish Bible, incompatible with their environments. When the Hounds of Hell run in the streets, there is clearly a discontinuity comparable to the mixing of bodily parts from two species. Animals, especially birds, have long been observed as harbingers of season and weather, so that the idea of the omen or the monstrum as a sign of radical disorder in the world is a traditional mythic sign.

The Post-Imprint Forms, animal species seen for the first time after one has become an adult, seem vaguely familiar but deformed. Manatees, dugongs, armadillos, or pangolins, encountered by the individual after the closure of the taxonomic mindset (at about age eleven), have no affiliation. So it was that European travelers in Australia named the animals and the birds as if they were bizarre combinations, such as the duck-billed platypus, the wren-tits, and the whole panoply of marsupial "cats," "wolves," "rats," and so on.

Farrago Forms are a special case of the Post-Imprint Forms. They are strictly human creations and appear to be made of mixtures of the parts of various animals and therefore confuse the observer. Is the dragon a reptile or a mammal? These, says E. H. Gombrich, exasperate and outrage our sense of order and meaning with their surprising, uncodable deformation.[7] While some of them are merely comic cartouches, sportive grotesques, or other decorative flourishes, Gombrich observes that, in most of the human record, they are supernatural and often protective. The recent secular freedom of the artist to play with such drolleries biases our modern view of these hybrid creations of other cultures whose function is not to release our puzzlement as humor but to lend imagery to diverse concepts and educate our acceptance of ambiguity as natural. These images do not speak easily to the literary mind. The literate man, says Edmund Carpenter, insists that ambiguity is intolerable, that we choose between them when visual contradictions are presented.[8]

THE CLOSE ASSOCIATION of animals and the names we give them produces a kind of carryover in language—in semantic realms. Ambiguous animals are like puns: they have identity in two ways that we normally consider exclusive. While puns are lighthearted wordplay in much the way a monkey is a pun on a man, the monkey's liminality can also seem threatening to our sense of ourselves, just as puns can become obscenity. In fact, obscenity dotes on an anatomical basis, just as species differences require anatomical comparisons. It has been widely observed among anthropologists that the passages of the body between inside and outside, being neither one nor the other, are a focus of prohibitions, as are the organs associated with them, their functions, and the substances they produce. Indeed, such terms as "swine," "cock," "cunt," "pussy," and "ass" combine such body parts with the names of ambiguous animals—the pig, rooster, cony (or rabbit), cat, and ass—in which their association with the human household is contradicted by their edibility, wildness, and servitude. Anxiety makes them tabooed and hence their linguistic use in taboo-breaking in verbal abuse. Obscenity is a violation of sacredness; being intermediary and double in sense, such terms and animals violate holy order or mythically mediate contradictions. Body openings, their exudates, and the borderline animals are sacred in some cultures, degraded in others. It is not that the semantic sets of terms and the organic entities to which they refer are intrinsically good or bad, but that they are stimulating because they verge categorically.

The Bible classifies animals, according to Genesis, as occupants of land with legs, air with wings, or water with fins, each known by the appendages suitable for its habitat according to the creation story. But shrimps, crabs, and shellfish, though creatures of the sea, lack fins; snakes, though land animals, lack legs. These forms, anatomically defective, are disordered or polluted beings and thus are prohibited as food, perhaps a reminder against the failure to ignore biblical injunctions of all kinds. The Bible also prohibits eating hare, hyrax, camel, and pig on the grounds that the definition of edibility must include cloven hoof and cud-chewing. One suspects that the abominated foods had less to do with diet and more with distinctions between themselves and the camel- or pig-eating tribes around them. The Hebrews' chosen incongruities in animal taxonomy were a symbol of their discord with other peoples. Leglessness in snakes could likewise be seen as a kind of defilement by categorical standards, associating the stresses of perceptual disorder with enemies who worshiped snakes in their temples. Snakes embody the corruption of sex as they do the semantics of identity. Judaism and Christianity found other zoological, categorical equivalents of evil, such as the twilight forms (owls and bats in the dusk), those between earth and water (toads and other amphibians at the stream's edge), and those who undergo transformation (larvae, nymphs, and moulting forms). They serve as tension-inducing signs that can be appropriated for instruction, understanding, or tainting by association. Owls are the demonic equivocator of day and night; the larvae of insects and amphibians are the deceivers of appearance. The image or call of each can be appropriated to signify disarray. Passage makers—the fetus at birth, the individual between health and sickness, novitiates facing initiation, the shaman between worlds, the spirit at death—may be imagined as riding an animal intermediary, which is held in especially high or low esteem.

Animals shift habitats and change anatomically during their lives. Because of growth and development, all humans and other animals are in some degree "marginal" at times. But to say that an adolescent is "in-between" is difficult to convey or ritualize without some more sensible expression. Animals who live at the margins between farmlands and wilderness, such as deer, or between the household and the garden, like monkeys or dogs, at the water's edge, like the beaver, above and below ground, like the bats who compound their status by flying and having hair and living at the dusky edge of day and night, serve these purposes. We ourselves are descended from forest-edge "brush apes," provoking reflection on finding ourselves in the ecotone between forest and grassland—the savanna. Ecologically, "edge effect" is rich not only because it includes passage by the creatures of adjacent habitats, but also its own fauna, and is therefore perceptually enriching as well.

Our human transitions are penetrations into new states, the way dragonfly nymphs climb up the reeds out of a pond. In a thousand other ways, animals emerge from nonbeing, go through life stages by breaking from an egg, coming up through tunnels, leaving nests in hollow trees, or passing through the skin of a host. Despite their fixed appearance, each is, in this sense, two or more things in one, models of orderly transformation, masters of something both mundane and astonishing, embodiments of acceptance in the face of alteration and uncertainty. Like them we are each immersed in a personal becoming, evident in the physical aging of our bodies, and must cross "a shore" or "a river" into death, an underworld grave or a heavenly home.

Like the inhabitants of the savanna—where forest and grassland intergrade—all border forms stitch antitheses with their tracks and trails. We are the seamstress who makes sense of the fox at the fringe of the woods. Edge animals ourselves, we are our own most difficult task. As apes, dog-heads, yetis, fallen angels, diverse races, and emergent androids lurk in the margins of our identity, our species is beset with a problem of the categorical imagination.

PART III

Identity

The question at the center of "myself" and "ourself" is "Who?"
Our identity is shaped not only by our differences from the
Others but in the alchemy by which assimilating them informs
the genesis of self. They are both the reciprocal and the
constituents of our consciousness. Their own belonging is like an
anatomy of membership: the image for us of our own social
union.

5 | The Self as Menagerie

We are Mosaics, forgive me, I think this wiser
  than the emulation of Zeus, or the harsh-axed
  Vikings in their Valhalla.
I have found animals in me when I stroll in the forest.
  I hesitate before a large dragonfly, I step
  like a cat in the night, I have felt something
    lift along my neck
    when a wolf howls . . .

LOREN EISELEY

MY FATHER USED to sing into the mirror as he shaved in the morning when I was small:

> Oh, I went to the Animal Fair,
>   The birds and the beasts were there,
> The old baboon, by the light of the moon,
>   Was combing her auburn hair.
> Oh, the monkey, he got drunk,

And sat on the elephant's trunk.
The elephant sneezed, and fell to his knees,
And that was the end of the monk . . .

I was enthralled by the bizarre figures of a coiffured baboon, a drunk monkey, and a sneezing elephant that were part of an animal fair. I felt the ambience of that fair, as if it were a playground or a zoo, the scratch of the comb and the warmth of "auburn," the comic dizziness of a boy's idea of drunkenness, and the momentary loss of balance when you sneeze. Much later I came to think of such images as mocking our certainty about our own identity, but I never completely lost the sense of having joined that circus of my father's song. Like the owl and the pussycat in their pea-green boat, birds, cats, people, and animals all play out the contradictions we feel in their human-like animality and our animal-like humanity. Each species seems whimsical, as if it were an increment of our personal, multiple self. Of each species we can say, "I am not that—and yet, just in this one respect, it is like a part of me," and so on, as though with every "I am not that one" we keep some bit of them. We take in the animal, disgorge part of it, discover who we are and are not.

Distanced for centuries from the wild world, we now speak of our identity as choices of "lifestyle" and tastes, distinctions of vocation, race, political loyalty, formal education, religious affiliation, national geography, and chosen memberships. By disdaining the beast in us, we grow away from the world instead of into it. Yet the ambiguity of kinship, likeness with a difference, presents itself with great force, inescapably revealed by bodily analogies.

Some societies still affirm such similarities, formally acknowledging that animals play a physical role in the sense of self. The Ajamaat Diola of southern Senegal, for example, believe that persons produce an animal from their feces, a kind of "double" that runs off and seeks shelter in the house of a female relative, as though a "birth" has taken place. This "fecal animal" might be a monkey, antelope, snake, leopard, or some other mammal, though it does not exactly duplicate the everyday wild form. It is bigger than normal, with a stubby tail, acts strangely, and lives unusually close to people, as though somewhat domestic. The person and his animal share the same soul. If the fecal animal, or *siwuum*, is injured or sick the human feels the effects. When the individual ceases to see his *siwuum* it is thought to have died and is replaced in a ceremony in which the person mimics copulation with a goat, thus reaffirming a mythic kinship with the animals based on a marriage with them at the beginning of time.[1]

Animals born of the body, such as the *siwuum*, suggest a preexisting presence there as aspects of the self. A modern therapist, Eligio Stephen Gallegos, has developed a form of meditation in which a conversation takes place with one's inner animals. The idea came to him when he was walking one day among the carved Tlingit totem poles in the park at Sitka, Alaska. Familiar with the Jungian concept of seven energy centers of the spine from the fundament to the top of the head and the parallel Indian theory of Tantric chakras, Gallegos was struck by the thought that the totem pole was a physical presentation of the mythic animal "speakers" positioned along a vertical axis in both concepts. To the American Indians the figures on the poles are those about whom traditional stories are told. They are not only components of a visible structure but players in a heard tale. Gallegos speculated that the animals associated with each chakra might correspond to Jung's symbolic animals of each neural center and to the totem-pole imagery. From this he developed a form of guided meditation in which the troubled subject visualizes entry into his own body along his central axis and imagines an encounter and conversation with an animal occupant of one or more of his own energy centers. In successive interviews, he invites the animal from each center to come forth for a conciliatory exchange. The creature is addressed in a friendly way with an offer of help if needed. Its concerns refer to its own domain—spirit at the top of the head, intellect behind the eyes, communication in the throat, compassion in the heart, power in the solar plexus, emotion in the gut, and connectedness at the base of the spine. The animal who "comes out" reveals some aspect of the center to which it belongs. The condition of the center is indicated by the particular kind of animal, its condition, and its circumstances. It may be robust and bold or small and shy. It may be caged, injured, or malnourished. Sometimes the strength and stability of one chakra is at the expense of the others. The animal is invited to tell of its concerns. The meditation continues in successive sessions, the patient addressing each center in turn under the therapist's guidance. When all the chakra animals have been reached they are brought together in a communal council.

As his technique developed, Gallegos went on to foster conversations between his patients and the animals of the senses: eyes, ears, smell, taste, and touch. At first testing it on himself with the aid of a friend, he says: "I closed my eyes and relaxed and following her guidance went to meet these animals.

In each of my eyes there was an eagle and these two eagles flew in tandem at high speeds. My ears were a single rabbit, sitting quietly, listening to all that went on around him. My animal for the sense of smell was an elephant probing the world around with his trunk."[2] Animals of the chakras, he decided, were modes of action and power; those of the senses were modes of reception. This meditation on animal speakers of the centers and organs of the body reveals wounds or issues which the person has not adequately faced, perhaps cannot solely on his own, but can do so when supported by an "animal" intermediary. As the dialogues continue the animals improve in health, change in form, and reconcile their differences: "The individual feels a growing harmony and centeredness in his or her own life and the inner support of the chakra animals is deeply and richly felt."

Gallegos' concept may seem at first to smell of New Age fantasies, but I see in it an approach to a firm foundation in the human perception of the self. There is no way of articulating our inmost circumstances—lively and vital—without reference to something concrete. Even when we understand our deepest thoughts as programs and circuits in memory banks, we refer to the external world of machines. But the animal is like an ambassador in a way that no clever machine can ever be, though they may correspond in some extended metaphor to the ionic and mineral aspect of the self. The animals—and perhaps the plants—within us are like the beings of a larger and older reality. They exist within us in much the same sense that our parents and our ancestors are in us, not as ghosts but as shared form, a continuum of which we are only the present expression.

DREAMT ANIMALS, FOR example, are about natural events otherwise unembodied or lacking the visible or audible features that we associate with coherence and intent, vague potentialities such as earthquakes, floods, and storms, or about the human drama in which the identities must remain hidden because it is too painful to face them.[3] Dreams, in the words of James Hillman, are "congruences in sets of associations within and across domains." In one domain are animals in children's dreams connecting the dreamer to the domain of society and family. The animals may be people or situations in disguise, masked participants in the daily lives of the dreamer. Among small children a self does not exist in animal dreams. David Foulkes observes: "Stories which name an animal as a character at the beginning may end by assuming that the child himself is the animal. . . . The essential continuity seems to lie in the action rather than in the nature of the character."

The child's absence from its own dream is part of normal unself-consciousness, without the "self," "you," or "they." The child is said sometimes to be frightened by such dreams but not frightened in them.[4] What can we make of a dream with animals, sometimes frightening, in which there is no self to participate or be a victim?

These dreamt animals may be, very early in life, the dispersed elements of the unknown self—the body's sounds, contractions, upheavals, secretions—and then also disguises of familiar people in stressful circumstances of ordinary experience. The evidence that the animals refer to people is from studies of the dreamers' socialization. At ages three to five, 61 percent of children's dreams contain animals, slightly more in boys than girls; at seven to nine, 36 percent; at twelve to fourteen, 20 percent; at sixteen, 9 percent; and beyond that they stabilize at about 7 percent.[5] As the child gets older, animal dreams diminish and a self emerges more frequently, as do familiar faces. By age thirteen, animal dreams are different in quality, more oriented to fantasy than behaviorally expressive. It may be concluded that animals are an eclipsed content having to do with something other than themselves—for instance, problems that children have with other people.

Dreaming is categorical work; the animals are a cast of surrogates and vehicles for riding out a problem. They substitute for actual humans in dreams, especially parents or other relatives, who are too necessary to the dreamer's well-being to accept as ambivalent or threatening. Perhaps the dream scenarios bring an issue to a crisis that can then be resolved. In this way the child works on relationships to playmates, siblings, parents, strangers, other kinfolk, and the endless procession of semifamiliar people as an internal narrative. Anxiety is thwarted in order to cope with uncertainties about others. That the animal disguise should screen the true situation in a friendly way seems odd in a society like ours with its mythology of dangerous wild beasts as the antithesis of human security. But the child's "work" does not require knowledge of the behavior of wild animals, only the comforting sense of order in categories themselves, already experienced in the naming of the animals in the development of speech. Normally the animals in dreams gradually disappear and the masks drop away. Children who lag behaviorally and socially, however, continue to have a high frequency of animal dreams as though stuck on social barriers. Throughout our lives animals in dreams may continue to signify unresolved concerns, intolerable truth, or interpersonal uncertainty.[6] They are not a random choice of symbolic vehicles of the unconscious mind but a continuation of the maturing processes of humankind. They are nurturant among small children because animals are already synonymous with the mind's drive to find order and the heart's desire to affirm given reality.

A "FAUNA," NO less internal, animates the fairy tale. Like the animal-masked people of dreams, here too is a theater of representations. The fairy tale dramatizes intrinsic childhood worries which the youthful listener unconsciously interprets as his own story and his own inner self. Bruno Bettelheim defines these tales of death, decapitation, monsters, and transformation as stories with happy endings, profoundly consoling despite their harsh details. Their message is that special skills, often the powers represented by different animal species, will come to the rescue, solve the problem, save the day, and guarantee a happy lifetime if we will but trust them.

Bettelheim believes the problems to be universal, having to do with protection from malicious relatives, the uncertain intentions of strangers, one's verbal or physical limitations such as the skills of speech or strength, the bodily changes and functions associated with growth, frightening dreams, fear of the dark, oedipal feelings, sibling rivalry, jealousy and envy, and the child's sense of limited intelligence, information, or techniques which adults already possess. The skills are often those that come with ordinary maturing: strength, coordination, size, understanding, a place in the world. Every story is a magic prophecy of personal transcendence, like a promise to the listener, who typically ruminates on its contents and then selects and fashions its meaning according to need. Various humans in fairy tales are allegorical with respect to the social milieu; the animals tend to represent aspects of mind and body, the inborn capacities that will unfold with age. The child is invited to place its faith in its own organic substratum and timetable. The quest in the fairy tale is for abilities foreshadowed in the organic world, personified as animals.[7] It is as if the story says to the child: "As sure as the tadpole will become a frog, you will grow into a strong, attractive, keen adult, provided you have the faith of the princess who marries the enchanted frog"—as though there were in us, as in life, rabbits, toads, doves, wolves, and eagles who, besides being themselves, are incarnations of kinesthetic and nebulous realities, both threatening and saving. "Both dangerous and helpful animals," says Bettelheim, "stand for our animal nature."[8] The different animals tend to be all-devouring or all-helpful, like feelings of love and hate or happiness and sadness. They portend threats to both a growing integration and, alternatively, a nascent maturity, sense of place, kinship to fellow-creatures, healing, the integration of one's personality, and the natural unfolding of one's own role in the world.

According to Bettelheim, fairy tales often help the listener sort out one's human status from the more general animal one. In "Hans, My Hedgehog," for instance, a half-hedgehog boy proves his humanity when treated kindly in bed by a princess. Likewise, the enchanted prince in "The Frog King" recovers his royal human status because the princess, who has been forced to marry him, endures his gross nature in bed and her own sexual experience grows from disgust to happiness. In another tale of this sort, "East of the Sun and West of the Moon," the groom, who looks like a bear, becomes a prince when the princess endures him. A slightly different interpretation may be made of "Beauty and the Beast," in which the monstrous groom, part human and part beast, is healed of the animal/mind duality by the unifying effects of the woman's patience and true love. Goldilocks, in "The Three Bears," tries out three familial roles as baby, mother, and father—then flees because she intruded on her own natural developmental schedule, which would in time resolve her childish confusion about being the baby, the mother, or the father. In the story of "Queen Bee," a simple and innocent brother wins the hand of a princess by the help of the ants, ducks, and bees he had befriended during his wandering, each representing a resource within himself. In "Brother and Sister" she dissuades him from drinking at streams which would have made him a were-tiger or werewolf. He drinks instead from waters which transform him into a deer and is captured by the king's hunters, a high achievement. A young man in "Two Brothers" spares the lives of the hare, fox, wolf, bear, and lion, who later work together for him at tasks which are analogous to the integration of his own personality. The "stupid" son in "Three Languages" can learn only "what the birds sing, the dogs bark, and the frogs croak," in contrast to his academically educated brothers. His exasperated father drives him from home for failing after three attempts at schooling. When a village is terrified by a pack of wild dogs, the boy is able to mediate peace between the animals and the people because, understanding the dogs' speech, he can explain their anger. The croaking of frogs later informs him that he should go to Rome, where he eventually is chosen pope, and the murmuring of doves gives him the words he must say at his first mass. In another of the "three sons" genre, the kind and honest son befriends a toad and later has his wishes granted by it, thus surpassing his mean siblings.[9]

Dogs are earthy, connected with practical necessity and friendship, matters of the ego. Frogs are aquatic, associated with evolution, transformation, and the unreflective magic of the id. The earthbound frogs, lizards, and snakes embody basic biological processes, while the birds connect with matters of mind and soul, with air and spirit, with high goals and the aura of the superego. The bird's purity and flight lend themselves to esthetic, intellectual, and religious ideas. The keen, watchful raven seems to be consciousness itself, the laconic owl is wisdom, the murmuring dove is love. Together these three birds represent aspects of the mature self in the story of "Snow White."

Together the animals represent the wisdom of the body and the necessary faith in our organic being in spite of its peculiar manifestations and temporary limitations. For children the stories are consoling; for adults they give meaning to their past and inspire their sense of purpose as storytellers.

ANOTHER ASPECT OF personal identity associated with animals is practiced by the Nuba of Africa. These part-time horticulturists were hunter/gatherers until recently and remain keen observers.[10] The Nuba paint natural forms on their bodies, mostly animals. Different kinds are not prescribed, but the design features follow rules. There are conventions of style, form, and color that define species, so that the animals are a kind of code. Paintings vary as one might expect in "an artistic tradition that is chiefly motivated by aesthetic and decorative factors."[11] The painting is purely body-enhancing, without connection to cosmology, totemism, particular powers, or special relationship to the animal depicted. The major factor in choosing which species to represent is whether the surface features of the animal appear esthetically appealing and worthy of imitation. Such designs are therefore highly individualized, reflecting the taste and skill of the painter and painted and thereby enhancing their presence in a society that values esthetic competence. The plastic surface of the human body is followed—as when the biceps become the bulge representing the tortoise's shell. In its most abstract form, only spare features such as eye stripes for antelope or leg stripes for jackal are employed, congruent with human anatomy, so that the stripes are, placed, respectively, over the eyes or on the legs. Sometimes whole body designs are matched to the human body. Although the examples of the Ajamaat Diola and the Nuba are very different—one profoundly spiritual, the other entirely esthetic—the animal abstractions arise in a creative insight, are employed in the distinctions made in human individuality, and commit the individuals throughout their lives to a studious, esthetic enjoyment and playful observation of the animals themselves.

Adornment of the human body is very widespread—not only stylizing animals but employing their bodies. Feathers, shells, teeth, skins, claws, bones, and other parts of small organisms are widely worn throughout the world, following fashion and custom but with individual touches. Like all cosmetics, coiffures, and jewelry, such embellishment "makes a statement" rather than simply denoting the individual, perhaps one not fully understood by the wearer who may never examine the impulse for animal signatures. It makes visible an equilibrium between the given and the made, the animal and the human, conformity and creativity, in which cultural practices are given a kind of natural vindication. From Papuan elders, wearing the arm bones of birds through their noses, to babies photographed on bear-skin rugs, to dapper men flaunting lizard-skin shoes, animals provide the distinctions by which we conceive and announce ourselves.

Along with the fur coat and gloves, the twentieth-century person adds "accessories" such as pearl earrings, animal brooches, pins, bracelets, barrettes, and wallets or belts from bone, ivory, or metal, and jewels in the form of an animal. Such "access" is an appeal to a larger world of living things even if it only appeals to an ancient and primal root of esthetics itself, though more often bearing half-forgotten associations and qualities. Tribal distinctions between the wearers of ermine and those of hare are like the modern distinction between rare and common furs, insignia of rarity and class, signs of rank, privilege, and success. Within each group the effigies or parts of animals are given personal touches in the pattern of feathers in the hair, beads on a string, and perfumes mixed from the musk of animals as signatures of our own rhythm and style, marking us as uniquely associated and differentiated by the way we compound the detached elements of a fauna. The self emerges from this extraordinary correspondence of inner and outer worlds, linking us to those unlike us in shared respect for our differences, reflected and magnified as species.

Perhaps putting on the mammalian fur and lizard boot not only defines us according to common models but truly incorporates us, a physical connection to the animals. Wearing the bits of animals as signs is a mnemonic, a recollection that they are already part of us. The human mind depends on a brain composed of layers of an evolutionary past which may recall itself in unknown ways. Perhaps that layering is why we "experience the real chaos of the brain,"[12] which is composed of a newer, outermost, primate neocortex that creates the visual consciousness of self, the middle mammalian stratum shared with all mammals, the heritage of a kind of smelly, antediluvian antiquity, and a still deeper core of tactile archaism of our forgotten lives as fish and reptiles. No wonder our thought travels among animal images, as the human brain represents a fair share of the animal kingdom.

"Internal animals" have appeared again and again in the foregoing: in dreams, in the spinal cord and organs, as the spokesmen of chakras or energy centers, in the emotions embodied in creatures in fairy tales, as the art we exercise in representing ourselves to the eye and nose. We are strangely composed of animals who flesh out our being, a diverse zoology of the self. They are more like indigenous inhabitants than casual symbols, more essential than decorative. To be conscious of our feelings does not require an image, but to name them does, and the first names of things are animals. The forms of liveliness of different species seem to correspond and be reflected in an equally rich inner experience. When they became objects in the external world—by being named—their existence in us was duplicated.

Identity precedes the usual social markers, skills, memberships, family name, age, and gender. It tickles me that each animal lives along the spine, where body posture and movement start, and I tend to imitate or to feel its look, in my neck the stretch of the giraffe, the tug on the arms of the swinging ape, the shade as the cool immersion of the diving turtle. Loren Eiseley, while watching the pigeons swirl up from the city streets at dawn, observed: "the muscles of my hands were already making little premonitory lunges."[13] Or as D. H. Lawrence puts it: "Nonhuman Nature is the outward and visible expression of the mystery which confronts us when we look into the depths of our own being."[14]

6 | Aping the Others

If neurosis is sin, and not disease, then the only thing which can "cure" it is a world-view, some kind of affirmative collective ideology in which the person can perform the living drama of his acceptance as a creature.

ERNEST BECKER

PLAY HAS MANY meanings: the frolic of children, performance on a stage, games, the whirl of cosmic energy, our use of a musical instrument, or easing fish on a line. A thread of common meaning among them is enactment—the joining of performer with a counterplayer according to rules in a rhythm of necessity and chance. Our play, whatever it may be, feels like an extension of some larger reality. It links a sequence of acts, a range of emotion or personality, or accidents and chance. The drama, the forces of nature, the games of childhood, the musician's and even the fisherman's experience, share the idea of a whole with many parts. In it there is a special place for the imitation of animals, which, however capricious, lays the groundwork for understanding ourselves as being: as actors.

Since we are "flesh and blood," it is not surprising that other players, the counterplayers, are lively, too, even when they are as impersonal as the gravity against whom we swing or teeter; it pulls. We are born prepared to assume that opposing forces are "others," and to feel the give and tug of a fish or the musical instrument quicken in our hands as if it were a conscious counterplayer.

Games with rules are among the earliest forms of patterned play, helping us know that whatever the play there are constraints. One of the most common features of children's games is the naming of games and players after familiar animals, which the children emulate in a kind of shorthand. There is something about the perception of animals which tells the child that, for a few minutes or forever, life is a game.

Despite our modernity, we are embedded in a venatic, evolutionary past with its foraging and the hunting of game. The "game" animals are those subject to the chase. In them we can see that to "be game" is to be willing to risk—to accept the possibility of losing in a play of forces with uncertain outcome. The success of the chase is not entirely a matter of chance, however, but also a question of each "side" understanding the ways of the Others. The lives of animals are ruled by what, to humans, appears to be a code of interaction with one another and with the landscapes in which they live. The "rules" of the game animals include those intractable forces and unbreakable bonds of their own natures and the natural world. Yet because the world is complex and creative, the outcome of the chase cannot be foreseen. The venatic faith is that knowledge improves one's chances, but also that losing is eventually inevitable and necessary, as the whole is greater than any of its players who are only temporary participants in something eternal. Like playing the cosmic and ecological game—participating in the nature of the game animals—playing human games is an access to an orderly world.

Beginning in infancy with gruff nuzzling and rowdy bearhugs by a snorting parent, tumble-bugging of somersaults and leapfrog, the child goes on to making silhouettes, or the finger games of "ten little squirrels" and "itsy-bitsy spider." There are crab walks, duck waddles, and song games like "farmer in the dell" and "old buzzard." There are "piggyback" rides and "chicken fights" and organized games such as "sharks" and "pom-pom-pullaway." "Horns" is played like "Simple Simon," along with "snail" and "sardine." Every country has its own children's games based on the imitation of animals: "hare and hounds" in England, "fighting cocks" in the Philippines, "cat that wanders by night" in China, "gecko gecko" in Australia, "badger the sun" in Japan, "wild horses" in the African Sahara, "lynx and rabbit" in boreal America, "follow the reindeer" in Lapland, and "fox is the warner" in Southern Europe.[1]

In snow country there are many forms of "fox and geese," in which two

large, concentric paths are tramped out in the snow and connected with ra-
dial lines, creating a mandala pattern upon whose paths the players must
stay. The game is a form of tag in which a single "fox" chases any "goose."
When caught the tagged goose becomes the fox and the fox becomes another goose. The metaphor of assimilation and reconstruction suggests a
balance of creation and destruction in which the fox seeks to catch (eat) and
become the goose, and the goose, being eaten, rises to its foxdom in an endless cycle. There is no escape from the closed universe of the circles, no fixed
duration, and the lone fox and many geese testify to the rule of numbers that
regulates the lines of flow in the world of life.

The representation of a universe in small may not be the game's only
function in the education of the child. Like the game itself we are individually composite. Each of us is an ocean of motives, emotions, and ideas. Each
animal in play reveals a certain trait or feeling exhibited in its behavior. Each
kind of animal gives concrete representation to an ephemeral and intangible
element of the human self such as assertion, intimidation, affection, doubt,
determination, kindness, anger, hope, irritation, yearning, wisdom, cunning, anticipation, fear, and initiative. Only when these feelings are discovered outside the self and then performed can such intense but elusive
"things" be made one's own.

The game, recounted later as a story, is composed of enactments of feelings in the name of the animal which gives them substance and thus allows
them to be affirmed in oneself—not taken in as though the feelings do not
already exist in us, but given shape. Playing the fox does not create my foxi-
ness. It helps me to discern it. It is endorsed by the social agreement of the
players and the natural being which it reflects. Of course, animals in nature
are also complex beings, not one-dimensional shadows of feeling, but the reality of their complex natural history comes later in our lives, after childhood, belonging to the adult's world of ambiguity, not the child's of definition.

The lively world of our emotions, fears, and responses is like a great forest
with its fauna. We experience those feelings as though they were wild animals bolting through the foliage of our thick being, timidly peering out in
alarm or slyly slinking and cunningly stalking, linking us and our unknown
selves, as though they were at home in an impenetrable wilderness, bearing
the gift of themselves as mitigators of our inchoateness.

Why must we see these aspects of ourselves in animals instead of other
people? Certainly the whole range of feelings is visible in other humans, but
in them the enactment is diffuse or sometimes concealed or deliberately dis-

torted. People, even our mothers, are contradictory models, plural, uncertain, undependably changeable. Being human is the problem to which aping the animals in play is a partial solution in the shared acceptance of brief enactments. Only as they get around to learning that human society has roles that we must play do children begin to play house or office or cowboy or nurse. Before that we apparently must perceive what we feel in a brief mime of the animal's behavior accompanied by the implicit declaration that I am "now" the fox. It is not the roles in life that are being learned but the more fundamental lesson that there are roles, and these, because we are a poetic species, are best grasped indirectly.

The miming of animals in play does not animalize the child, since it does not teach us to live in holes, run on all fours, or catch mice in our mouths. Nor does it humanize the fox, as even the child knows in its heart that the whole matter is "just a story." It is important as a joint adventure by children, part of the biology and culture of childhood, in "capturing the Other" in order to constitute a self.[2]

ANIMAL GAMES ENRICH our inner life. We watch the animals as cues to our own feelings, even when we are not engaged in formal play. Transforming them into an inner plenitude encourages our curiosity, watchfulness, wonder, and admiration. The child's spontaneous interest in animals arises in encounters with easily identifiable, external expressions of our impulses to beg, present, invite, appease, conciliate, displace, facilitate, appeal, threaten, defend, search, and avoid. Once these actions have been associated with specific animals, they may be represented by the names of the animals or by horns, antlers, feathers, beaks, teeth, face marks, rump patches, wings, hooves, fins, songs, and calls.

Accounts in the media of yesterday's athletic events remind us that games translate easily into narrative. Dramatic play-by-play accounts heighten the event to a degree that the purely visual media lose—an excitement that even the original event may not have had for those who participated. Against the grain of our picture-book culture, stories without pictures prosper. Play is drama before it is told, or a story coming into being.

The kitten chasing the ball prefigures something central to the adult life of cats, something of which the kitten has no actual experience. Chase games under the auspices of predation are a "capture," each of the other. Separation merges into identity in the assertion that I am at this moment the fox, in mimesis of cunning pursuit. I am "it." I pursue and take the prey; the goose now becomes the fox and I become the goose. In action, in my foxing, identification is realized; merely latent in the noun, it is truly realized in the fox as verb. Since I am finally not a fox, the assertion that "I am the fox" is helpful to my selfhood only if I then become something else. Changing roles safeguards against too close a unity. The same is true of "being" the goose. I am that honking, panicked being, in flight, only until replaced by the fox.

At the game's end, both vanish, assimilated in the abyssal self, each having become some part of what I am becoming. In games the child is engaged in a thrilling and somewhat frightening shift among identities, from which he or she may withdraw at any time, like swimming underwater. It is this flexible back and forth between self and asserted identity that gives confidence in a world in which people and things move and change, the turbulent, mercurial nature of being human. As anthropologist James Fernandez puts it: "The reality for all of us is that we live with a variety of categories, converging upon us, to which we must relate and in which we must find some identification. . . . We are all of us in constant passage, in reality or in imagination, between categories."[3]

The declaration that "I am a fox" or that "you are a goose" is the predication of an animal on a pronoun which is more or less amorphous and helps to teach the art of metaphor. Just as I may be foxy in strategy I can be a tree in my rootedness or a rock in stolidity. Such multiple ritual assertions are a kaleidoscope of successive, shared domains that define me ever more precisely. My identity is not simply human as opposed to animal. It is a series of nested categories. These shifts in domain are like the alternation of solo instruments in a piece of music. Paradoxically, the difference between me and the fox begins with a claim of unity. But that claim really does allude to common ground which remains even after the separation—when I become a goose and, finally, when I become a whole zoo. I am foxy in part because the fox and I, both mammals, share a common heritage, and because our ecology converges in our hunter's kenning. Children do not understand the statement "I am a fox" as a metaphor. One does not explain it to them. The pretending of play is an act of mimicry, not one of theoretical comprehension.

Like the other primates we humans are obsessed with the relentless question of status, the pronoun inquiry: who are you, we, they, and I myself? An adult might say that I am a lion, Clara is a parrot, you are a fox, we are sheep, and they are snakes. We know that physical resemblance is not intended: Clara does not look at all like a bird. This shifting of Clara toward the parrot domain is a leap in the work of language and image.[4]

Playfully stalking each other, the lion cubs participate in an unspoken pretense that "I am an adult lion hunting gemsbok" and "I am a gemsbok." Cubs and adult lions seem to know that the serious hunt and the gemsboks are a fiction. The child's assertion that "I am a lion" in play is an untruth with perhaps some of the same function of the lion's play—as practice—with the additional purpose of laying the foundation for a succession of ritual avowals of identity that coalesce in an inner nature.

Ritual acts are mimic pretensions based on the suspension of disbelief, performed "as if." They are enactments of metaphoric assertions. Childhood play is predication in anticipation of any conscious awareness of such metaphors, the first step in an individual "ontogeny of ritualization."[5] Mimic play at being animals is an inherent activity, keyed to the calendar of individual development, essential as a precursor to the adult capacity to engage in rites and ceremonies. Drama as a stage play emerges historically from Greek and Oriental ritual performance. In an evolutionary perspective, ritual has its roots in the imitative play of immature human mammals, goes on to the declared pretense of interspecies impersonation in child's play, and ultimately makes the play the thing. That both ritual and the earliest drama are sprinkled with animal masks suggests a link in this sequence of connections, a sequence in which we never quite lose the original animal idiom. Nor should it be lost, because it infuses our all-too-serious roles with humor and the mellowing ingredient of ironic participation in life as a fiction. It reassures us that roles are natural to us and provides the infinite complexity of the multitude of species and the inexhaustible subtleties of each— stereotyped and one-dimensional at first but inexhaustible as we study the animals all our lives.

Personal identity is not so much a matter of disentangling the self or "the human" from nature as it is a farrago of selected correspondences in which aspects of the self are projected into the dense, external world where they are discovered among a variety of animals who are both similar and different from us. Aspects of the animal are then reintrojected into our psyches by a wonderful chemistry of imitation. When we observe this unlikely agency at a distance, animals seem like mediators, appearing in music, story, song, narration, dance, and mime as participants in the narrative.

The emphasis on what the animal does has influenced language. Speech itself may have emerged in concert with sounds for actions borrowed from the names of animals. We verb the animals not only in games but in ordinary speech. We duck our heads, crane our necks, clam up, crab at one another, carp, rat, crow, or grouse vocally. We cow, quail, toady, lionize, and fawn in servility, admiration, and fear. We fish for compliments, hog what should be shared, wolf it down, skunk others in total defeat, and hawk our wares. We outfox and buffalo those whom we dupe; we bug and badger in harassment. We hound or dog in pursuit, bear our burdens, lark and horse around in frolic. We bull, ram, or worm our way, monkey with things, weasel, and chicken out. We know loan sharks, possum players, and bullshitters.

If indeed such animal verbing derives from the earliest emergence of human speech, perhaps the distinction between noun and verb did not exist—as it still does not in some languages such as Hopi, lacking separation of the thing and its action. What is now a convention, absorbed in the individual acquisition of speech in childhood, may once have been the remarkable discovery of the most erudite and mature: the discovery that one might convey the intention of a certain action or a done task by vocal reference to an animal not present. Even though this use of animal names as verbs has become one of the routines of learning to talk, modern writers, poets, storytellers, dancers, and other artists continue to plumb their creative resources for such images. Perhaps the reason why there are no more incisive or efficient ways of describing action is because the human mind was organized this way in its genesis.

KONRAD LORENZ HAS observed that all mammals learn by rote memorizing. A mouse learns a maze by running it over and over, because memory and enactment are combined. Human learning is commonly said to have three stages or "modes of representation": enactive, iconic, and symbolic. Children are committed to the first stage, acting out what they feel; they learn to walk like mice learning to run a maze. By first hiding an object and then looking repeatedly they are able, at a certain age, to know that it continues to exist while hidden beneath a blanket. Manual skills are not taught but learned by watching and imitating. Other large mammals do this also. The human repertoire seems peculiar, though, in that children imitate not only other people but other species, a characteristic of our own kind of consciousness. It has long been a truism among humanists that individuals start life as animal-like and graduate slowly into the full human condition. But being an animal and acting the animal are not the same. Enactments are transformative traits of our humanity. And we do not rise above them; the skills of categorical thought by means of the taxonomy of other species is never completely put by, and we continue as adults to extend our knowledge of the nomenclature of plants and animals. The emulation of animals in childhood games is translated into dance and ceremonial performance in adult life.

The second "mode of representation," the iconic stage, blooms in middle childhood, when many children are charmed with drawing and pictures. Iconic refers to images in the mind's eye and in art as simulations of things tangible. It includes the personification of animals in ways much wider than games and is profoundly related to mythic narration and performance. Some scientists, unaware of the functional role of pretending that animals talk and wear clothes, argue that "anthropomorphism," being fictitious, makes it more difficult for a youth to understand the true lives of animals. The moralistic idea that it reduces humanity to animality and rationality to instinct, or elevates brutes to human status, is equally shortsighted. The stories of talking animals are joyfully accepted not only by small children but also by juveniles—indulged by parents who sense that the illusion has its own ring of truth. The mistake made by the critics of anthropomorphism is the supposition that it is an end in itself; they judge it without regard to its function in childhood as a precursor to poetry and other metaphor. Bruno Bettelheim remarks: "If we do not understand what rocks and trees and animals have to tell us, the reason is that we are not sufficiently attuned to them. . . . A child is convinced that the animal understands and feels with him, even though it does not show it openly."[6] Despite the pretense of games and stories, children know that the different species are not a society with a single language, traditions, roles, or ceremonies. Like kittens in a play mode or chimpanzees who are said to "put on a play face," we are conscious of the pretense. Yet it is this make-believe that keeps before the child an unspecified meaningfulness of animals.

Fantasies of animals wearing clothes and building houses, what may seem a too-close similarity between them and people, create excesses of common ground, stored for late adolescence when disjunction almost swamps the ego. The imaginary continuity between animals' lives and our own reinforces a profound and enduring metonymy, a lifelong shield against alienation. Especially at the end of puberty, the end of innocence, we begin a lifelong work of differentiating ourselves from them. But this grows from an earlier, unbreakable foundation of contiguity. Alternatively, a rigorous insistence of ourselves simply as different denies the shared underpinnings and destroys a deeper sense of cohesion that sustains our sanity and keeps our world from disintegrating. Anthropomorphism binds our continuity with the rest of the natural world. It generates our desire to identify them and learn their natural history, even though it is motivated by a fantasy that they are no different from ourselves. We enjoy animals so much because we are laying up the basis for a language of analogy, the terms for the abstractions of cosmology and poetry—though the purpose is no more revealed to us as children than the outcome of a healthy diet.

The second form of representation—iconic—alludes to images and the older child's capacity to imagine. Juvenile life is a potential trap in platitudes of literalness because its purpose, from the standpoint of human evolution, is to become fully acquainted with real forms. Beginning in adolescence and continuing throughout life, the third form of representation—metaphor and symbol in poetry and song as well as all other arts and myth—results in the intellectual realization that things have more than a face value. Guiding the young adult in this work is the cultus, with all of its exercises in tutorial, myth, ceremony, and test, traditionally employed to open the doors into maturity. A rich, literal knowledge of animal life is fundamental to this process, generating a respect for the natural community as a higher language, as clues toward wisdom in the immense panoply of nonhuman life, to which mature adults will look for the terms with which to describe a cosmology. Failure to nurture childhood enthusiasm for animals produces adults bereft of diverse living forms as the metaphorical basis of religious conceptions and values.[7] Such a society fails to honor the underlying unity of diversity because it lacks experience of it, is blind to the humor and irony of the human/ animal juxtaposition, and projects human cruelty, murder, tyranny, and enslavement into the ecology of animals, thereby justifying concentration on "higher" things and the reconstruction of animal life according to an ethics suited to the worst aspects of our own species. That poetry should have its roots in anthropomorphism's illusory consanguinity and juvenile emphasis on iconic representation at a time—childhood—when the taxonomy of nature is still a major preoccupation may seem strange. Elizabeth Sewell, arguing for the primacy of organic metaphor, writes: "Morphology and taxonomy are postlogical and consequently nearly related to poetry, which is in its turn morphological and taxonomic in character."[8]

Florence Krall suggests that there is a fourth type of representation beyond psychology's conventional three of enactive, iconic, and symbolic. After the symbolic divisions, she says, where things at first thought to be the same are discovered to be analogous, there should be a culmination in the recovery of literal connection where we and the animals meet as kin.[9] In this sense the early play at being an animal, with its interiorizing of a morphology of feelings—giving species names to acts and emotions—leads to a clearer self-definition. The biological kingdom, Animalia, is composed of many species, but psychologically it is we who are composed of animals. Despite modern usage, as in "humans or animals," we are not part of a binary duo nor elements of a simple juxtaposition with them. However we may define ourselves as a species, the final act is to recover our animalhood, to see all taxonomy for what it is: a means, not an end, to thought.

7 | The Ecology of Narration

*The world of animals forms a backdrop against which humans
continually formulate and reformulate, in the language of
metaphor, their ideology of themselves, the particular
characteristics of their own forms of social interaction,
and their personal and social histories.*

GARY URTON

STORIES WITH ANIMALS are older than history and better than
philosophy. History tries to describe the world as if it began with writing
and only humans mattered; philosophy attempts to abstract truth as if it
were defined only by discursive thought and experience of the natural world
were unimportant. Speech (by which we have clamorously and erroneously
declared ourselves as unique among living things) permits us to recapitulate
experience to one another in ways that make sense of a world where the others communicate—tell their stories—differently.

As protagonists and characters in stories, animals are sometimes surrogates in the collective unconscious of humanity. But they are not there out
of fear and disdain. Animal imagery was not burned into the human heart by
terror, in a long, prehuman darkness, traumatizing generations of "cave-
men" who cowered in their holes, scarred even now by half-remembered

struggles against snakes and lions, who return as monsters in our dreams and stories, as the sculpted demons on Gothic cathedrals, or as "beasts" in the imagination of historians.[1] One art historian has theorized that our present "security" from real, wild beasts results in their being relegated to comic decorations such as gargoyles. This jungle view of the past is the bad retrofit of history and progress onto the past.[2]

The symbolic use of animal forms as narrative does figure in our deep past. It is witnessed in worldwide rock art whose painted shapes and etched outlines of large animals constitute a 30,000-year tradition. Not simply "subject matter," these animals on the walls of the great Paleolithic sanctuaries from Spain to Russia, and hundreds of portable carved objects, resemble dream images or fixed stills in the imagination. But they were probably not used either to terrify and subdue a human audience or to gain "magic" control of the animals they represented. To the Paleolithic observers of that art, the subterranean world was briefly illuminated by torches—the way we experience visualized images in our individual heads. The elephants, bison, horses, aurochs, and other wild, hoofed forms seem to externalize the content of the painter's mind as though it were the interior of the earth. They are the eidetic, surreal fragments of a transformed reality, a shared thought.[3] Their lack of scale, absence of linear perspective, and timelessness speak to a mythic cosmos. Species are often paired in complementary sets, analogous to the two divisions or moieties of tribal peoples, or grouped in triples or double pairs that seem to paraphrase other social distinctions. The location of the different species in the cave corridors, galleries, and bays follows a pattern, as though the cave were the habitat of the universe. Some animals seem to be emerging from the rock itself, their muscles, bony contours, or profiles given relief as though affirming the shaping of life in the earth. One feels before these astonishing figures in their solitude that their ineffable silence was probably balanced in human song or performance, intended to tell a timeless story, like the figure of Osiris on Egyptian sarcophagi or the manger scene in the Christian tradition of the birth of Christ.

The participants in those ceremonies went inside the body of the earth to attend to whole animals—hunters who in ordinary life had learned that the insides of the animals they butchered could speak of the conditions of life, hard times and good, disease and injury, age and births, and the passage of food though corridors suggestive of underground caverns. The telling of the hunt was an adventure in tracking and interpreting signs inside the landscape, in reading interiors of places, an insider's view.

The cave ceremonies distanced participants from events just enough to recreate them without their becoming alienated. Encountering these representations as vicarious expressions of themselves, the participants were plunged into the pronominal enigmas that enliven our conceptions of who we, and they, "really" are. Octavio Paz comments that "pronouns, like nouns, are masks, and there is no one behind them—except, perhaps, an instantaneous we which is a twinkling of an equally fleeting it." Here Paz glimpses that moment in which we see ourselves between subject and object, the moment when we can be an "it" to ourselves, without losing our place in the world, and the animals become a significant "they." We have always known that the animals were the great "between" forms, like us yet different, whose serene bonds with the world are like and unlike our own.

If stories were indeed told before the illuminated walls and around nearly a million years of campfires, it is likely that they include especially a recapitulation of events of the day and traditional wisdom stories or myths. Apart from a recounting of recent adventures, two kinds of narratives occur among traditional peoples, the first suggestive of the immense past of the Pleistocene caves and the second of the life of farms and villages which succeeded it. These are myths and folktales, alike in evoking animals in ways that go beyond their physical definitions.

UNLIKE FAIRY TALES (inventions of urban societies which require happy endings because their hearers are so distanced from their roots and awash in dismay and doubt), myths are intended to enlarge our sense of connections in a story of the cosmos—that is, of origins. They are simultaneously about the inner life of the individual and the gross forces of the world at large. As in other tales their characters often travel through landscapes, but myths are not set in abstract noplaces, nor merely in "a forest," but at a unique mountain, lake, seashore, spring, or locale with its plant, animal, and spiritual inhabitants that echo in some way their own inner landscapes. Their events do not occur "once upon a time" but "in the beginning." They do not ignore specific place and time but make it transcendent.

Psychologically, the audiences are different—the child of the fairy tale, the juvenile of the folktale, the adolescent of the myth. Fairy tales are addressed to personal growth, folktales to the social game, while myths have an inclusive sweep combined with unique reference, both exemplary and cosmological. Myths seem to meander, with superfluous details and oblique content, a waywardness unlike the story line of the fairy tale, yet rhetorically declaimed as though the teller's phrases were active in it. The protagonists are often people but also may be the sky, a huge snake, human genitals, fire, a giant bird, or any other aspect of the world. Monsters appear, things come once and vanish, some encounters happen inexplicably. Whims shake reality, tremors distort the universe, magic is common. Supernatural beings intervene, and animals play central or gratuitous roles. The motives of the players are often unknown, the scenes inexplicable. Transformations and confrontations take place. Skills are tested, direction is occasionally seen, but the passages and encounters are fortuitous. Improbabilities are linked in a play of accident and irrational decisions like a game of chance.

Children are fascinated by the telling of myths because of their strangeness, but the gist is about a multiverse for young adult ears. Some anthropologists see in myth a universal, binary framework, as when fire and water, raw and cooked, sweet and bitter, balance each other, conveying the idea of movement around a stabilized axis. Perhaps an even more prevalent theme is that of chance or "luck," which is always out of personal control and yet always influenced by formalities and attitudes. The enigma of the myth is that luck is never truly fortuitous. We live in a world more complex than anyone can fully understand. Despite the strange turns, meaning underlies events, not quite available to consciousness. The broken shadow of structure teases the imagination of the listener, who has learned that no event is insignificant and none mean only what they at first seem.

The animals of myth are like the composite souls of species. A bear is all bears or the powers of the bear. Many mythic animals, like griffins, are hybrids—unifications of disparate ideas or domains, made "creatures," anatomical amalgams, prefigured by the body. The animals of a myth are avatars of cosmic skills who speak to one another and intermarry—an "as if" society sharing customs, a necessary fiction in the perception of the living stream as purposeful, sentient, communicative, and interdependent. Just as nature itself is given, that community of different species is a model of the givenness of culture, justifying customs as "natural."

It has become fashionable to describe myths as though they were entirely about the psychic life of humans. The cosmos is exhausted as a projection of the unconscious life. Joseph Campbell, for instance, interprets the bull, snake, or eagle as aspects of the inner world, neglecting everything beyond the personal and social. By contrast, the Navajos, for example, never overlooked a larger context in their coyote stories, a world in which people are surrounded by real coyotes that catch mice, scratch fleas, and defecate, coyotes which may symbolize aspects of the human psyche but never became such exclusive inhabitants of the personality that they disappeared from the whole community of life. It is difficult for a naturalist to believe that the wisdom and necessity conveyed in myths are directed only to understanding the inner human self. There is a kind of arrogance in the idea that all the animals of the natural world are there only as mirrors of a Jungian interior world. We conclude a little too easily that the nonhuman others are mere background and that the "real" world will smile on whatever concoctions we formulate as stories and rites. Myths may indeed illuminate unconscious processes, but the context in which that inner world came into being is *ecological*. Surely myths, then, reflect a wisdom about that environmental world—in which the rabbit (in addition to being a timorous or amorous aspect of the self) is an aspect of some larger organic reality of which we too are part.

Myths often deal with cosmic dance or play, the rules of which are sometimes beyond human understanding. Mythic events, therefore, seem fortuitous and, as play, correspond to games of chance, reminding us that in small-scale, independent human cultures gambling is especially popular.

THE FOLKTALE DOES not deal with an inner structure or a cosmos so much as with a human society in which the weak survive by their wits or are perpetually duped and exploited. Official power is often the villain. Strategy and scheming are the weapons against greedy, cruel overlords or narrow laws and customs. Folktales address conflict between knaves and fools, patrons and peons, age groups, men and women, children and parents, buyers and sellers. While myths humanize nature, folktales animalize human society. Animals in folktales are the principal characters of stories which are not so full of bizarre events or creatures as myths. Often they resemble adult nursery rhymes or burlesques, as in the song:

> All around the chicken coop
> The monkey chased the weasel,
> The monkey thought 'twas all in fun,
> Pop goes the weasel!

The chicken is the victim or scarce resource, sought by two competitors, the naive monkey and the sly weasel, who, lulling the monkey into a playful scamper, pops into the coop. To a peasant the scene has almost unlimited analogies to daily life. The setting of folktales is a world in which all goods are limited, where your advance is my loss, my gain your misfortune, where power can be opposed only indirectly and conniving is central. There are chickens who will be taken and weasels who will take them by disguising the real game from the poor monkey (who merely apes his betters). The sneaking, petty, seducing cheats, the tricksters, amuse us because they oppose suffocating laws, rules, and mores. The conniver defeats a more powerful landlord or sheriff by his wit, avoiding retribution. But success is seldom final and it is possible to be too smart. The trickster is often the rascal who ends up in the soup.

The animals in these stories are often caricatures of status and role types, travesties of real bailiffs, landowners, burgermeisters, woodcutters, and bureaucrats. As animal equivalents they acquire the conventionalized features of their natural prototypes, noble arrogance in the lion or hysterical fear in the hen. The folktale is an internal dialogue by which a society inculcates its members in a language where animals parody not only classes and values but personality traits, hierarchical notions, and moral principles.[4] Folktales are the narratives that correspond to games of strategy, deceit, design, maneuver, bluff, feint, and entrapment. There are equivalents played on gameboards as checkers or chess, comparable in small to military tactics.[5]

While myths emphasize chance, fantastic events, archetypal images, and cosmic motifs among foragers such as hunter/gatherers, who often live in a more generous world than farmers, folktales look to the struggle for limited resources in more centralized worlds of shortages and taxes, where the clever seek to thwart the rules, frustrate competitors, or trick the inept. The tales are told as entertainment but convey the attitudes and means by which the weak deflect the greed of the powerful. They center on the tension between obedience and authority in village life, or between the powers and the vassals at court, with skirmishes and alliances at every level in the hierarchy, betrayals in the interest of self-survival, where cunning can protect one from those above, deceive one's competitors, keep one's advantage, and otherwise bend the system without defeating it.

The rabbit who outwits the fox who tricks the dog who catches the rabbit reveal analogies in a world where the participants, especially the oppressors, ever vigilant against slander, are disguised as animals. Tanzanian tales tell, for example, how the ostrich got its long neck. A hungry crocodile, pretending to have a toothache, persuades the helpful but naive ostrich to have a look. The crocodile snaps its jaws shut but the strong, young ostrich escapes—with a stretched neck. The story warns against being duped as well as admiring resilience in escaping from danger. What Tanzanian would not be reminded of the wisdom of this story every time he saw an ostrich?

G. K. Chesterton, writing of fable and folktale, observes that the persons in them must be impersonal, predictable, and familiar. Unlike those in myth, "they must be like abstractions in algebra, or like pieces in chess. The lion must always be stronger than the wolf, just as four is always double of two. The fox in a fable must move crooked, as the knight in chess must move crooked. . . . Everything is itself, and will in any case speak for itself. . . . This is the immortal justification of the Fable: that we could not teach the plainest truths so simply without turning men into chessmen. . . . In this language, like a large animal alphabet, are written some of the first philosophic certainties of men. As the child learns A for Ass or B for Bull or C for Cow, so man has learnt here to connect the simpler and stronger creatures with the simpler and stronger truths . . . that a mouse is too weak to fight a lion, but too strong for the cords that can hold a lion; that a fox who gets most out of a flat dish may easily get least out of a deep dish; that the crow whom the gods forbid to sing, the gods nevertheless provide with cheese. . . . All these are deep truths deeply graven on the rocks wherever men have passed."[6]

The folktale offers much for rumination. Its rambling nature gives the impression of unfinished business. *The History of Reynard the Fox* is said to have had 100,000 verses. Even if Reynard, the unsentimental, lying, cynical underdog, along with Noble the lion, Isingrim the greedy wolf, the Old King, the pompous bear, and thug cat become trite, the story is true as a distant mirror of the class struggle, hardships, and provincial thought in the age of rural towns.

Folktales in the hands of moralizers become fables and parables, cautionary tales whose one-dimensional animals stereotype the human possibilities. In Aesop's fables the moral and its animal are a cliché. In modern hands fables have become quaint and prettified subjects for the gift book trade. The medieval bestiary is entertaining in a scientific age for its quaintness and medieval flavor. Modern literati read Aesop's fables for their amusing logic, and they are published with elegant bindings and baroque drawings, to be given as gifts to children.

ALLOWING THAT DANDLING tales and lullabies for infants may have come first, the ancestors of stories are probably narrations of foraging adventure, the hunt or the gathering, the essential form of the quest. Most of the memorable events of all forays have to do with animals. "Game" animals are those with whom we are engaged in a perpetual but episodic seesaw in which the "rules" of the game are in fact the habits of the animals, learned primarily from the observation of important animals. Perhaps, as humankind became increasingly reflective, the whole of the living world seemed encompassed by such rules. Instead of such rules applying to the animal only, our lives seem refracted by play with rules, which are metaphors of the habits of animals. These rules could be guides in miming the animals in fun. Games as play are reenactments of the quest, with its equilibrium between opposing sides. Play miniaturizes life, and in recollection it is narrated in a spirit of excitement and instruction, whether heard and enacted in the mind's eye or mimicked by actors.

Because they center on action, the narrative forms with animals have "game" equivalents connected to particular animals. Games progress like stories. In general, most societies tell all three kinds of stories—fairy tales, myths, and folktales—and play their game equivalents. But there are discernible differences. Egalitarian small-scale societies seem to emphasize games of chance and mythic narratives; hierarchic rural or village cultures prefer strategy games and folktales; urban peoples tell fairy tales and play games of skill. Skill and fairy tale deal mainly with physical and emotional aspects of the self; strategy and folktale deal with social circumstances; chance and myth deal with the larger cosmos.[7] Within all human groups, social concerns are narrated as folktales, personal concerns as fairy tales, and spiritual matters as myths.

Games are "played out" and then recounted in story. Each narrative/ game form has its animals. Typical among them, the frog is the master in the fairy tale and the skills of life—a transformational icon with a kind of organic givenness, rising as it does from the muck of the earth at winter's end, going from the water to the land in summer, emphasizing the mysterious, inborn potential for growth and change that is like an enchantment in its power to transform life and our lives.

The fox and coyote are the famous rascals of the strategies of the folktale. Pigs and hens define gluttonous bureaucrats, hens the gullible or foolish, and lions the aristocracy or military. The environment is the social milieu, characterized as divided and contending. The animals are the avatars of flux and negotiations among people, where words are used not only to inform but to trick.

The snake is the supreme animal of myth. Unlike frog or fox, it is beyond calculation. It is the coiled serpent upon which creation rests: the keeper of the underworld and of time. Grasping its own tail, it is everything contained in cyclic duration, hence all the play of uncountable cosmic factors. Being ineffable, these factors interact to produce what we experience as chance and are spoken of in myths with multiple-leveled meaning.

If the three types of story, game, and animal are envisaged as having specific objectives in human life, a table of relationships can be constructed:

story	*game*	*object*	*life stage*	*capacity*	*focus*	*animal*	*economy*
fairy tale	tag	skill	child	motor	self	frog	city
folktale	checkers	strategy	juvenile	image	society	fox	village
myth	dice	chance	adult	symbol	cosmos	snake	band

This table associates the life cycle of the individual with widening horizons on the world. Despite the levels of meaning and sequential ordering, the three types of narration and game continue to figure throughout the individual's lifetime. None of the categories in the table, either vertically or horizontally, is intended to be exclusive or necessarily to wholly define a way of life. Their relative emphasis depends on the culture's attitudes toward the natural world in different economic and ideological systems (which in reality are often hybrids). Eclectic, modern urban society tends to blend all three and blur their distinctions. Although the animals in modern stories remain, they burlesque more often and are marginalized wherever cultures are dominated by otherworldly religions or metaphysics of subatomic microevents and celestial matters described in light-years. But in all cases, the scramble for certainty is more frenetic in a world without the assurance and mediation of the Others and their examples of repose and resolution in an otherwise baffling universe. The table, moreover, has an overarching implication: as a philosophy the myth line is clearly associated with maturity, the folktale line with thwarted development along juvenile lines of literal absolutes, and the fairy-tale line with therapy in societies where an elemental confidence in the self as an organism has gone awry in childhood.

Narratives in which animals are protagonists are not necessarily better than other stories, but they occur in all kinds of societies and in different forms at three stages of the life of the individual. Humans intuit the essential wisdom that animal figures are necessary to thought and communication (just as they are physically necessary to the health of the natural community or ecosystem), that their efficacy is related to the accessibility of humans to wild animals, and that even in societies stuck in the juvenile absolutes of folklore animal images are shields against madness and despair.

8 | Membership

People are intensely interested in the animals that live around them, and along with that interest and fascination comes the desire to make other living things participants in what the humans are doing . . . through the metaphoric process.

J. CHRISTOPHER CROCKER

THAT TO WHICH we belong, of which we are part, is as important to our sense of self-identity as our personal beliefs and attitudes. We "play out" our lives creating a self as though driven by a restless demon whose whisper is the unfinished "I am ——." Its closure is postponed all our lives by accretion from two directions: an inner or subjective meditation on one's feelings and choices and an outer positioning by belonging to something larger than ourselves. From one perspective, membership is part of a taxonomy of selfhood. As we saw in Chapter 3 on the skills of cognition, animals were among the first objects of classificatory thinking. It follows that interspecies concepts became the model for our social definitions.

Perhaps it all began with a distinctive primate intensity. Anyone watching a troop of terrestrial monkeys notices that each individual occupies a place in a social constellation and changes places in that group, aware of its own position as it moves through perennial patterns of social structure which themselves are as impersonal as the stars overhead. How important is this to the monkeys? Status is charged with emotional furor, the abrasions of capricious mateship, contingent status in peck orders, and conflicts of kinship and precedence, all reflected in the animal's behavior as the troop travels and its members orient to food and danger, genuflect to tyrants, punish transgressors, and escape from danger.

Then there is the distinction between one group and another—that shared identity despite personal differences by which the members know themselves as apart from the others. Membership has this dual quality. There is the role within with its microdifferentiations—the leaders, the future leaders, the new mothers, the virgins, the babies, the deposed elders, and so on—and then the macrodifferentiation between groups one, two, and three and between this group of groups and that other association of groups.

Our deep needs are demonstrated by our simian cousins. Their social slew of attachment, confrontation, affiliation, subordination, and the random play of daily circumstances is acted out in monkeys, as it is among ourselves, while in us also it is given names which signify belonging, symbols of our membership. Since animals seem to demonstrate intragroup roles and intergroup differences, they provide categorical models—the names and imagery—for thinking about and referring to ourselves, totems by which we individually are conscious of this aspect of our own identity.

Such membership distinctions have a visual prehistory. Appearance is everything in these matters, because primates depend mostly on sight—especially on body size and proportions and patterns of facial hair. The role of color is crucial if we are to understand ourselves. Primates have little color variation in their bodies except for muzzles and rear ends. Color vision, lost among ancient, nocturnal mammals, was recovered in tropical jungles by our arboreal monkey forebears as part of their sunlit search for insects, flowers, and fruits. As though by the hand of the god of impressionism, they regained what their whisker-, nose-, and ear-oriented ancestors had forsaken.

But it was not reinvented for the sake of art or for discerning differences among ourselves by color as do butterflies and fish. Mammalian body colors have never caught up with our chromatic vision and social desire. We inherited that primate eye which allows us jolts of pleasure in the presence of rainbows and sunsets. Most mammals with their tan, tawny, buff, and brown pelages are disappointing to our color-livened minds. In our own sight we too are lackluster, less beautiful than flowers or lizards. Even allowing for blushing, paling, tanning, and the variations of skin, hair, and iris pigmentation, we are dull compared to the fishes of a coral reef.[1] Our eye's versatility, disappointed by human uniformity, quickens our envy of iridescent scales and gaudy plumage. Birds, butterflies, tropical fish, and the rest of the vibrant world all seem splendid in ways that we are not. Their superiority is twice blessed—making them not only more pleasing but also more easily identifiable.

This cleavage between the pallid and the prismatic would be unimportant were it not that culture and consciousness have escalated our inherent primate concern about identity. Differences of dialect and custom are responses to an old furor: the unrequited need for personal and social particulars that could be signaled and recognized as effectively as we differentiate by a glance the parrots at a water hole.

To recreate the origins of this disappointment we may imagine our distant forebears as they first ventured into open country, emigrants onto the ground from a sea of foliage, vicarious newcomers in the edges of the savannas, peeking into the neighbor's world. What they saw there, as information, was as fragmented as if they were commuters glancing at others' newspapers. In time, as they entered more seriously in the game of savanna ecology, they awoke to a world of signs unlike anything in the canopy of the old leafy world from which they had descended. Yet what they brought from their forest background was a social existence in which all signals from the world at large were translated into the provincial vanity of the group obsession with itself.

It seems probable that our kind of self-consciousness began with the conceit of the monkey mind that everything shapes itself about "us"—that is, whatever we see and hear is assumed to be a sign. This assumption took a great shock with the ambiguity of our lives after we left the social womb, the deep forest. On the ground, hairless, bipedal, we were like an ape and yet not one, carnivore and yet also herbivore, edge creature between forest and grassland. Our sense of ourselves as a species became immensely more difficult. There would be overlap everywhere, making us composites in our own minds. In open country it could all be seen at once, all the others in motion, as though the landscape itself, peopled by a great spectacle of contesting and cooperating life-forms, were somehow what we had become.

As our ancestors became grassland omnivores and their attention sharpened, both to the variety of life and to intragroup confrontations, they watched the social displays of other species. They saw ritual among the different animals, signals to each other by means of stylized movement, the parade of horns and antlers, the flourish of brilliant feathers and corresponding calls. Early humans would graft onto their own poorly differentiated bodies and groups paraphrases of these displays of other species. The mating foot-drumming of grouse could be mimicked as a dance of courtship, the ritualized combat of horned animals as a language of conflict.[2] Particular calls and animal skins could be appropriated to identify human subgroups denoted by age, sex, and marital status. Fragments of animals became insignia, ritually assigned in ceremony, that established or heralded human membership by translations from the zoological realm.

Claude Lévi-Strauss has observed: "The diversity of species provides the most available intuitive picture and direct manifestation of ultimate discontinuity of reality: the sensed expression of objective coding."[3] Nature is not only a storehouse of names and a logical model of differentiation but an encyclopedia of behavior. He says: "The animal world and that of plant life are not utilized merely because they are there, but because they suggest a mode of thought. The connection between the relation of man to nature and the characterization of social groups, which Boas thought to be contingent and arbitrary, only seems so because the real link between the two orders is indirect, passing through the mind. This postulates a homology, not so much within the system of denotation, but between differential features existing, on the one hand, between species and, on the other, between clan a and clan b."[4] And he adds: "An individual, in his social capacity, combines multiple roles, each of which corresponds to an aspect of a function of the society. . . . He is continually confronted by problems of orientation and selection. . . . As a member of a large clan a man is related to common and distant ancestors, symbolized by sacred animals; as a member of a lineage, to closer ancestors symbolized by a totem; and lastly, as an individual, he is connected with particular ancestors who reveal his personal fate and who may appear to him through an intermediary such as a domestic animal or certain wild game."[5] Such animal eponyms of social entities are therefore not based on a similarity between animal types and human types but rather on an idea by which two sets of relationships are bound. The key to this kind of thinking is the similarity of the systems of differences.[6] "Totemism," Lévi-Strauss says, "is a particular way of formulating a general problem: how to make opposition serve integration."[7]

When we think about it, social place or rank is an abstraction that cannot be easily imagined. Animals are concrete; their system is a categorical grid on which human roles can be fitted with name and ensign. Human identities are elaborated and performed—sung and danced—by the human participants who may also be costumed, masked, tonsured, tattooed, and painted.

Unlike the physical changes early in life, as their social roles and their ages change, adults lack dramatic differences that mark these phases of life. Many cultures make the most of the physical changes of childhood—such as tooth eruption and loss, hair growth, voice pitch, breast enlargement, and menstruation—as keys to the timing of new designations and formal attention to new roles and responsibilities or separation from parents. But as they get older, changes in individual roles within groups overtake their changes in appearance. Individuals are repositioned with such "bodily" changes as hairdress, costume, and tattooing, all coded by a variety of animal monograms which reflect categorical types.

People perceive themselves not only in terms of things but actions. "Being," as in "human being," is a gerund. It encompasses the donning of badges. The blurring of human difference is remedied by signs that tell of "my being" and "our membership." Labeling human goups with natural forms is evidence of the intuition that a great, invisible mind—Nature, the Cosmos, the Creator—really knows who we are or what we are "meant" to be, and that we will find signals in the natural world which can be adapted to each new becoming by narration and performance.

To be urban—to live in mass society at a distance from wild diversity—is to share a heightened angst about the pronominal enigma: the identity of I, we, you, it, and they. As if to deny our poverty of wild things, we declare a cultural superiority over such "primitive" reference. It was said by "civilized" people until recently that "savages" suffered from "congregational unself-consciousness," a kind of communal daze. Our schizoid alienation from the animals has led us to project the frightening confusion of our urban grayness upon them. Personhood among tribal peoples is no less volatile and relentlessly scrutinized than it is among ourselves, though with less inverted Christian and Jewish anxiety, less psychoanalytic guilt and introspective soul-searching, which seem to aggravate our concerns.

Individuals in primal societies, maturing into an ever-richer wild world of metaphoric signs, seem to become more confident rather than less. Their repose reflects the sensibility of the fully individuated person. Wisdom for them is comprised of a lifelong expansion in the perception of natural signs, while for us it is the abstract extension of a limited number of signs into an alphabet. Primal thought opens out into a frontier of new experience in the exploration of the natural world, as opposed to the mirrored hall of urban cognition. The loss has haunted civilized humankind since the first cities replaced the natural plenitude of the ancient river valleys of the Euphrates, Indus, and Nile.

ROCK ART ON the walls of Old World caves may mark the emergence of animal signifiers in a system of human social grouping. The great Pleistocene mammals depicted were central to human life and thought. As signs their figures are etched, sculpted, and painted in a lost language, a legacy of thirty thousand years of the imagination in signifying something human and social by animal species, perhaps as tribal totem animals from which came the widespread binary division of local human groups and membership in linked clans.

Why does the savage, asks Jane Ellen Harrison, "persistently reaffirm that he is a bear, an opossum, a witchity grub, when he quite well knows that he is not? Because . . . to know is first and foremost to distinguish, to note differences and classify."[8] Alfred Radcliffe-Brown adds: "The resemblances and differences of animal species are translated into terms of friendship and conflict, solidarity and opposition. In other words the world of animal life is represented in terms of social relations similar to those of human society."[9]

No single species could model all the guises needed for the human repertoire. What was required was a reading of ecological diversity as if it were a society. Skins and feathers could be used in mixtures, as though the variety of pelage and plumage were made for human use in a perpetual drama of self-becoming and the juxtaposition of ideas in mythic tales. The choice of certain plumes to mark a particular social group was probably made at first entirely by esthetic chance—the way chimpanzees decorate themselves with bits of colored material—then later given its logic and mythic explanations. In time the decoration would be accompanied by narration or performance, fantastic mystery plays or dances in which the analogy, from ecological to social, was justified by myths averring that, in the early days of the world, the natural world was social, plants and animals sharing the customs and language of a single culture.

Most human societies are composed of subgroups such as clans. Dividing society into clans in many tribal societies includes the use of animal species' traits as clan emblems and then explores the symbiotic relations between species as logical analogies of human social subgroups. The discovery of this metaphorical leap was one of the profound ideas of Claude Lévi-Strauss. It requires a good knowledge of natural history, especially the relationships between species. The primary bond is the food chain—perceived in the mythic sense as a dialogue between eater and eaten about optimizing sunshine and its trickle through the living community. Studying nature has a practical aspect: the whole idea of stalking game may have come from watching the great cats as they hunt; the notion of digging edible roots may have come from observing the wild pig. But the cultural translation is unpredictable: are the men in a "lion" clan those who will stalk or those who will be prohibited from stalking, those who will teach neophytes, or those who will pacify the souls of lions? Badges of identity such as buffalo horns and eagle pinions were appropriated in a logic of distance rather than replication. It is as though we accepted a code provided by a thoughtful cosmos under the condition that we not confuse it with a literal assertion.

When the Swazi of southern Africa commemorate the annual, national *ncwala* festival with rite and dance, marking male, female, and the hierarchic ranks of king, priests, princes, warriors, queen, king's mother, married daughters, and so on, in costumed regalia with distinct kinds and numbers of feathers and pelts, the dress also refers to wild/domestic, predator/prey, land/water, or other binary sets of the larger natural system. Pelt differences define human roles—as when the Swazi chiefs alone wear lion skins and the women wear only the leathers of domestic animals, warriors only those of wild. Although the celebration has astrological, seasonal, and mythic aspects, its sophistication depends on a detailed knowledge of the animals, "concretizing intangibles" which correspond to the range of cosmic powers.[10] Other peoples, in interior New Guinea, prefer plumes to skins. Most mammal pelts are not brightly colored and the hair eventually falls off. Feathers last longer, have more color, and can be individually arranged in a way that the fur cannot. They can be used in their original patterns or they can be separately mixed and matched. Indeed, feathers have probably exceeded any other adornment in defining ourselves as richly and beautifully as the most esthetic group of beings in the world. As codes for social difference, or as the means of personal expression, feathers have no limit.

A SPECIES KEY to social rules is made more concrete where the animals are also classed by their edibility and distribution. In rural Thailand, for example, the use of animals as food and their place in the farmstead correspond to degrees of marriageability of individual humans. As the anthropologist S. J. Tambiah says in a now classic paper: "The system of marriage rules made themselves felt in the system of rules about touching and eating animals."[11] Animals are lodged or range at distances that correspond to social distance. Those living in the house, such as dogs and cats, are inedible. Oxen and buffalos, housed under certain rooms of the stilted structure, are eaten only ceremonially. Pigs and ducks, sheltered away from the house, are ideal foods. Boars, civets, deer, and wolves, out in the forest, are normally palatable. Ele-

phants, tigers, and bears, inhabitants of the deep wilderness, are not eaten.
In a parallel series, the marriage of brother and sister is not tolerated any
more than the eating of a household cat. Cousins are conditionally marriageable just as oxen and buffalo may be eaten. Marriage between members
of nearby villages, often second cousins, are ideal like the meal of pig and
duck. Inhabitants of more distant villages have about the same limited mat-
rimonial valence as deer and wild dog have as cuisine. Finally, marriage with
the absolute stranger is regarded with the same revulsion as eating an elephant. It is stipulated, moreover, that siblings have separate sleeping quarters, marriageable kin may stay only in the guest room, first cousins may enter but not sleep in the house, more distantly related people with whom
marriage is possible must cleanse their feet before entering the house, and
outsiders are not invited in. There are yet more subtle distinctions—ambivalence about the habits of the household pets, edibility in terms of ceremonial feasts or daily meals, sacrificial rites, taboos, power species, mythic
forms, and so on.

This system of references and equivalences seems to embrace a universe
of possibilities, going full circle from prohibition to ideal to prohibition.
When it comes to monkeys, however, the analogical system crumbles in
horror. As both an animal and a degenerate human, the monkey is both too
intimate and too remote; its edibility borders on cannibalism and marriage
with it is unthinkable.[12] In brief the marriage/edibility/spatial code can be
seen in a table:

distance	kinship	animal	marriage/food
intimate	sister	dog	prohibited
close	cousin	castrated ox	qualified
adjacent	distant kin	pig	ideal
distant	known	deer	acceptable
remote	stranger	elephant	prohibited

This looks like a complicated way of doing something otherwise simple.
Why link habitat, species, and edibility to marriage rules? Apparently marriage categories are too important and yet too abstract to be left to language
alone. The correspondence of human groups to animals, their location and
edibility, is both a tangible and a taxonomic means of memorializing rules
among people who do not have printed statutes. Books enable one to forget

such mnemonic signs (and the intellectual discipline which they imply) or to reduce them to alphabetic ciphers. But reference to the visible world of daily life is a way of teaching and remembering. It is more than a convenience among illiterate peoples. It reflects a larger philosophy of compliance and subordination to systems beyond those created by humankind—not in the sense of a dumb bowing down but as a stimulus to inquiry and observation. The actual use of marriage rules, like all conventions, is frequently tested and bent in individual cases. Animals are a constantly novel presence whose natural history fertilizes the conventional wisdom. Social rules may work in the ordinary course of things. But in complex situations or punishment for violations, the reservoir of natural history provides an unlimited tool for deliberating on the means and consequences for resolving such situations by creative, analogical extension.

The correlation of food, sex, and space exploits an appropriate parallel of scale. Embedded in the code is the fundamental dyad of sex and death—marrying and eating (by killing something). Both are consummations in the cycle of life: assimilation is linked to renewal. The connection of the two appeals to our sense of symmetry and order, to prelogical understanding, and to the poetry at the heart of meaning, justified by old stories, sometimes forgotten or lost, that refer to common origins, as though both the people and the animals shared a single rule in the beginning.

ANCHORING THE CODE in a distant time is sometimes made explicit. The Eastern Tukanoan Indians of Colombia speak of their ancestors as "tapirs" (a large, native, South American jungle pig). They call the tapir the "old man" of the forest or "father-in-law," using the word *behkara*.[13] Tukanoans refer to the nearby Arawakan Indians as "tapirs." This reveals ambivalence about the similarity and yet difference between themselves and the Arawakans and ambivalence about humankind as like and yet not like animals. Tukanoans call their own ancestors "tapir people" and "deer people." A myth tells that the deer and tapir once intermarried to produce the Tukanoans, who have deer-mothers-in-law and tapir-fathers-in-law. This tradition is elaborated in star constellations, seasonal festivals, and food laws.

Individual differences among the animals are noticed: their color, sex, age, marks, wounds, or parasites. The forest animals are perceived as members of a household of regeneration from which all the different game animals are sent by a powerful master, as food for humans, and to which the spirits of the dead game return, circulating as both food animals and souls.[14]

The tapir and the deer are said each to be like a different human personality trait, and therefore rather odd mirrors of humanity. The tapir is large, heavy, and fast, sometimes appearing clumsy as it rushes in panic through dense underbrush. With its conspicuous genitals it is regarded as sexually very active and an eater of foods which increase fertility. Like other pigs it gives birth to large litters. Tukanoans ridicule the tapir as "big testicle" and "big ear," but there is envy in their remarks. To this end they have a horn ceremony simulating the tapir's voice, which is said to make pollen fall, fertilizing the blossoms of forest fruits.

Any tapir encountered in the forest may be a transformed person or simply a tapir. Though good to eat, tapirs are clearly souls as well, and therefore not to be treated casually. They are hunted with cautious respect during an annual hunt in which the death of tapirs is said to balance the loss of human souls by infant death during the hard, dry summer, making the hunt "an ecological stock-taking and a balancing of books." Food prohibitions at the time of the annual hunt and their own human renewal and descent from these sexual animals draw together the two forms of consummation, sex and death, toward a dream of that moment in the beginning when the terms of a metaphor lose their separation. Behind metaphor is the still more profound state of consanguinity, as though the distancing necessary to the rational analogy were merely a temporary fabrication.

THE DECLARATION THAT a person "is" such and such an animal need not, however, be rationalized as an origin myth. Men of the Bororo Indians of central Brazil say "We are red macaws," which sounds to outsiders as if they minimize their humanity or oversimplify their circumstances by calling themselves parrots—that they hope to become parrots after death or think of themselves as descended from parrots. Instead, they are making a familiar and ironic comment on their own social circumstances by means of reference to the macaw's situation.

Red macaws are the only pets kept by the Bororos. There are other domestic animals—chickens, dogs, and pigs—but the people show no affection for them. The macaws are the property of women. They are admired, well fed, groomed, given proper names from the owner's matrilineage, taken on trips, and become part of an estate. Allowed to wander freely in the village, they are regarded with indulgent pleasure and protective care. They are never punished or eaten, seldom sold, and mourned when they die. The macaws are "serious personal property," however free they may seem.

As wild species living in the penumbra between forest and human settlement, their status is dual. They are regarded as part soul and part body, liminal beings in whom are joined radically distinct physical and mystical parts.

As pets they embody opposite principles, since the birds are free yet captive, wild yet tame. In no sense is any macaw, wild or tame, seen as "merely" an animal in the sense of being an object. All macaws are believed to be sentient and sometimes inhabited by spirits who cannot enjoy sex and food unless embodied. These spirits, the *aroe*, temporarily occupy bodies in order to be incarnate and participate in physical pleasure. People also perform these spirits in rituals having to do with naming. In the ritualized presence of the *aroe*, opposing moieties or other social entities convene, bringing together what is normally separate.

The Bororo reference to their "macaw" identity grows from the custom that newly married men move from their mother's house to that of the wife's uncle, presided over by females of the wife's family. Each man feels himself to be an intruder in this female-dominated household—to be free yet constrained. By controlling food and procreation, the women bind masculine loyalties and fetter their freedom just as surely as they domesticate macaws. Yet the man does escape in the course of the ritual enactment of the totemic spirits, when he wanders from his body, just as the *aroe* or spirit can leave the body of the parrot. During these ceremonies the spirit can also enter the men. But the existential dilemma they share returns in ordinary life. Indeed, the conflicting constraints of their birth and marriage are seen as analogous to the captive parrots.

The nuances of this "parrot code" sift through the details of parrot life, their behavior, taxonomy, natural history, and appearance. To "be" such a bird is to say: "I, my mother's brother, my sister's sons, and all other members of the *aroe* (the spirit of the macaw) clan share a cosmological and social status, which is reflected in the red-breasted macaw's pure form."

Life for Bororo men who call themselves "red macaws" has many more subtleties that bear on the traditional binary division of the people into two groups or moieties: the ceremonies in which the men may become the *aroe*, the presence in the forest of three species of macaw, the shifting details of the relationship of men to women as they grow older, analogies to the macaws' mythic relationships to other animals, and variations on the theme resulting from individual circumstances and personalities. The implications carry over into daily life in diet, clan membership, decorative styles, roles in ritual sacrifice, and other connotations.

SMALL-SCALE, EGALITARIAN societies, like the Bororo, tend to explore single-species metaphors with greater freedom than do more centralized regimes. An odd but persistent association of political structure and the use of animal metonyms seems to separate authoritarian from egalitarian systems.

And it results in a subversion similar to that just mentioned. In highly centralized, dictatorial societies a one-dimensional allusion controls the metaphor. The elephant offers a good example.

In Africa the elephant has been a major focus of cosmic and social meditation for many centuries. In a recent traveling exhibition, arranged by modern art museums, a large number of very beautiful objects, made from ivory or other parts of elephants or representations of them, show their extraordinary prominence in tribal cultures. But examined in the context of their purposes in the cultures, rather than their quality in the eyes of the connoisseur, the objects reveal a radical difference of meanings, depending on whether the society from which they come is a political state shaped by "obedience" or a less centralized people. In small-scale, egalitarian groups the sacred and totemic place of the elephant is paramount. Here the object's allusions tend to be cosmic—to illustrate mythic roots in the whole community of life and place. By contrast, among groups centered on hierarchic chiefdoms and monarchies, with their chains of command and the privileges of royal lineage, such objects seem constructed to celebrate and transfer power among the human ranks. Here the elephant symbolizes force and superiority. Instead of a cosmic avatar it has become one more ministerial device, conscripting all the elephant's awful integrity as a free and spectacular being into rites of subordination. Rather than invoking the spirit of the elephant as a holy presence, the head of state sits in his throne of tusks to hold sway, controlling the people just as the elephant is said to dominate the other animals. As in Mogul India, where absolute rulers and generals rode elephants over their enemies, the animal is subordinated to the human dictator, although African elephants were not tamed and ridden.

IN PATRIARCHAL SOCIETIES the sexual roles of men and women have a rich animal reference along similarly ranked lines. Such a typology requires a legacy of stories and widespread common observation and beliefs about the sexuality of the different animals.[15] In India an elaborate sexual code denotes individual men as hare, bull, or stallion and women as deer, mare, and elephant. Partners whose sexuality is balanced are the stallion and elephant, bull and mare, and hare and deer. According to the *Kamasutra*, other mixtures throw sexual relations out of kilter.

Although such metaphors are highly durable, their edges can wear off in time and they become crude, or turn into literary tropes. In the Mediterranean region, for example, the contrast between goats and sheep has long been employed among peoples from one end of the sea to the other. In ancient Greece rams were sacrificed to gods or goddesses to show gratitude for a bounteous season, whereas a goat was sacrificed to appease wrath—the original scapegoat. The distinction is based on differences in the animals: the sheep docile, pure, enduring, and noble, associated with verdant pasture; the goat devilish, sensual, cunning, stinking, noisy, making pasture unfit for sheep. The ram is virile, strong, and fierce, defending its harem and its territory against rivals. The goat, its inverse, tolerates access to females by other males in its domain. The ram gathers and keeps its wives by formal combat and never permits betrayal and promiscuity; the billy is both betrayed and betrayer. The difference can be reduced to right and left, honor and shame. From Spain to Persia, women have been regarded, like both ewes and nannies, as morally weak, more vulnerable than men to uncontrolled sensuality. Because of woman's supposed "natural" carnality, it was assumed they would yield to temptation. In rural Italy their honor is traditionally placed in the hands of their husbands, brothers, and fathers. In some places only men care for the sheep, only women the goats, as if to say that women are naturally associated with promiscuous animals, being of a similar degree of carnality. In Spain and Portugal a man who tolerates the seduction or adultery of his wife by another is a billygoat or *cornute*, the wearer of horns. It is implied that such a man who tacitly consents to his wife's looseness is himself without honor. The cuckold is dominated and defeated by another man through the seduction of his wife. He is not only deceived and dishonored but of dubious virility. He is equivalent to a castrate among the goats. Unable to face others (suffering a "loss of face"), he finally cannot even appear in public.[16]

Analogies to the distinction between sheep and goats arise in pastoral, patriarchal societies where there is probably an underlying admiration of the goat's sly "horniness" by men, although, in its more puritanical expression, the distinction is absolute, as in a 1662 poem by Michael Wigglesworth:

> At Christ's right hand the Sheep do stand,
>   his holy Martyrs, who
> For his dear Name suffering shame,
>   calamity and woe,
> At Christ's left hand the Goats do stand,
>   all whining hypocrites
> Who for self-ends did seem Christ's friends,
>   but fostered guileful sprites.[17]

In more recent times, class distinctions have assumed the convenience and logic of animal symbolism. As the bourgeoisie set out to define itself from both peasants and nobility, it employed animal associates among domestic forms that had long signified status. In seventeenth-century Sweden, a family horse was handsomely kept for going to church in order to make a good public impression. Kindness to animals became part of this distinction. Except for pets, who were given personal names and were not eaten, "animals were something you read about or looked at, rather than things you handled in everyday life."[18] The new middle class separated its groomed and housebroken pets from farm animals in the same way they saw themselves separated from peasants and laborers, controlling impulse and instinct in their animals and themselves, and, in the process, the animals came to represent the classes with which they were associated.

In the bourgeois world birds were clean, refined in their sexuality, devoted as mates and parents, homebodies, singing, industrious. They were felt to be refined "paragons of bourgeois virtues," members of nuclear families of their own, natural symbols of middle-class virtues. Bad birds, predators and scavengers, were like the occasional criminal; the grubby sparrows in the streets were a lower class. The birds' instinctive behavior was one side of a tension profoundly experienced in proper society between indulgence and restraint. The clean and virtuous side of birds was a Victorian moral example, while their defects confirmed the superiority of ourselves to animals, or at least our opportunity to become so. Birds became a veritable catalog of human traits within the new mercantile economy, represented in the vernacular by projections of role and personality onto persons as vultures, old hens, cocks of the walk, parrots, hawks, lovebirds, geese, magpies, turtle doves, or cuckoos. Cheating in business or marriage, avarice, the hidden vices of a decorous society, were evidence of the natural self hidden beneath the plumage of acceptable housebroken animality.

In the twentieth century pets continued to signify status and class, evident in advertisements in which high-bred dogs and horses, or exotic lions and eagles, connect social status with choice brands of automobiles and other commodities.[19] Abstract categories in human thought continue to require concrete reference. The species model implies an imaginary, ecosystem-like superstructure. Such names embrace activity and color and are a vivid counter to the reduction and stylization of categorical signs in museum art, the tendency to become hieroglyphs. Groups tend not only to equal species taxonomy but to borrow it for the names of clubs, genders, corporations, political parties, age groups, or teams (bulls, timberwolves, hornets, bucks, broncos, dolphins, cardinals, rams, eagles, buffalos, tigers, panthers, blue jays, hawks, orioles, sharks). There are Lions Clubs and lodges of the Elks and Moose, loan sharks, hawkers of goods, militant hawks and pacific doves, colonels wearing eagles on their shoulders, squadrons of flying tigers, an immense, self-typing of pet-keepers, boy scouts belonging to wolf and beaver dens. All inherit the idea of making groups visible by an identity with species of animals. Thousands of athletic teams are named for animals—not only because the "panthers" are synonyms for ferocity, but because the taxonomic system creates the cognitive facade of difference. That there are more "hornets" than "banana slugs" and more "lions" than "anteaters" attests to the poverty of social/ecological analogies and to the deracinated, modern appeal to the raw strength of animals.

The modern world of work roles, professions, and mix of religious, political, psychological, and economic classes opens unlimited possibilities for the cloak of animal species—extending on the one hand the skin-wearing distinctions among Swazi aristocrats and Renaissance European merchants and, on the other, the totalitarian obsession with sheer animal power. "Does not society turn man," wrote Balzac, "according to the settings in which he deploys his activity, into as many different men as there are varieties in zoology? The differences between a soldier, a workman, an administrator, a lawyer, an idler, a scholar, a statesman, a merchant, a sailor, a poet, a pauper and a priest, are just as great—although more difficult to grasp—as those between a wolf, a lion, a donkey, a raven, a shark, a sea-cow, a sheep, etc. There always have been, and there always will be, Social species as there are Zoological species."[20]

Clothing still signifies social or economic status, sometimes with leather, feathers, and skins as well as animal pins and jewelry, as it has perhaps from the beginning of human society. Until recently the sable collar marked the bishop among the clergy as it now separates the affluent from the poor. Costume adds to our own native pelage—or our naked absence of it. The derivation from the animal is indicated in fashion's "coordinate" sense. Only in a society rebelling against traditional signification does one mix the parts of one symbolic attire with parts of another—beaver-skin top hats with bull-skin sandals or the lion's pelt with the belt of cowhide. Yet the chosen act of disarray emphasizes the power of the signs. Worn by humans, the pelt or feather ceases to resemble the animal from which it came and persists as an abbreviation, as in the yellow and black football uniforms of the "tigers" and red of "cardinals," or as logotypes, stylized forms, ideograms, and, finally, letters of the alphabet. The eventual form of the badge is not important. What matters is that it represents a species idiom, a means of belonging in the world. Clothing and other adornment based on animals became the mark of social difference because it is the most efficacious way ever discovered of representing the gaps so obsessively desired by society with its bone-deep primate heritage. Nothing else in the given world lends itself so perfectly to stressing disjunctions in a series and representing the perfect embodiment of typology.

The true richness of these social/natural metaphors depends on knowledge of the natural behavior of the animals themselves, on the diversity and complexity of those animals, a habit of watching we have lost by stages. We disavowed the animal mind and soul three centuries ago, reduced the beasts to representations of the appetites twenty centuries ago, replaced the wild forms in our environment with simpler domestic animals of fewer species sixty centuries ago.

Like wild herbs in cooking and medicine, notions of tapir ancestors and parrot paradigms have the quaint air of rustic stories. The species system of animals as images of their own social rules may seem far indeed from our own habits of thought.[21] Yet in a time of deeply felt loss of "nature," which we mistakenly think of as a kind of pastoral anodyne, we may wonder about the necessity of "thinking" animals. Terence Turner observes: "That human (cultural) beings should . . . represent social and cultural phenomena to themselves through the symbolic medium of animals . . . raises a number of questions. . . . The 'nature' incarnated in animal symbols is not simply the biological domain of animal species, adopted as a convenient metaphor for human social patterns. . . . It consists of aspects of human society that are rendered inaccessible to social consciousness as a result of their incompatibility with the dominant social framework. These alienated aspects of the human (social) being, which may include the most fundamental principles of social and personal existence, are therefore mediated by symbols of an ostensibly asocial, or 'natural,' character."[22]

It seems to follow that there are political and social adjustments which would make us less alienated from the "most fundamental principles" and reduce the need for animal metaphors. But there is another interpretation: that the alienation of which he speaks is a result of the burden of self-consciousness, of the human condition, our being what Neil Everenden has called "the natural alien." If this is true, no philosophy will end it, and we may freely accept and affirm our peculiar use of animal terms for social membership as part of the ecology of mind. It is one small bit of cement that helps keep us connected to our animal kin while other aspects of our culture drive us away.

PART IV

Change

As vicarious models of our own inner state, and ourselves, the Others are the concrete expression of a dialogue between the internal and the external. But more, they represent the art of becoming. In their metamorphoses and their movement through the penumbra of places and form, we see them as masters of transformation. In dancing to the rhythms of change they become our own guides, healing and mediating our own passages.

9 | The Masters of Transformation

The fact that the animal species, in contrast to man, are comparatively stereotyped in their innate releasing mechanisms has made them excellent representatives of the mystery of permanence in change.

JOSEPH CAMPBELL

SOME LIVE BY affirming time and change, in the knowledge that all forms give way and physical states are provisional; others seek to escape the transient appearance of things by emphasizing essence and eternity. The first path may be identified with "primal" culture, the second with "civilized" culture, a distinction obviously artificial, merely a useful fiction. The difference is best understood as a matter of degree, for all people are "primal" in that they live in a world of incessant alteration. For us all, the idea of transformation, in all of its meanings, is made flesh in fellow creatures.

Those for whom change is the way of the world are centered on traditional story and recited genealogy. Voices—human, wind, and animal—signify brevity and religious imagery rich in beings who turn into other beings at the moment of a cry or the end of a song. Sound resonates from inside and refers to the quick and vital passage of breath and thought. Theirs is a world of manifold, internalized otherness, a place of crossings, transformations, incarnations, and metamorphoses, where there is no dishonor in becoming an animal or returning to dust. Their shelters of skins and wood, like their own tracks or themselves, are expected to vanish into the earth.

Such societies differ from those for whom the story of existence is the "novel," a text progressing away from the past rather than recovering its timeless presence. Modern people, amid an architecture of steel and concrete, pursuing High Culture and political stability, look to the fixity and logic of print as visual truth, as if it were a stone edifice, and are suspicious of the transience of sound—of mere oral tradition—and the fragility of living forms which produce it.[1] When Romeo pledges his eternal love "by yonder sacred moon," Juliet objects, "Oh swear not by the inconstant moon." Romeo, who is already plagued by contradictions, his family reviled by hers, his own feelings flipping from hatred to love, claims his devotion in mineral constancy. Against mutability and dissolving ties, looking beyond permutation, like him we want a true state, an essential, transcendent self, a final "I am."[2] With us, he belongs to a mindset placing hope in stone tombs, shunning impermanence, fearing the corruption of death, avoiding the twilight zone between human and beast.

There are many literary expressions of this fear of losing one's humanity in the form of a fateful permutation: Odysseus' transformation into a pig, the alterations in Ovid's *Golden Ass*, Marie de France's werewolf, the punishing, shifting incarnations in Dante, the *Metempsychosis* of John Donne, Kafka's hero changed into a beetle.[3] Kafka did not invent the fear of human/insect transformation; he merely used it to express the terror to which our culture is heir. Caught in such passages or in contemplating death itself we wonder whether one retains one's own consciousness, carrying it between incarnations like a little egg in a basket, or is it (we fear) lost with the body? The historical/literary ego prefers a subjective survival of bodily reincarnations, an "I" preserved, as if one had merely changed clothes, keeping the little homunculus soul who drives the body like a captain at the helm of a ship. It is the hope of "world" religions, "classic" literacy, and computer "culture" that individual memory can be immortal, even downloaded into a machine if necessary. If the inner self is an inviolate entity, different kinds of beings cannot truly coalesce. Altered states, as in werewolves, digestion, metamorphosis, and rot, are the unraveling of order, marring, obscuring, and thwarting the true self.[4] The merely organic body and its changing appearance are not to be trusted.

Yet we all "came" from invertebrate animals, whose presence in us is coded in our chromosomes and mitochondrial DNA, in cells immersed like coral animals in fluids as old as the sea. A whole fauna is in us still, tacitly. The human body is unlike and yet not opposite to those Others it incorporates. This Möbius-strip quality of being—the simultaneity of two states, of transformation and kinship—reveals how tentative is the idea of a fixed self/body separation or unlikely an absolute distinction between ourselves and animals.

THE HUMAN LIFE cycle, our *ontogeny*, is a series of changes and part of that animal heritage. The passages of life in traditional societies are formalized in ceremonies and stories that give them concrete expression in performance and imagery. Change is made objective and then becomes internalized, true to the self by the skillful and poetic use of external forms such as the weaving of animal changelings into thought by means of art, ceremony, and myth. Being born with all of these transitions from infant to adult (and their animal equivalents) within us implies that we are actually composite beings. Without traditional affirmation of our recurring emergence, we learn to fear our plural nature and despise the creatures who signify its stages.

Such stories tell of humans arising from the marriage of the first people to animals and convey images of our multiple nature. At El Juyo, a 14,000-year-old cave sanctuary in northern Spain, there are mounds constructed of alternating layers of colored earth, vegetation, animal bones, and spear points, one-ton slabs with smoothed edges, circular fireplaces, trenches with antlers in them, colored earth in alternating bands, pits with shells, needles, and ochre, all carefully positioned and covered with clay. "Presiding" over all of this mysterious mélange is a large, head-shaped piece of sandstone, modified by human hands into a face whose left and right sides differ. The right side is human, with lines for head hair, beard, and mustache. The left is feline with cat whiskers. This dual face is thought by anthropologists to express "shared symbolic behavior," balancing opposition. The theme of the whole shrine, with its many objects and forms, is that life is essentially transformative, as reflected most emphatially in the double face. A large carnivore who mediates opposed principles is a traditional divinity in many initiation rites; the face represents that unity and the cave is the archaic site of its rites.[5]

Systems of names seem static. Yet among tribal cultures, movement appears to be the pervasive theme in their taxonomy. Different species indicate divisions between earth and water, between travelers by day and night, and between flying, walking, and crawling sacred beings. Hunter/gatherers tend to see themselves as subject to these life-stage changes, as the eggs of wild birds are to hatching, and to formally affirm the inescapable passages. Movement denotes shifts in our lives, beginning with the birth that gives us a first identity. Later we are reborn, passing through stages, each a "leap" between domains marked by thresholds, corridors, barriers, gates, and doors. Change as movement is stylized in procession and dance. The rhythms of seasons and climate, the cycles of plant life, the migration of animals, and the pulses of the celestial universe are continuous with those of the person, the phases of growth and age, health and work echoing the larger cycles. Links between the cosmic and the human are made by animal mediators who are expert in gestation and hatching, in larval and adult stages, in moulting, fledging, and shedding skin. Ceremonies that formalize human passages, the catalysts that transmute our being, borrowed from such images, are administered by a mentoring group in proper time with the psychological readiness of the individual. As reflections in small of a cosmic scheme, growth, marriage, aging, sickness, recovery, gestation, birth, and death can be celebrated without fear or psychosis.[6]

In societies in which the gods are like humans and control the world as humans control domesticated plants and animals, such rites include animal sacrifice. Transformation requires that the end of one form be marked as a death. The Akikuyu of East Africa say that the child is "born of a goat" which is killed as a surrogate for the death that must precede "the second birth of early childhood." The sacrificed animal is itself immature: a heifer, foal, piglet, kid, or lamb. Thus do ceremonies of transformation take shape from the cosmic realm of the sacrificial engine that moves the wheel of change. Christ, the Lamb of God, was "given" in a sacrifice derived from the larger context of animal sacrifice, and at baptism, a drowning death and rebirth, the Christian child becomes a lamb of Christ.

THE VILLAGERS OF rural Peru, thirty kilometers north of Cuzco, refer to the lion or puma, an occasional killer of their dogs and pigs, as "Machu Compadre" (male) or "Paya Compadre" (female), the same terms they use when referring to the grandparents of first-born children. The grandparents select godparents for the infant, which is said to be "taken" by the new godparents. Young men, hunting or "taking" deer in the high country, sometimes "find" or "take" newborn sheep or cattle, which are said to be born from the mountain springs. Both the hunters and the godparents are responsible to Machu Compadre, the puma spirit and mature "advisor" who communicates with the earth forces just as grandparents communicate with their children, the infant's parents. New parents are likened to the fox, Ni-nucha, or "young father." In the spring, as the people go up the mountain to plant crops, the fox may be heard howling as part of its mating behavior. The farmers must protect their crops against foxes by using slings, the foxes like themselves having "come up" from below. Even so, the howls foretell a fruitful season. They are regarded as the mythical spreaders of cultivated plants from heaven to earth as a result of gorging in a heavenly garden until bursting and dropping the seeds on earth. The fox crosses these vertical realms like the people ascending from the lowlands or adults moving up in their life stages, having their first child. Prior to their foxhood as parents, the young adults are "bears." The spectacled bear belongs to this cast of charac-ters, portrayed by Andean dancers as a boisterous, somewhat aggressive, sar-castic, sexual flummox with a typically adolescent falsetto voice.

These transitions center in the planted fields of potatoes and corn as the scene of the domestic responsibilities of young parenthood in which the young couple matures like the crops. Yet even while protecting the maize against depredations by wild animals and birds, the young men wear a fox skin. Full adult status is seen in the distinction between fox and puma. Al-though the human group and its fields are the principal concern, the wild spirits of bear, fox, and puma are those which most fully embody and em-power, by a kind of parallel logic, the seasonal events and the human life cycle. Perhaps the wildness of the animals reflects an unconscious human awareness of the stages of the life cycle as beyond their control in a way that domestic plants and animals are not. The social practices relating infant, ad-olescent, parents, grandparents, and spiritual sponsors are equated with a story of baby animals, bears, foxes, and pumas who become instruments of understanding and representation of their own society and economy. The different animal species correspond to human social groups, membership or *ayllus* categories, the individual's age grade, adult and elder status, each of which has specific social and ceremonial duties.

The animals are markers in the human life cycle: the dependent infant is analogous to a deer or newborn lamb or calf, the unruly adolescent to a bear, the gluttonous young adult to the fox, the monitoring grandparent to the puma, and the godfather or spiritual keeper in both realms to humans. Oth-erwise, the human passage through life—actually a series of levels or pla-teaus—progresses without "sufficient" natural divisions and without con-nection to larger phenomena such as the grand model of seasonality. The bodily changes—size, weaning, toothing, menarche, and hair growth—are concentrated in the early years. The rest, the other phases in the progress of the individual, are flagged by comparison to the gaps between the different species and their ecologies. The true focus is on emergence and transformation. The fox and the other animals are "movers" by saltation, as we humans are by analogy to them. No mere travelers, they leap through boundaries in the mixed pastoral and wild landscapes of submontaine Peru.

AMONG THE OTHER crucial transformations of life are those between sickness and health. A master among the transformers is the honeybee. Apart from milk, honey is the only bodily substance produced by animals and used as food by people. But its nutritional value is second to its role as a medium of metamorphosis. Bees live as a community, making a precise "architectural" structure, the comb. Hives contain an extraordinary sequence of life-forms in a highly organized and ritualized society with an enduring queen and an ephemeral king. Nothing quite so spectacular and suggestive occurs in the lives of cattle, even through their milk may have had greater impact on human life than honey. The qualities of the bee's *elixir vitae* are associated with genesis, renewal, and repair, and it is a substance used in human birth, marriage, and death rites. Some of the oldest known libations to the dead were honey, older than oil or wine or milk, although honey was later mixed with milk, perhaps as a complementary substance, causing the milk to ferment. Rhea's sixth child was Zeus, whom she saved from being swallowed by King Kronos by hiding him in a cave on Mount Dicte on Crete, a cave inhabited by sacred bees who fed him on honey and protected him. Zeus was called Meliasaios, "bee man," at first, and he had, by a nymph, a son who was named "Meliteus." Sir Arthur Evans found the remains of a libation table which served that purpose in the Dictaean cave. Actual bee larvae are "nymphs." In ancient Greek myth, goddesses called nymphs— Thriae, or bee-maidens—reared Hermes in a glade of Parnassos as if he were a larva and taught him soothsaying.[7] The nurses of Apollo were *thriae*, and Dionysus too was reared in a cave on Mount Nysa by nymphs, his sacred drink made from honey.

All the ancient Great Mother goddesses were served by *melissae* ("bees"). Between 1200 and 700 B.C. the worshipers of the Greek goddess Artemis ("bear") of Ephesus were Essenes, or "king bees," who got their name because they were bound, like larvae, to chastity for a year. The connection of worker bees, which are asexual, with Artemis, who was chaste, is clear. The Cretan name for Artemis is Britomartis, or "honey-taker." The bee is the source of the widespread figure of the winged Artemis, a motif of female head and bee body found on the jewelry and coins of Ephesus and Greek islands. Artemis was not a mother goddess but a *kaurotrophos*, a caregiver and protector. Her substance was not, therefore, milk but honey, in keeping with her nonreproductive role.

Any "land of milk and honey" would seem, in this light, to be one combining two different but complementary life-sustaining substances, one procreative and the other sustaining and transforming. One bonds the infant to its mother; the other engages the infant with a wider world of needs and care. The myth of Zeus and the *melissae* further affirms the protective power of honey. According to the historian Pausanias, chthonian or rock-cleft bees were associated with the worship of Rhea, Demeter, Persephone ("Melitodes"), and "Black Demeter" in Arcadia by means of a honey offering. No doubt caves were the actual homes of bee swarms as they are in Africa today. Such stories connected caves, hollow trees, bees, and netherworld with the goddesses.

In metaphors based on hibernation, the bear is the primordial traveler in the underworld and its bones are used in funerary rites. Offerings to it were made in the form of its favorite food. Thus honeycake offerings were widely used in netherworld entrance rites. Spartan women offered honeycakes in a cave near Mount Tagetus. In Lucan's *Charon*, Hermes speaks of such libations, and there are funerary uses of honey in the *Iliad* and *Odyssey*—for instance, Aeneas near Lethe encountered "bee-souls." Like birds, the bees could incarnate the souls of the dead. Honey was used also to preserve the body after death. A votive offering of honey in Socrates' prison was seen taking place as late as the nineteenth century.

The Oracle at Delphi was called the "Delphic Bee." Not only oracles, the bees were also weather prophets. A swarm led the Delphic envoys to the cave of Trophonius at Labadea, where the men sought relief from the heat and made offerings of barley cakes kneaded with honey. The chief upperworld divinity to whom honey was offered was Helios. There is an old story that honey fell from the sky and was gathered by the bee—an idea consistent with the observation by anyone whose automobile, parked beneath the flight path of worker bees, is speckled with tiny drops of spilled nectar.

From ancient times the honey-colored lion and bee were connected, perhaps because they shared habitats, as forests diminished, in isolated rocky outcrops, the last stronghold of prides of Mediterranean and Asiatic lions and the cave and crevice sites of hives. The sun, gold, lions, and honey all had royal associations. The Greek philosopher Porphyry says, "bees are begotten of bulls." As lions disappeared, the association was shifted to bull and bee. Bees were said to be ox-born in Egypt, probably attracted to the dead bodies by moisture. In lands where forests were depleted and cattle were common, the weathered skulls of bulls may also have served as hives. New swarms, lacking tree hollows and rock clefts or caves, may have occupied the bleached crania of the dead oxen.

Among the Finns the "Little Mother of Honey" is a forest spirit invited to the "wedding" or ceremony of the slain bear. Maybe the Finns had observed bears being pursued by bees whose hives they had raided. The connection between bee and bear was not only the latter's appetite for honey and their casual association in hollow trees. Both were major chthonic spirits, as caves and tree holes are the dens of both bears and bees. Both disappear and reappear seasonally into the earth. The northern hemisphere story of the bear grandmother of humankind is paralleled by the myth, described by Mircea Eliade, in which bees were widely held to be human ancestors. Those animals which undergo dramatic transformation are the best representations and masters of change. Honey, the supreme substance of nurturance, healing, and metamorphosis, was produced by the masters of the season and of prophecy, the models of the radical alteration of the immature to adult. Honey was the food of larvae and larvae became food for thought.

Honey is a mix of nectar and pollen from different blooms which are as seasonally predictable as the monthly full moon is predictive of werewolves. The bees advise each other on the direction and distance to the flowers, and their honey is the spice of life to the preeminent holy mammal, the bear, the forecaster of equinoxes, seasons, herb sprouting, salmon runs, berry ripenings, and elk births. So it was for centuries to us, who, watching the bee and bear, found them exemplary.

When beer was first made among the Finns and Saxons it was mixed with honey, which caused it to ferment. Beer (from the same root as "bear" and "bier," the slab on which the dead were borne to the grave, just as the beer itself was fermented underground) was originally mead-beer, "Saxon bee," or "beo." Such mead-beer preceded modern beer, which was unknown before the ninth century. "*Med*" (as in "medicine") is the Mordvin (Finnish-Ugrian) word for honey. In the ancient Finnish *Kalevala*, honey is a healer of the sick and injured and restorer of the dead. It is the basis of salves and ointments. In Wittenberg, Germany, before the Reformation, some churches used 35,000 pounds of wax a year. On Candlemass Eve, hives were decked with ribbon and a song sung as people carried wax candles, beginning, "Bees awake . . ." The thousands of tons of beeswax used in medieval churches not only illuminated them as candles and gave the sweet smell of incorruption but continued an older tradition associated with earth temples and the language of renewal long before Christianity.

AMONG THE RENEWERS the frog is the most familiar. Of all the great enchantments by which we conceive of the mystery of change, the frog is chief. The wonder of its life in ponds as a fishlike tadpole, followed by the growth of legs, enormous eyes and mouth, and life on land, is matched only by insects such as the dragonfly which climb out of the same ponds and fly away— and are justly celebrated in some societies. Everyone for whom frogs played a part in their childhood can remember the fleets of pollywogs, the vicarious joy of hopping, the fun of catching the frogs in the algae or pondweeds in shallow water, chasing them in the sedges and grasses, watching them swim and dive, holding the muscular, cool bodies, the retractable eyes with their protective lids, their grasping hands and webbed feet, their chorusing, the tree frogs that are colored like bark and the toads who "come out" at night. What would fairy tales be without the enchanted forms locked up in the bodies of frogs and toads? Beyond all bees, butterflies, and dragonflies, the frog is the preeminent manifestation of the transformational principle as it hatches in the water and then climbs out, changed totally, just as we ourselves begin life in the amniotic pond and come ashore at birth. Similar principles apply to our shared life cycle and deep history, as they are representatives and we are inheritors of the evolutionary emergence of vertebrate animals from water to land.

Images of frogs reverberate in human culture in those leaps that carry us between domains. Folktales abound in frogs and toads, their poisons, their favors, their fantastic demands and miracles. Larval amphibians or tadpoles, legless, armless, and with fused head and body, are like free-swimming embryos. Sensitive and vulnerable to the temperature and ambient air, frogs remind us of our own vulnerability, while toads, intrinsically interesting if not repulsive, suggest imperviousness; we are in awe of the grotesque endurance we see in them. But they too, in the end, must return to the water to breed or reproduce, contracting the transitions from egg to larva to maturity into just a few days.

At the end of the twentieth century the amphibians are interlocutors between humanity and the planet. Their reduction in numbers and the loss of species is catastrophic. In their worldwide demise they are signs of the earth's malady. The frogs, toads, and salamanders, presiding over the shoreline between tailfins and legs, gills and lungs, land and water, seem to be making some final gesture as a symptom of planetary shock.

Recently there has been a recovery of the old idea of Gaia: the Greek mother earth as a living being. Such an image bears the corollary that she, too, is subject to change, not only in the normal seasonal cycle but in health and sickness. Perhaps she is the princess who needs the enchanting frogs in order to be well. The living community has been deformed by our irresponsible pouring out of poison in a mesmerized illusion of autonomy and scorn for the juicy pullulation and shameless, naked vitality of swamp life. Our only hope, according to the message of the prince of change, performed by his own death, is to affirm and protect this elemental earth and restore the frog to the princess.

THE REALM OF animal cures for human maladies is immense. And cures are but one of the transformations that all humans must face, from birth to death. Ancient peoples believed themselves to be participants in the rising of the sun and the change of seasons. If that seems like a quaint fancy now, perhaps it is because of a lapse of participatory understanding—of the sense of our own lives as corresponding to and part of the larger events of the universe, an understanding informed by the ambassadorial presence of other species.

In this chapter, I have described only three examples of the ways in which certain peoples have perceived those connections and transformations by means of an animal idiom: the movement through life stages as corresponding to a series of wild species who are the actors in a single story from Peruvian farmers; the bee as maker of medicine and the ultimate model of the nurse who monitors growth from nymph to adult, from illness to health, from life to death; and the frog who demonstrates that we have the potential to become fully human in an environment of patience and kindness as well as the potential to destroy our Mother Earth. These animals, and many others, are, as James Hillman has said, bearers of our souls across boundaries and borders.

The primal human world from which we all come believes in these animal mediators, for they show that we are not actually shipwrecked on a desert island planet, but that our needs, especially the transformations of ourselves, have been provided as if in advance, in the form of fellow beings who constitute a great congress of shamans. If modern culture is broken from its own primal nature in the fear of time and its changes, perhaps it has lost faith in those masters and interlocutors who give form to our understanding and reassure us of the wisdom of transformation.

10 | Heads, Faces, and Masks

The faces of the beasts show what truly IS to us.

RAINIER MARIA RILKE

STANDING BEFORE THE mounted animal heads on his wall, the gentleman in the smoking jacket in the cartoon says to them, "I suppose you're wondering why I have brought you all here." We, who no longer serve whole pigs at Christmas with apples in their mouths, laugh at this grotesque spectacle of the sportsman and the stuffed heads. But such a macabre scene leaves us uneasy. In a world in which pavement and machines are replacing wild things it has become fashionable to scorn the hunter with his trophies. Yet even those who see the heads as evidence of human cruelty are moved in their presence, not only by their sympathy for the animals but by something more gnomic, less certain.

The presentment of heads has a long and strange history, an enigma in which the face and the mask are keys, and the mounted head looks into the room as though its body extended through the wall and the animal had paused in the act of moving through it. No matter if the hunter himself is a vile drunkard or a heartless squire, a roomful of such heads is like a council, as though they were bringing different potencies together in some kind of arcane consultation.

There is an archaeology of collected heads which has little to do with the measure of a person's skill in the chase. In England alone the evidence is suggestive of ancient, widespread customs. During the repair of St. Paul's Cathedral in London from bombs in World War II, an inner wall was discovered in which the heads of oxen were entombed. Celts and other Germanic peoples "hung the heads of sacrificial animals on trees."[1] Goats', pigs', and horses' heads were similarly treated. An old Anglo-Saxon ritual sacrifice of oxen had required the separation of the head from the body and its special preservation. At Harrow Hill in Sussex more than a thousand animal skulls were buried together. Place-names are associated with the heads of stags, sheep, cats, ravens, badgers, snakes, and eagles, evidence of "animal-head ceremonies." The Anglo-Saxon "Mother's Night," December 25, and the *guili* or Yule celebrations of December and January were marked by feasts centered upon a boar's head. The long scrutiny of the characteristics of heads is indicated by 168 forms of "headedness" in Webster's *New International Dictionary*, from "addle-headed" to "yellow-headed."

Doting on heads has aspects far older than iron-age Europe. At a biological level, carnivores such as hawks and cats eat the brains of their prey first. Perhaps there was a time when we, too, preferred the heads of the killed game. Holes drilled in the fossil crania of African baboons and other primates are evidence that prehumans may have extracted the brains. Neanderthal peoples left caches of bear skulls in German caves. At the culmination of a Winnebago Indian feast at which I was a guest, a deer cranium was handed around, opened through the palate from the underside, into which each participant dipped his fingers to scoop out a bite of brain. Brains are widely relished by farm and village peoples in central and southern Europe and on American ranches and farms.

Perhaps because our own personal consciousness seems to be in our heads, the head is widely regarded as the place of sentience, or of the spirit, and therefore its last residence at death. It follows that if the spirit of a dead animal or person persists then the head deserves particular respect. In the circumpolar ceremonies of the slain bear the head is placed on poles or scaffolds or in trees, and the watchful soul of the dead bear is thought to hover about the skull, aware of everything as if the animal were alive. If the corpse, especially the head, was accorded due honor, the soul would eventually return as a new bear in another season.

The history of reverence for heads is speculative, but, given the foregoing facts from recent observations of tribal peoples, the head hung in the tree, on the pole, or on the wall might sometimes have been brought to festivities rather than the reverse. As a portable object it could be carried in procession or in a dance, and in being danced its characteristic movements in life restored. Hollowed out, it could be worn. Putting on the severed head of an animal in order to vivify its sacred meaning was apparently customary for millennia, to judge from artistic evidence, myths, and the anthropological records of ethnic peoples, even continuing into the present century. It remained only to make a substitute head to achieve the effect. The carving of heads and the widespread apotheosis of the *caput*—from bronze-age herm, classical bust, and sculpted relief in the architecture of churches to the heads on coins, modern commemorative profiles, and the photographer's portraits—all seem rooted in this prehistoric practice.

Although the spirit is thought to be associated with the whole head, the face is its primary aspect. Indeed, "aspect" means "the act of looking" or facing—looking not only by the observer, who is transfixed by the stare of the severed head, but looking by the head, as if there were a beam from the eye. (In all the hundreds of painted and carved figures in twenty thousand years of the cave art of Pleistocene France and Spain, only lions and figures believed to be shamans look directly at the observer; the rest are profiles. Apparently the function of the painted and etched figures in those labyrinthine tunnels was not the same as the preserved head: the animals represented were intended not as encounters with an observer but as objects of meditation.)

The face has its own biology. The differences in faces among all the vertebrate animals are simply a matter of skewing the same parts—configurations among mammals of jaw and cranium, forehead, nose, ears, lips, eyelids, lashes, brows, and hair, feathers, and scales. Faces include a wet or dry nose, chin and cheeks with or without whiskers, eye details such as iris color, the shape, mobility, and position of the pupil, mouth and tooth patterns, and contour of bones of the cheek, brow, and chin. The human face is a relic that tracks our long evolutionary journey, a reworking of a basic set of bones and muscles in different planes, each diminished or emphasized according to the needs of the jaw, eye, braincase, and breathing apparatus.[2] "Countenance," a synonym for face, has its roots in a term for "behavior" and also means "approval," which is to say that we, like all higher primates, study the face to ascertain intentions or responses by grimace. We combine playful, sad, happy, sleepy, and frightened faces with gesture and voice, watching people speak in a way that listening to the bugling of an elk or the call of an owl does not require.

The ears, lips, teeth, and position of the eyes on the head reflect food habits. Facial muscles are not only for chewing but eye and mouth positions, for squinting, vocalizing, and sniffing. The faces of animals are therefore a key to their ecology, and a good paleontologist can recreate a whole biological realm from a few fossil faces. They condense information which is a clue to appearance, not only of themselves, but of the world in which they lived.

Members of an eye-oriented and face-watching order of mammals, we have long been interested in espial and, on the part of the observed, the strategies of appearance. To the naive eye, the things of the world are what they appear. Twigs on a branch and stones on the creek bottom look like what they are, just as stinging wasps and bees are flagged with banded colors, poisonous caterpillars have warning marks, and dangerous owls have mesmerizing eyes. But the clever seer can be tricked by harmless mimics who boldly go in daylight with wasplike bands and poison marks, as though wearing masks to frighten us. Insects and birds look like twigs and branches, are striped to resemble grass and its shadows. Fish are shaped like seaweed; moth and caterpillar wings have spots like staring eyes; the motionless turtle gapes its mouth as a trap baited by the tip of the tongue wriggling like a tasty worm. These mimes and the mimed play out a vast game of statistical probabilities in which the wary predator and hesitant prey regularly test the legitimacy of the signals, making choices when confronted by an animal who seems to say, "Am I or am I not what I seem?"[3] The visual world is composed of signs and de-signs, cues and signals. I once saw a blue grosbeak and an indigo bunting together on a branch in an oak tree in the Ozarks, two closely related blue birds. They ceased singing and were briefly immobile, looking at each other. I felt as though I had witnessed a momentary double miscue, as though each were successfully masked to resemble the other.

To identify another being we look first at its head, which, with its countenance, are like the signatures of the different animals. Eyes are not only the windows but the lamps of their souls. The mounted animals' faces are fixed in death much as they are in life, narrow in their range of expression when compared to ourselves, as though keeping some special knowledge. It is this masklike characteristic which is uncanny. We learn that things are not always what they appear to be. And we know this, too, from our own human, mobile faces, which may reveal feelings or misrepresent them, like the wasp and its imitator. We can deceive by means of our expression in ways that the animals cannot. So we ask: Does the fixed visage of the animal signify a singularity, or does it hide its true state, like us who put on a happy face? How can we reconcile the masklike deception on the one hand and the tacit revelation of the face on the other? Deeply committed to the play of facial features and the power of expression, we find the immobile faces of other animals to be suspiciously concealing, or to be the guileless mind of pure, untroubled divinity—transcendent, serene, detached, innocent, knowing. Fish, amphibians, reptiles, and birds are even less expressive and the most uncanny. Even in the fox's "mask" we sense a being essentially like us and yet beyond us, in the guise of a special wisdom that denies the ambiguity of our own fluid look. Sheep, bat, weasel—each has a fixed species aspect, as if contemplating its own monstrous or wonderful secret, an idea made perfect, as if for our attention.

The range of animal masks is wide, as though connected to the play of sacred powers. But in our time we may misunderstand "play." Joseph Campbell argues that pretending—the game of "as if"—frees us from "the banal actualities of life's meager possibilities." He thinks that this special play is embodied in the mask.[4] But the idea that the mask is merely a little tweak in an otherwise tedious world is too easy; it may reveal more of the academic's limited sense of the diversity, complexity, and beauty of the natural world than of that world's banality. Campbell seems to dismiss the mask as childhood frolic, just as we turn equinoctial festival fires into wiener roasts and Halloween into sartorial games of social roles and disguises. In this view, masks are commonly associated with clowns, marginalized tricksters, and worn as "false faces" for "costume" parties. Small children, before about five years of age, have not learned to finesse arbitrary appearance and are frightened by masks, just as six-month-old infants cry when they confront a strange face too close. Older children learn to enjoy this fear as the enticement of safe danger and to play at it—to lay the foundation for adult amusement or lawless concealment or the connoisseurship of grotesque exaggerations that anticipate the arts of caricature and satire.

THE PLURALITY WHICH the masks represent collectively seems to apply to us; every species appears to manifest an aspect of ourselves. The changes in our human faces with mood may be read as evidence that we individually are a community of feelings that surface independently, each of which may be represented in dramatic performance and its own mask. We have access to these vehicles of singularity, just as each animal has its own strength—the power of flight, contact with the earth, transformation, healing, foreknowledge, whatever our stories say of it.

For the bearer of the mask or the watcher at a ceremony, masks are a visible means of representing and easing the transit between alternate states—between now and later, body and spirit, this world and the other world. It teaches that reality is contradictory; it reconciles antithesis and impermanence. It both denies and affirms the wearer as "himself." Putting it on and taking it off is a becoming. The range of masks represents our mutability as an echo of cosmic diversity, asserts that transformation is central to reality, and helps it along. Such masks often represent the faces of animals who are marginal in habitat or masters of metamorphosis, as the fox and the frog. The principle which they embody is like the assertion "I am a young man, but I am also a frog" or "I am a maiden, yet I am also a bird," a way of saying "I am both physical and spiritual, animal and human, good and bad."

In this way the mask addresses categorical ambiguity—the question how anything can be two or more things at once—by subordinating the two to change, giving visibility to intrinsic multiplicity by enacting alteration, resolving paradox by accommodating it as transitional. Mimetic rites and dances with masks represent such movement. Drums and other percussive instruments accompany these performances because staccato rhythm invokes leaping better than does the flow of a melody. Certain animal figures refer to this process of the mask itself, to "crossover bearing." The effect on the observer is a leap in imagination. Stridulation reminds us of frog and insect calls—those who "know" metamorphosis and who are, therefore, two in the one. For the human psyche all marginal areas create anxiety, and passages are especially dangerous, but the lesson of the mask is that they are normal, needing only discipline and access.

As James Fernandez puts it: "The first problem is not how animals take human shape, but how humans take animal shape and enact nature."[5] "Metaphor" is from the Indo-European root *bher*, the same root for "bear"—the bearer. The mythology of the bear emphasizes its transforming and nurturing power as manifest in the hibernatory death and awakening new birth, the she-bear's monumental caregiving, weaning, discipline, and the skirmish of young adults in making a place for themselves in the social system, all boundary phenomena, all to be given metaphoric use in myth. The wearing of the bear robe by the bear-doctor-healer, anointing with its grease and the talisman of the paw, are agents through which we came to see the bear in ourselves—given its epitome in the face of the bear.

The many "faces" of sacred power represent not only the human personality but the plurality of sacred power. On one side of the "seated human figure bowls" used in puberty rites of the Salish Indians of the British Columbia coast is painted a seated female figure; on the other side is an owl, snake, or frog. Their faces have bulging eyes, large mouths, and heavy brows.[6] Such arresting and cautionary figures seem to anticipate those Northwest Indian masks in which a human face looks out of the mouth of a raven or bear, or the Eurasian figure of Herakles, the hero wearing the skin of the lion pulled over his head so that his face appears in its mouth. Here is the ambiguity of the hunter/protector, killer/preserver, eater/eaten, inside the animal, in its guise, yet a man. In the Christian/Hebrew/Islamic cultures of the West the man in the mouth of the animal is a fearful image, humanity trapped in beastliness and the jaws of hell, swallowed by a demon.

The mask which we make in the guise of the animal concentrates that essence—as if it were given to us in order to perceive the independent features of the cosmos. The animal masks invoke their authority. Neither a toy nor a disguise, the mask is an artifact of what Lévi-Strauss calls "the science of the concrete," and its meaning is exhausted in its features. David Napier writes: "Because of this special way that a mask pantheon combines religious belief with the cathartic value of mimesis, it embodies a type of order and a means of accommodating religious and perceptual paradoxes that legitimize appearance in a potentially constructive and educational manner.... The recognition of illusion is the single prerequisite for understanding something that seems self-contradictory—in other words, for recognizing paradox—and the recognition of change is possible only with this understanding: that something may appear to be something else."[7]

So we are faced with the nature of deception, or what Napier calls illusion, when we reexamine the apparent division between those primal cultures for whom the mask is central to their rites as the means of access to the truth. Otherworld, monotheistic religions beatify essence as superior to appearance, perceive the mask as evidence of evil intentions, and abandon its use in religious rite and secular drama. When we compare the two approaches, the question is not whether the mask invokes illusion, but whether such illusion reconciles apparently contradictory qualities as both true, or whether it is evidence of a world in which all appearance signifies chaos and heresy. Perhaps the answer is yes to both questions, since, in the aggregate, the masks demonstrate that no single conclusion constitutes the whole.

In ritual the animal mask reveals the limits of taxonomy, of thinking of ourselves as only one thing. It embodies the concept of transition or change, making it a visible aspect of ourselves, and it demonstrates the division of the powers of the world into many strongholds. Altogether, masks address the world as dense with incongruity and contradiction by showing that both sides of the apparent contradiction are true, as paradox.

Historically, scorn for masks reveals a cultural attitude toward the nature of evil in a world populated with fallen angels, in which disguise serves to mislead or abuse, where all appearances are suspect. In this view, the physical world is itself tainted as a false reality, resulting from the conflict between the physical and spiritual, or the opposition of the "natural" and the "super-natural," and the idea that latency is a tool of the devil. This is the Gnostic and Essene side of Christianity, a world besieged by veiled devils, the atti-tude of the Jains, hating the physical world, or of the Buddhists who find it insignificant. In this we have come full turn from an ancient widespread be-lief in the inherent spirituality of ensouled heads with their sacred faces. We seem caught between this and that: the choice between the purity and the deception of the illusion of the mask. Perhaps we ourselves are like the ani-mals whose heads hang on the wall and who seem to be sticking through it, occupying both sides at once, as though in passage.

ANIMAL MASKS ARE at the primal roots of drama. In the modern theater and cinema facial expressions convey a typology of the emotions. This hu-manization of plurality celebrates diversity in ourselves and omits the older tradition which sought authority in masks as the source of our diversity of the community of life as a whole. In ancient Greek and Japanese drama, masks provide for disjunction and mutability at the heart of a larger unity: the community of life. Religious masked performances are iconic conven-tions that elaborate classificatory changes as essential aspects of the uni-verse. Between these two extremes—the modern, mobile verisimilitude and the folkway of the mask—is a range of intermediate performance, such as Hindu classical dance, in which only certain stylized expressions disturb the painted human visage. In any case, the absence of the mask in modern drama is one more sign of the charter of humanism, in which we declare that we can be all things to ourselves and no longer look at human diversity as shadowed forth in nature.

What difference does this make? In polytheistic performances appear-ance changes without threatening the sense of self. Roy Willis observes: "The distinctive peculiarity of animals is that, being at once close to man and strange to him, both akin to him and unalterably not-man, they are able to alternate as objects of human thought, between the contiguity of the met-onymic mode and the distanced, analogical mode of the metaphor. This means that, as symbols, animals have the fortunate faculty of representing both existential and normative aspects of human experience."[8] The diver-sity of animals lubricates our notion of difference-with-similarity, extend-ing our meditation on human kinship, those others-but-not-strangers, whose faces are subtly distorted reflections of our own.

Such "polytheism" is misunderstood simply as disjoined "many gods." In fact it affirms an organic whole: communities, ecosystems, ponds, or forests.

Its organic equivalent is the community of tissues and organs within the individual. Belief in a single creator does not preclude belief in multiple, sacred, animal-represented powers, because the creator often remains as distant as the sky itself, a shadowy deity who does not intervene like Jehovah opening the Red Sea. Between that remote demiurge and everyday life are multiple deities, eccentric specialists, natural and supernatural at once, each limited and sometimes inept and vulnerable. Some singular animals, such as the bear, can represent the whole, as well, counterbalancing the disintegrating effect of equating different powers with different species.

LIONS, NOT BEARS, give us the best lesson of the history of the head, face, and mask, as the old world of egalitarian cultures gave way to that of the modern state. Having spent a million years in the cosmos of the animal powers, humans did not readily give up zoological images with the rise of the state. Centralized political power distorted the pantheon, so that it decayed slowly, finding certain animals more appropriate to the tendencies of Western Civilization toward empire, monotheism, and humanism.

In this process, the preeminent bear was replaced by the lion. Until recently lions were found from Greece eastward as far as India, south into Africa, and west to the Atlantic shore. Golden as the sun's rays, fierce in predation and sex, the female a passionate protector of its young, fiery in temper, intelligent and mysterious in its movements and moods, regal, sometimes dangerous to people and cattle, its ecological position is at the peak of the pyramid of predators, a clear basis for royal or divine analogies. The Assyrian kings kept lions in their hunting parks and proved their valor by hunting them from horseback or chariot. Despots depicted themselves as lions, the beast of kings, representing both monarchs and those against whom the emperors defended their people as a shepherd would defend his flock. They were guardians, as in Egypt at the Ptolemaic temple of Horus at Edfu. Door and chest locks were made in lion shape. The story of Daniel in the lion's den, the pet lion of Saint Anthony, the fables of the lion and the mouse, the removal of a thorn from the lion's paw by the Roman slave Androcles, the lion as sacred to several goddesses—all reveal its imperial and divine connections. Alternatively, the Gnostic cosmos, is "a den of lions and dragons." Howard M. Jackson adds: "For generations of martyrs the 'lions' of the Psalms had been the standard symbol for a despot's persecution and his instruments of torture."[9]

As in archaic traditions of the head as the soul's seat, the lion's face condensed its qualities, with the result that sculpted representations are abundant in the archaeological remains of antiquity as well as pre-twentieth-century architecture in Europe and Asia. Short-muzzled and binocular, with prominent teeth, eyes, and superciliary tufts, the lion has more of a face than most mammals, a face reproduced in art for so long that it became stylized to the point of abbreviation. Its features and the ease with which they can be dissociated produced an iconography of ears, whiskers, eyes, jowls, and brows.

Some lionesque features contributed to the most startling face in the history of masks: the Medusa, a Gorgon with large, bulging eyes, fearful snake hair, protruding tongue, tusks, pierced ears, beard, and forehead marks based on the lion's superciliary tufts with their clump of long, whiskerlike hairs. Howard Jackson notes: "That they occur exclusively on lions and Gorgons suggests indeed that there is a direct connection between the lion and the Gorgon especially—that, for example, the protective function of the Gorgon as Athena's aegis is analogous to the heraldic function of lions as gate protectors throughout Greece and the Near East."[10]

In art the superciliary tufts could be combined as a single central mark or third eye. In ancient India the third eye was associated with the effect of sacred drinks such as soma and mead that produced insight or union with the divinity in one's self. The lion's symbolic solar connection seems to have been epitomized in the third eye as *laksana*, which reconciles the ambiguity of enlightenment with the power of destruction. Aleene Neilson observes: "The creative and destructive powers of the third eye illustrate that knowledge has both good and evil qualities, that self-knowledge also implies the necessity of self-discipline and an awareness that meanings change when sequence or context changes. To most Western minds, conditioned by monotheism and rationalism, such ambivalence may be psychologically devastating. In polytheistic cultures, however, the idea of appearance is a powerful intellectual tool that allows one to assume a mask or discard a role in order to deal with paradox or effect transformative change."[11]

Medusa, the mortal of three Gorgon sisters, was so terrible that to look at her turned one to stone, perhaps the remembered fear from the petrifying encounter with a real lion. She was connected to horses and was sometimes represented as a centaur, the mother of Chrysaor and Pegasus, the winged horses which sprang into the sky from Medusa's neck after she was decapitated. Her story may be related to the Vedic myth of an equine figure who mates with a wrathful mother. She is certainly related to a fantastic lion-faced protector, the *kittimurka*, who is so fierce that it consumes its own body. In Indian mythology Vishnu sends a Gorgon from Siva's third eye: "Here we see the permanent establishment of the Gorgon as an apotropaic aegis—i.e., the ambivalent Gorgoneion, at once horrific and deadly, but at the same time capable, in the right hands or in the proper location, of meting out justice."[12] The Medusa is associated also with other Asian creator/destroyer deities such as Kali and the Babylonian Humbaba, a cedar forest guardian who gets her name from the Sanskrit *kumbha*, a superciliary protuberance of rutting elephants when they are most dangerous. What could be more frightening and protective than the combined third eye of lion and elephant? Humbaba's frenzied visage is abundant in the Greek art of 2,700 years ago, when East and West were linked by Phoenician trade and by the incursions of Indo-Europeans into the Mediterranean region.

Our spontaneous fear of the faces of guardian figures represents two deep streams of human experience. The first is the psychic imprint of the fierce faces of carnivores with their threatening teeth, protruding tongue, and large noses and ears—those occasional crunchers of human bones and the strongest of the animals. The second is the correspondence of the *kittimurka* to a "sensory homunculus"—the bizarre image of the body when its parts are represented in proportion to their relative sensory areas in the cerebral cortex of the brain. The homunculus, at once like a visible projection of the brain's emphasis on mouth and eye and like the representation of a primate scream, rivets us in terror. Its grimace combines the contradictory expressions of fear and ferocity.

The exceptionally wide eye aperture of the Medusa and her counterparts is arresting, perhaps because we primates signal with the whites of our eyes and the size of the pupil. The guardian face holds us suspended by its expression between laugh and frown, fear and aggression, a focal tension stopping us at thresholds, dangerous because of the uncertainty of its intentions. It makes the moving intruder and the initiate pause to respect the precincts as they cross from one domain to another. As guardians at the doorways and peaks of temples these figures are "warding off evil." The Greek story of the decapitation of Medusa may have begun with an archaic, massive stone guardian figure; it became a portable head by means of the sword of Perseus, a son of Zeus, who was preserved from looking directly at the Gorgon's deadly eye by using Athena's borrowed shield as a mirror. The head becomes a decoration on the shield thereafter, helping to protect Athena. The Medusa face represents the Gorgon figure and spirit, just as the head of any animal preserves its spirit life.

AS HUMANIZED DIVINITIES replaced the ancient animal-masked performers, the Medusa and all other apotropaic or guardian figures receded
into decorative stone pediments, roof-line embellishments, joining the
other gargoyles as architecture. Masks receded into the Greek chorus, then
decayed as an iconography of festival and carnival license, emblems of chaos
and hell, eccentric decorations on manuscripts, as toys, and as representations of madness and other irrational human behavior. The Roman gods,
having gone through the anthropomorphizing process to emerge in human
form, kept the power to mask themselves as animals. When the giants
warred against them, the Roman gods were said to have fled to Egypt and
taken animal shape: Jupiter the ram, Apollo the crow, Bacchus the goat, Di-
ana the cat, Juno the cow, Venus the fish, and Mercury the bird. It was a
handy healing myth that helped the Romans to understand and incorporate
the Egyptian pantheon. Such psychological, mnemonic devices, the memory of our transition from animal to anthropomorph, gives our unconscious
also a healing capacity.

Christianity swept most of it away and made the remainder into the visible work of the devil. Christianity is hostile to the theory of biological evolution, not simply because it clashes with the creed of special creation and
monotheism, but because classical and Christian ideas of essence bled the
significance out of outward form. An Alexandrine church father warns
against putting on "other images instead of that of the savior. Instead . . . we
assume the form of the Devil, so that we may be called 'serpents.'. . . And we
also put on the masks of a lion, of a dragon, and a fox when we are full of poison, violent, crafty, and of goats as well when we are too inclined to lust."[13]
The Western mind distrusts appearances and change in nature, all signs of
overlapping and confused identity such as man and ape. The uncanny similarity of higher primates to ourselves triggers cognitive dissonance and challenges the moral, God-ordained discreteness of his order. Our culture denies the value of paradox (or trivializes it as riddles) and censures anomaly as
evil. The ape imitates man just as man confuses himself with God. Outward
signs of an intermediary nature and the ambivalence of multiple, simultaneous identities are threatening in a world where identities are fixed. Without the resolution of confusing identity by masked mimesis, mutability is
demonic, erroneous, and evil.

WE BEGAN WITH the modern reaction to the amputated head as a kind of
black humor. But humor conceals deep concern. At a subconscious level we

intuit the head as sacrosanct, the domain of the living spirit. As primates we are aficionados of facial nuance, mediators of intentions and feelings. As a signpost, door jamb, threshold guardian, temple protector, or territorial herm, the guardian face is the keeper of entrances and passages between realms. Taking the human, mobile, communicative face as a standard, the unchanging face or mask of the animal is either the manifestation of singular wisdom or a demonic deception. Worn by a human performer, a sequence of masks is the spectral imagery of transformation itself, the plural reality that is truer than our momentary state, the ritual instrument of movement between states and across boundaries.

When the head or face is worn as a mask, its double meaning proclaims and affirms the simultaneous reality that we think of as paradox. The mask presents in its most inescapable form the antithesis of unity and plurality. If we are determined that multiplicity is illusory and that the ultimate monism is an essence behind undependable appearance, then the mask becomes a child's plaything or a deceiver's cover. But if one believes that the spirit may be revealed in appearance, that time may contain all things in the present, that other forms of being may be latent in the one, and that spiritual power is dispersed in forms seen and unseen, then masking is a language of understanding. In this view, natural masks, worn by the animals, tutor us in the resolution of the two most difficult philosophical problems: the unification of antitheses and the nature of a singularity that contains multiplicity. In the spirit of play that marks the best of philosophy, we may find an explanation for the man in the cartoon who says to the mounted heads in his parlor, "I suppose you're wondering why I have brought you all here."

11 | The Pet World

*The Autonomous Self feels so Alone in the Cosmos that it will go
to any Length to talk to Chimpanzees, Dolphins, and Humpback
Whales. . . . Why do people in general want to believe that
chimps and dolphins and whales can speak . . . to Establish
Communication with an Extraterrestrial Intelligence?*

WALKER PERCY

AT AN AQUARIUM in California the fish swims indolently along
the glass. The man and two children watch. He waves his hand next to the
pane, taps it. The fish does not respond. At first it seems to be looking,
pauses, but then swims on, clearly oblivious of the man and children. He
turns away with the boy and girl.

At the crocodile sanctuary south of Madras, India, boys watch the immobile reptiles in the sun. Finally one boy, then others, throw pebbles at the animals.

I tell a famous ornithologist at Crater Lake National Park where we both
work as seasonal ranger-naturalists about my experience on a high slope,
where two hawks played overhead with a wad of lichen—the kind that hangs
in festoons from conifer branches in wet forests. One hawk, higher up,
would drop the lichen, to be caught in its talons by the other. It would then

climb above and drop the lichen to the first. After several rounds of this, the lower bird, having caught the lichen, sailed over my head, "and," I say to the scientist, *"dropped the lichen to me."* At this he snorts, turns abruptly, and leaves.

All of the foregoing scenes are glimpses of a widespread yearning for a sign from the Others and of skepticism that it is possible. Usually we wait in vain and in frustration. We seem frantic to contact some intelligence more assured than ourselves, to be blessed in their witness of our mutual presence, to be given their surety that life is real and purposeful, even if the purposes lie beyond our grasp. As human numbers increase and the Others recede, it becomes our last passionate desire and we feel like calling, as the birds fly away, "Come back! . . . Hear me! . . . Look!" We tap on the aquarium glass in the hope of the slightest acknowledgment, wait at the zoo where we are no longer allowed to feed the animals, or bang on the cages hoping for a crust of attention. We await a reunion with absent beings on a crowded yet increasingly lonely planet. But the zoo animals turn their glazed eyes away. "Those arrogant animal bastards!" we think . . . and throw a rock. Even when they react we are skeptical of their capacity to think and feel, having been taught that wild animals are little more than mechanical toys. While most people still watch for attention from them, my biologist friend has been conditioned by his education to disdain that possibility. He had perfected his dissociation.

Against the indifference of the wild animals, the impetuous affection of our pets seems like an enormous boon. In a world so full of problems and suffering, only the worst curmudgeonly cynic would sneer at our indulgence, their simple pleasure in us, and our joy in them. Something, however, is profoundly wrong with the human/animal pet relationship at its most basic level. Given the obvious benefits of that affiliation, one has to poke very carefully into its psychology and ecology before its fragile core can be exposed.

We believed ourselves until recent centuries to be continually in the proximity of a multitude of wise animal elders—avatars of kenning—who filled human life with excitement and strange associations for so long that our species continues to anticipate their reassurance. The artifacts of industry and media, all the human mob and its distractions and therapies do not make up for the loss. Even the shambling domestic forms that pulled wagons, laid eggs, or turned our garbage into sausage have been removed from sight. Only pets remain, a glimmer of that animal ambience, sacredness, otherness.

Pets are abundant throughout the world except in Africa. In addition to Western peoples, the Australian Aborigines, indigenous Southeast Asians, North American Indians, Polynesians, Brazilian Kalapalo Indians, and many others keep animals "perceived as subjects or quasi-persons—hence the tendency to indulge and fondle them."[1] With few exceptions people feel a strong sense of their comforting presence. An abundant literature tells of the personal renewal mediated by them. They are a buffer against loneliness and "negative life effects." City people especially, without access to wild animals and without the communality of small-scale life, enjoy pets as an amenity, but there is evidence that their role goes beyond the pleasures afforded by affluence.

The pets kept by Kalapalo Indians and many other tribal peoples are merely captives trapped from the wild. They are not the domesticated and genetically altered descendents of certain wild ancestors, the products of centuries of controlled breeding, the original form of "genetic engineering." Indeed, the breeders chose individuals that were the least wild in behavior, the most responsive to humans, and the most productive, making them over the generations increasingly dependent on people by thwarting and redirecting their social drives and crippling their intelligence in the interest of pacification, the constraints and needs of the household, and sometimes appearance.

During most of the five thousand years of pet keeping, people also had easy access to livestock and wild animals. Towns or their margins were occupied by an abundance of small wild animals: birds, insects, fish, and amphibians. People were seldom more than walking distance from a still richer fauna in nearby streams, fields, and forests. Domestic animals have gradually become surrogate companions, siblings, lovers, victims, workers, parents, competitors, deities, oracles, enemies, kinfolk, caretaker-watchmen, and so on.[2] In seventy centuries of human cohabitation with animals, the makers of manure, milk, meat, and skins, as well as sacrificial offerings and symbolic and esthetic objects, have not always been separated from their pet function. In the past a tiny bullock might be cared for with familial warmth and attachment in the household, exchanged as currency, kept as a fertilizer machine, admired for its strength and beauty, or sacrificed on an altar and then eaten, all the while talked to, touched, and loved as a member of the family. From subsistence farmers who occasionally keep "familiars" in the household to royalty and heads of ancient states with their private zoos and exotic pets, we come to modern, relatively affluent classes among whom animal companions seem ubiquitous.

Even in cities, until the twentieth century, rabbits, chickens, ducks, and geese were still kept in backyards, local fairs and markets had large livestock sections, draft animals were still abundant, farmers drove pigs and cattle to market down the streets, and knackers butchered them in alleys. There were few laws against keeping birds and other wild captives. Dogs and cats ran freely in the streets.

During the rise of this biological void in urban existence in the industrial world from 1850 to 1950, the middle classes began to have fewer children, for whom the household could be a lonely place. In fiction, such a child who represented childhood without siblings and without easy access to street friends was Christopher Robin, for whom Winnie the Pooh was the substitute. Pooh Bear is an animated and storied teddy bear. In this way the Anglo-American concept of animal friends is prefigured in childhood, with the aid of "bedtime" stories, by pretending that one's stuffed toys are alive. This scenario creates a very different childhood orientation toward the living pets than in earlier times, when household animals like rabbits might be eaten, cats caught mice, and dogs served as guards or hunters. From Pooh Bear it is not very far to the doggy friend and but a step from the doggy friend to an imaginary relationship with or among wild animals—animals which, for the first time in history, are almost completely lacking in the child's experience. This sequence is a drama in five acts.

Act I is an outer circle of wild animal life which was a major focus of human attention, establishing the expectation of a rich, surprising, meaningful, and beautiful diversity of life around us. Some animals were sacred. All were conscious, unique, and different in spiritual power.

In Act II people took certain animals into captivity, manipulated their reproduction, and altered their biological natures to conform to human dominance, reconstructing them as members of the household. These became the domestic animals. The wild forms, reduced in number and diversity, literally receded.

Act III begins not with animals but with a class of things called "transitional objects." These are toted around by anxious three-year-old children who are having difficulty becoming independent from their mothers and who are comforted by a soft object that is subjectively intermediate between themselves and the outside world. The children who do not seem to require the security of such objects are those who are surrounded by abundant other forms of life. The exact reason for this is not entirely clear, but apparently animals in their diversity model a world of likeness and difference which makes the child's impending separation less frightening and also resonates with internal, psychic structures which can best be described metaphorically as a fauna. The stuffed toys are simultaneously huggable, transitional objects and "animals." They appeared in large numbers in the industrial, nuclear-family era, compensating children for their lonesome social and ecological situations and preparing them for lifelong pet keeping. The mode of this preparation is pretend-play, the self-dramatization of all life as a happy playground.

Act IV is the transfer of this affection from effigies to dogs and cats. As the toys had been pets, the pets became toys. Even "wild" animal manikins, such as stuffed bears and lions, are little people in the imagination, who participate in a household society, who have expectations, reasons, worries, expressions, voices, tastes, and complicated affinities and antipathies toward each other and their human companions.

Act V extends the equivalence of the living domestic pet and the stuffed wild toy to living nature. If the domestic forms have all along been substitutes for the wild and the latter have become unavailable and unknown, it is easy to fuse the domestic and wild. The wild are simply those potential pets who do not happen to live with us. They are each other's pets, or perhaps creatures whose friendship we have lost. Zoos seem to affirm this identity. The zoo has "toy verisimilitude," foreshadowing the modern child's menagerie of stuffed animals and friendly pets, each zoo creature enduring in blind lethargy, withdrawn except in moments of hyperactivity when the feeder comes, like the puppets waiting in the closet to be flung into tea parties or wagons by a child.

Yet the zoo does not fulfill its promise. "Why do zoos disappoint?" John Berger asks.[3] The cells are like frames around animal pictures, the zoo an art gallery. Like the oeuvre of a dead painter, the animals serve "as living monuments to their own disappearance." The animals are rendered marginal by being out of place, having been stripped of the ecologies which defined and energized them. Wild habitats are complex environments to which the new animal-park zoo, as distinct from a steel cell, can only be another simulation. One may look at animals at the zoo, Berger concludes, but the "look of animals" has vanished, for they stare blindly beyond the zoo visitor who, wanting more than an image, finds himself alone.

Adults stand apart from this childhood fantasyland, amused and indulgent, but when the imaginary life of the toy, projected upon nature, fails, we are ripe for denial and dissociation. Our expectations begin with the sugar-plum roles of the stuffed toys and move on to the surreality of telephoto film. The remarkable close-up films of animals' lives no longer suggest ordinary access. The inside view of a termite nest, the eagle gently feeding its downy young, the gasp of the gazelle in the cheetah's jaws—these are as remote from our lives as the "friendship" of the kangaroo, donkey, tiger, and Pooh Bear. As a result, Berger says, these arresting images only emphasize the animals' normal invisibility.[4]

To PET IS to touch. The desire to hold the wild as we do our pets is acute. In the past quarter-century of television, Marlin Perkins, Jacques Cousteau, and others, in the name of science, captured manatees, lassoed gazelles, grappled with crocodiles, or anesthetized lions—ostensibly to treat a disease, restock an area, mark for further study, or rescue from rising waters or industrial development. Perkins was an American zoo director, Cousteau a French media entrepreneur. In every sequence men clutched, held, or reached out and touched the animals. They fostered our yearning to recover a lost world, to be once again the trusted friend of all beings. They put the animals vicariously into our hands, where we wait for them to tell us something.

At home Rover is told to sit up and speak. Polly is asked if she wants a cracker. The news is full of the language of apes and dolphins, who, it is speculated, may be smarter than people. The scientists claim to be doing studies in communication, but the rest of us know what's afoot. We just want the whales and chimps to talk—not just about hunting shrimp or fruit but to speak to our mutual situation, claim our shared purposes, mend what Lévi-Strauss calls "the ultimate discontinuity of reality."

A *New Yorker* writer gives the uptown lament. Unable to speak to or understand the animals in a dream, she says it became a nightmare. She refers to the "wistful regret that all this life was hidden from us."[5] Reviewing the research on chimpanzees reared in psychologists' homes and taught to signal with hands or word processors, she speaks from the depths of that community of desolation in which we cheer each time an ape signals "I want a banana."

In his book *Lost in the Cosmos*, Walker Percy addresses "The Lonely Self" asking, "Why Carl Sagan is so Anxious to Establish Communication with an ETI (Extraterrestrial Intelligence)."[6] Percy concludes that it is because we recently lost our sense of uniqueness and special destiny at the center of the universe which had protected us from the effects of alienation from other animals. Humans are "like Robinson Crusoe, marooned on an island populated by goats," who, failing to have a man Friday, would like to "at least teach goats to talk." Not only the search for friendly beings from other planets but the whole UFO phenomenon is, he says, a manifestation of our having found that "chimpanzees don't talk, dolphins don't talk, humpback whales sing only to other humpback whales." The would-be conversants with chimpanzees have perceived truly that nature is a language, that animals are its principal actors, that reading the world was the first function of the human mind. Humans may be the "talking animal," but ordinary conversation raises blood pressure whereas "speaking" to animals lowers it. The *New Yorker* writer wants a discourse that might only aggravate the burden of babble. "One of the most appreciated qualities of animals is that they are beyond language," says Mary Allen in a treatise on animals in American literature. "That they feel but do not require conversation is a great relief to most people."[7]

In the wild world pets are, of course, alien. "Aliens," in the homespeak of modern children, are either hideous demons or grotesque but possible friends who need companionship and are lost like us. We understand their loneliness (and fear) because we are becoming alien to our home planet, distanced from the earth's most fundamental properties, including its wildness.[8] The animals in science fiction include otherworld monsters or lonely wayfarers from outer space, the void that resonates with us as though it were an extension of our inner emptiness. In a 1982 film, the space visitor E.T. ("Extra-Terrestrial") was warmly welcomed by its earthbound hosts in the story and in the audience, whose chronic estrangement is not social but ecological.

E.T. was pathetic. Like some disfigured child, it responded to the affection it generated in children who saw themselves reflected, sharing their loneliness with their celestial friend, as if they could see themselves as stranded on the wrong planet. A major myth of the civilized world is associated with messiahs, the arrival of beings from celestial space, sometimes superior to ourselves, often susceptible in their own way to feelings of being lost, or vulnerable to our own evil. Angels replaced the sacred cranes who brought springtime for a million years and seemed to share with humans an immense confidence in the round of season and life. As we lose a world in which every hour once had its animal voices, smells, and signs, the UFOs and flying saucers and friendly visitors from space are sops to our anxiety, saviors who not only share our solitude but mark our uncertainty about ourselves and all nonhuman otherness.

NOW PETS HAVE become part of the pharmacology of medicine, ten thousand years after people first took in the dog, sheep, and goat. Indeed, pets have recently taken a new leap into institutional respectability, becoming "companion animals" as part of an integrated treatment. The planning committee for a conference sponsored by the Institute of Animal Assisted Therapy in 1989 included two psychology Ph.D.'s, seven registered nurses, three clinicians with master's degrees, four doctors of veterinary medicine, one M.D., two foundation directors, one person from the Society for the Prevention of Cruelty to Animals, and an "animal resource manager." Topics included the human/animal bond, animal therapy in illness (AIDS, organ transplants, cardiovascular therapy, eye patients, and chronic illness), grief management, families and children, worker stress reduction, and national programs for schools, clinics, and hospitals.

"Animal-facilitated therapy," based on the "human/companion-animal bond," began about two centuries ago among the Quakers at York Retreat, an insane asylum in England where the inmates seemed better off with their gardens, rabbits, and poultry than shut in cells. Florence Nightingale is said to have advocated pets for the chronically ill. Since that time a number of other institutions have also developed programs. A hospital at Bethel, Germany, initially for epileptics, with about five thousand patients with multiple physical and mental handicaps, has systematically incorporated animals in the daily regimen. In 1942 the Pawling Army Air Force Convalescent Hospital in New York treated traumatized veterans on a farm with livestock and poultry. Beitostolen, a rehabilitation center in Norway for the handicapped, began using dogs and horses. In New York, Boris Levinson pioneered this approach among children who were autistic, withdrawn, inhibited, obsessive-compulsive, or culturally disadvantaged. He found the "pet-reared child" to be stronger in ego, more gratified with its environment, better able to handle ambivalence in life, less anxious, and more responsible. Videotapes documenting the interaction between the patient and the animal show the "Giaconda smile" of gentle relaxation, the calming "displacement gestures" and the serene "looking away," the closeness of heads of patient and animal, the quiet speaking to, voice pitch lowering and rising inflection, the touching, patting, and stroking. The benefits are confirmed by heart rate, lowered blood pressure, better general health, and psychological and social improvement.

In the presence of pets the sufferers from Alzheimer's disease and autism are inclined to speak. Incarcerated incompetents, handicapped outpatients, plain folks who are just getting old, impoverished or stressed executives and their lonely children—all are happier or live longer in the regular presence of friendly animals. There is also less suicide or aggression among the crimi-

nally insane, calming among the bereaved, quicker rehabilitation by alco-
holics, improved self-esteem among the elderly, increased longevity by car-
diac patients and cancer victims, improved emotional states among
disturbed children, better morale of the blind or deaf, more cheer among the
mental and physically handicapped, faster learning in the retarded, solace
for the terminally ill, and general facilitation of social relationships. (Dog-
walkers in London parks speak more often to strangers, especially other
dog-walkers, than do non-dog-walkers.) In a study of hospitalized heart patients, three of fifty with pets died within a year—and eleven of twenty-eight
without. Hearing dogs accompany the deaf, guide dogs lead the blind, hos-
pice pets give unqualified cheer, animals help retarded children, monkeys
have "hands" for the handicapped. In general pets ease the negative effects
of solitude in nursing homes, induce caring, stimulate comfort in touch and
fondling, provide something interesting to watch, make the keepers feel
safe, are nonjudgmental and nondemanding friends, integrate a routine, and
stimulate exercise. They are used to help abused children and accident victims get through their trauma. There are more than five hundred therapeutic riding programs in North America alone. Goldfish reduce the stress of
waiting in dentists' offices and noticeably enliven school classes and confer-
ences. Many institutions now attach animals to outpatients such as pension-
ers or the homebound.

The corporate use of animals in health care requires specialized veterinary services, technical training in counseling in the selection and placement of animals, support groups to console those whose animal dies, and
special programs in hospitals and schools for the nursing and administration
of pet therapy. There is enthusiastic participation by the animal food manu-
facturers. There are new university programs. Companion animal treatment affects the institution's liability when the dog bites, the cat scratches,
or the horse steps on a palsied rider, as well as the legal standing of the animals themselves, who must be protected from abuse by the patients or clinicians. Someone must determine when the animals are overburdened with
work and need to take a vacation or be retired.[9] New breeding programs are
required, as are new techniques for selecting and producing the right animals for specific therapies. A corollary of this development is the conversion
of the old city dog pound into a refuge and educational center for the public,
perhaps where pets may be prescribed and rented.

Professor Leo Bustad of Western Washington University's Department
of Veterinary Medicine wants aquaria in all dental and medical offices and in
waiting rooms, conference rooms, classrooms, lunchrooms. He thinks there

should be wards in all hospitals where patients may keep pets, animal refugia from which the terminally ill may get animals on loan, professional referral systems, visitor animal friends in nursing homes, and provision for them in government housing, penal institutions, and community service agencies. Echoing the depression era cry for a "chicken in every pot," he advocates two or more companion animals for every household. Bustad believes that regulations by landlords against keeping pets should be made illegal, that community service agencies in every town should include pet-therapy branches, that disaster plans should be made to include both the fate and the availability of animals, that government should subsidize pet-care programs, that all building design should take the housing of companion animals into account, that courses should be offered in public schools on "The Human/Animal Bond," and finally that every household should have two or more animals. "We believe," he says, "that the health of our society is dependent on the nature and the extent of the association between people and other animals and their environment."[10] This view would no doubt be endorsed (for other reasons) by Eskimos with their dog teams, by eighteenth-century furriers and fur trappers, the butchers and meatpackers of the world, all the pastoralists since the horse was domesticated, most of the personnel of circuses, the guilds of leatherworkers and shoemakers, and so on. Of course that is not what Professor Bustad means, but underneath there remains the shadow of utility: the animal commodity dressed out as medical treatment instead of pulling sleds or growing furs. In return for the work of a well-kept slave the animals in this bond get "friendship," Old Testament responsibility toward livestock, legal protection, food, and shelter.

In the writing of animal companionists there is frequent reference to "roots," cave art, myths, rites, and the bonds generated by the keepers of flocks with their dogs, to the imagery of lambs, goose girls, and milkmaids. If the Neolithic barnyard was the root, then from it sprouts the compassion, caring, and kindness necessary in all the hospitals, hospices, prisons, and medical offices where we are to recreate in new form these traditional relationships. But "bonding" in the sense of an extended custody is not from the Pleistocene, from Paleolithic cave art, or from primal human ecology. We falsify our relationship to wild animals with our husbandman's eye, social worker's agenda, veterinary tools, and breeder's agendas. Animals were present at the center of human life for thousands of centuries before anyone thought of taking them captive, making them companions, forming the "friendship loops" of which animal-facilitation therapists and ethicists speak.

Even a prominent philosopher, Mary Midgley, refers to "the human/animal bond," as if it were an ancient demonstration and right of amity. To the contrary, our primal relationship with animals is interaction in terms of utility and symbolism, our rapt attention to a diverse, dangerous, and beautiful Other with whom we are ecologically and spiritually affiliated on many levels. There is indeed a heritage of wisdom in the form of an ecological conscience, an attitude of humility toward the natural world, a sense of human limits and participation from a thousand centuries of the Pleistocene. But it has nothing to do with a display of entente cordiale nor an affectionate compact of familiarity or helping, centered on keeping, as if by a shepherd, nor a modern sentiment of gushing over kittens in a basket.

Pets are not part of human evolution or the biological context out of which our ecology comes. They are civilized paraphernalia whose characteristic combination of accompaniment and accommodation is tangled in an ambiguous tyranny. Constance Perrin, an anthropologist, calls it "attachment theory." The animal triggers nurturant behavior and serves as a kind of intermediate object between the owner and a more or less alien world, but at the same time it is dragged about like a tattered security blanket. Indeed, the domestication of animals has never ensured their tender care. In recent Anglo-American tradition the dog is "man's best friend," but it is abhorred in the Bible. In Muslim tradition the dog's saliva is noxious, and contact between people and dogs requires ritual cleansing. Over most of the planet the dog is a cur and mongrel scavenger, feral, half-starved, the target of the kick and thrown rock, often cruelly exploited as a slave. Although looked upon with affection, even modern pets are property that is bought, sold, "put down," and neutered. Pets are deliberately abandoned by the millions and necessitate city-run slaughterhouses, shelters, and "placement" services. This paradox of frenetic emotion and casual dismissal reveals our deep disappointment in the pet's ability to do something, be something, that we cannot quite identify. Yi-Fu Tuan considers our behavior to be exercises in casual domination that symbolize human control of nature.[11] In an earlier book I argued that pets were unacknowledged surrogates for human companionships or substitutes for the resolution of interpersonal social problems, and therefore impaired normal human sociality by enabling people to avoid mending, maturing, or otherwise dealing with their personal relationships.[12] Pets can cause family conflict, even divorce, and may become bridges of unhealthy transference relationships and regression to infantile human behavior.

Even so, I now see that the pet may be more than a human replacement.

"Pet-facilitated therapy," casual or institutionalized, reduces human suffering. It is truly an astonishing solace. The "companion animal" is a medical miracle to which we should be kind and grateful. But like all psychotherapy its presence is not a true healing. It cheers, modulates pain, and helps the owner/patient to cope.

Domestic animals were "created" by humans by empirical genetic engineering over the past ten thousand years. They are vestiges and fragments from a time of deep human respect for animals, whose abundance dazzled us in their many renditions of life, helping us to know ourselves by showing all that we had not become. The pet cannot restore us to that wholeness any more than an artificial limb renews the original; nor can it do more than simulate the Others among whom our ancestors lived for so long, the Others that constituted for them a cosmos. They and all captive animals are like organ transplants: healthy for us but cut out of their own organic fabric.

What is wrong at the heart of the keeping of pets is that they are deficient animals in whom we have invested the momentum of two million years of love of the Others. They are monsters of the order invented by Frankenstein except that they are engineered to conform to our wishes, biological slaves who cringe and fawn or perform or whatever we wish. As embodiments of trust, dependence, companionship, esthetic beauty, vicarious power, innocence, or action by command, they are wholly unlike the wild world. In effect, they are organic machines conforming to our needs.

No one now doubts that pets can be therapeutic. But they are not a glorious bonus on life; rather they are compensations for something desperately missing, minimal replacements for friendship in all of its meanings. Mass society isolates us in ways and degrees that seem to contradict our population density. Pets occupy by default an equally great human need for others who are not part of our personal lives. The diversity that nourishes the mind extends to the whole realm of life and nature. Pets, being our own creations, do not replace that wild universe. But as living animals they confuse our perception and hide the lack of a wild, nonhuman comity of players on a grand scale—a spectacular drama of life to which our human natures commit our need and expectation.

Wild animals are not our friends. They are uncompromisingly not us nor mindful of us, just as they differ among themselves. They are the last undevoured riches of the planet, what novelist Romain Gary called "the roots of heaven." We cannot comprehend the world as it is experienced by a bat, a termite, or a squid; we cannot force them into barnyard conviviality or household banality without destroying them. More than bearded prophets and great goddesses they are the mediators between us and plants, the rock and suns around us, the rest of the universe. Wild animals connote the wildness in us which cannot be equated to our domestic affairs or reconciled with the petty tyrannies of "dwellers in houses," domesticates, from the same root word that gives "constrain" or "subdue." As a fauna only the wild are a mirror of the multifold strangeness of the human self. We know this. It is why we scrutinize and inspect and remark on them, make them the subject of our art and thought, and sometimes kill and eat them with mindful formality, being in place with our own otherness.[13]

The Gift of Music

12

*We know that Orpheus did keep his paths of association open
because the animals understood everything he played, and he
was always drawn with the animals around him, listening.*

ROBERT BLY

WHY AND HOW did we first dance, sing, and make instrumental
music? What was the idea behind it? Perhaps we did not invent music at all
but appropriated it as an expression of our own felt rhythms. Like much else
in the human repertoire, music may have been already there when we arrived and its performance was everywhere audible and visible. The classical
Greeks told of Orpheus, who created music and taught it to the astonished
beasts. But the Orphic myth had it backward. It was the animals who first
made music and Orpheus' ancestors who listened. Hearing and mimicking
the natural world gave us our ear. When humans uttered their first words,
birds, frogs, and insects were already whistling, dancing, drumming,
trilling.

Long before that we had inherited the capacity for music from the old
sound brain of little ancestral mammals who attended to footfalls, wing-
beats, and calls in the dark, clutching the intervals and each note as they
would take bites from a beetle, keeping in mind the second in relation to the
first, and the third to the first two. Memory made sense of the sounds in

deep, wet forests seventy million years ago. In evolutionary terms: "Since the typical mammalian distance-sensing system at that time was an encephalized auditory and olfactory system, with capacities for storing information over time . . . a newly visual species of mammal would have its new visual system . . . modeled after the long established mammalian auditory system with its major central nervous system representation."[1] The mental structures for recognizing patterned sound provided us with the basis for managing patterned sight, as our visual world is composed of remembered visual fields. All spatial meaning has this shared functional origin and anatomical basis in sequence or temporal experience. Thus are melody and landscape related and terrain perceived as a kind of music.

And which did our immediate forebears do first: hum, strum, or jig? For peoples in the deep past, many creatures modeled aspects of ourselves, as though they were our tutors, or reminders of our own natures and the ways of giving voice. It was widely believed in the northern hemisphere, for example, that the bear taught us those paeans that must be chanted at the festival of its death at the end of the hunt. Do bears themselves sing? They who seem only to growl and roar have their musical moments. Cubs, before they have emerged from their natal dens, hum a singsong as they nurse, a contented "up and down the scale."

But our primate ancestors may not have needed bears to trigger their music. The quiet gorilla beats a tattoo on its own chest. Chimpanzees have spectacular group caterwauling and shagging for storms and sunrises and sometimes apparently just for the pleasure of sharing cadenced whooping. The gibbon is a sunrise singer of great repute. Some anthropologists, debating the origin of speech, have suggested that those ancient proto-grandparents of ours, who came out of Africa and could have known no bears, sang before they talked. In fact, the idea of song at sunrise suggests the spring madrigals of the birds with whom the simians shared the treetops.

Until recently we lived in a world of neighs, howls, bleats, cackles, barks, chirrups, buzzes, and bellows, of coyotes and wolves howling, whales doing their opera, many in chorus or in dialogue, such as the bugling of elk and moose, the synchronized trill of crickets and alternating rattle of grasshoppers, the competitive braying of the jackass, roaring and screaming among the cats, the bawling hubbub of the wild cattle, the trumpeting of elephants. But the more subtle calls are lost in the human rumpus, the clicking of exoskeletons, the drone of the flying beetle, the tinkle of bats, the tattoo of millipedes on leaves, the knock of the termite's head and lisp of the earthworm, the scuffles and squeaks of native mice, and the millions of muffled thumps, chirps, and sibilance of which we know little more than we do of the music of the sea. In their unison, like the clamor of a million kittiwakes or the unbroken round of chirring of a hundred tree frogs, we may feel ourselves suspended in sound.

Among the most ubiquitous, melodic, and contrapuntal of singer-musicians are the frogs. All of us at one time were in the same boat as the Indians of the Rio Icana in northwestern Brazil, who say that ghostly frog people pass in their canoes at night, the sound of their paddles made by the paddle frog. To the natives of Hispaniola, in the West Indies, the frogs' *"toa, toa"* sounds like infants asking for the breast, while others are more like calves calling for their mothers. Among the Pima Indians of Arizona a rattle used in the rain dance sounds like frogs, who are associated with rain, new growth, and fertility. The trill of a great Aztec earth goddess was like that of a certain giant toad.[2] Frogs are seldom solitary singers. It is as though they needed the chorus, like cicadas.

Somehow, as we listened, all of this came to us as performances of rhythms in sound and motion which we felt in ourselves as music. And some of our songs were about them, emulating their piping, treble, trumpeting, and modulation. We found that we could extend our song by blowing through their pithed wing bones and amputated horns, striking their stretched skins, plucking and scraping their dried guts, or shaking their skulls full of loose teeth. We incorporated their bodies with "singing" as part of our own.

Like gibbons and cicadas, we humans have our own distinctive musicality. But something happens to our idea of it as we hear it and think about it and make it a part of our conscious selves. Then we know something that the birds may not have thought of—translating other modes into our own. Perhaps in this way instrumental music was originally an extension of human song by the mimicry of animals, the rustle of plants, the consonants of wind and water. In traditional societies musical instruments are made from and named for animals.[3] The oldest known six-holed flute is the same vintage as the earliest human art, nearly 100,000 years old. Instrumental music already existed when the first pieces of mobile art were made in concert with a sharper consciousness of the rhythms of life, a "creative imitation."[4] Flutes, whistles, and pipes, like horns or reed instruments, are hollow bones or branches, the ocarina a dried cucurbit. Horns are made from seashells, but the originals are the horns of hoofed animals or of conch shell, whose calls are suggestive of horns blowing. The vibration of reeds still makes the whine and wheeze of clarinet and oboe. Percussive music is made from stretched skins, turtle shells, and bones, stringed instruments from hollow wood with the dried sinews and gut-strings of animals, bagpipes from stomachs and bladders, the keys of instruments from horn and bone. Musical notes are like the sounds of wind or rain, the rhythms of the heartbeat or hoofbeat, the pulse of frogs and crickets, the whir of cicadas and innumerable stridulations. Museum collections of musical instruments are instructive: the older instruments are not only made from organic parts but made to look like animals—the body of the viola, the whorl of the horn, the torso of the rattle.

Fugues of the night arise from moving wind in leaves and the polyphony of running water, earth disturbed and reverberating beneath tumbling rock, the voice of thunder and silent interval. Melody is widespread in the songs of passerine birds, and others clack their bills and vibrate their feathers in flight. Of the recorded growl of a snow leopard, the poet Michael McClure has written that it is "the purest, most perfect music I have ever heard . . . more beautiful than any composition of Mozart,"[5] and the *New York Times* music editor has reviewed the complex harmonics in the songs of wolves. Animals became "good to think" just as they were good to sing and dance.

Different animals are heard against the background of the larger hum of the universe. In India, for example, the moon is a visual cosmic sound, *"Om,"* the primal word that precedes giving names. Tempo is leaf against leaf in the wind, the cadence of the waterfall, the sonorous whine of insects. Rattle and drum evoke the pummel of rain, the patter of seeds falling or rattling in pods, simulated in the gourd, turtle shell, bone tambourine, and skull rattles. Tempo is recreated as the throbbing heartbeat and hoofbeat by striking bones on skins stretched over wood. As the old Dutch word *trom* ("imitate") suggests, the drum recreates a sound made by members of a family of fishes called "drum." Percussion is the woodblock knock of the woodpecker, the log-thumping grouse, the tattle of the kingfisher; the castanet is the vibration of the snake's tail against dead leaves or the clicking of beetles, shrimp, and crabs.

ACCORDING TO ONE analyst, music "enters into association" with repressed material and raises it just below consciousness, so that the feeling originally connected with it is reexperienced and yet remains alien to the ego.[6] We are charmed by its effect in a strangely final way outside both our logic and our dreams. Preserving sanity, assuring us of meaning, our stories, sung and danced, are built on that harmonic coherence, confirmed by all the other singers and dancers around us. It is because we are all musicians in a world of music makers that music reveals its true value. Life does not progress in stately measure. It jerks in bits and pieces. Our confidence in it as a whole is a function of music, shared with us by all the animals who are and were its instruments.

Time itself, as Albert Einstein reminds us, is elastic—not a river but puddles in an erratic stream. It needs added tempo. Robin Riddington says that American Indian medicine bundles contain the symbolic objects of the mystery of transformation, which are also represented in "the songs of the medicine animals within the man . . . always in his inner ear. . . . [These] carry a person's mind up and down the abysses of his subjectivity" and "reach out horizontally to touch the subjectivities of others. . . . The ultimate source of the dreamer's songs is the animal world for they are the prayers that animals sing when they have hard times. The dreamers in heaven have heard the animals dancing and singing and sent the songs down into the dreams of the seventh shaman, who then gives them to the people."[7]

The role of the shaman in music, in its unwritten prehistory, may have been intermediate in more ways than one. Shamans everywhere chant, drum, and dance their access to that realm beyond mundane experience where the healing and protecting spirits live. The *"sha-"* in shaman is similar to the Tibetan word for the urial, a wild sheep, to a variation on the word "she," and to a term for a Japanese Shinto temple. The associations of *"sha"* may have no etymological significance, but we are reminded that before shamans there were sacred animals, who were their own musical instruments in their song and who were the first dancers. In his ritual attire, the shaman is half-animal and half-human, a kind of sheep-man. Human healers may initially have been she-men, or female humans, whose performance took place on sacred ground, like the precincts of Shinto temples. The typical shaman's headdress or other costume continues today to include significant, symbolic parts of the animal masters, signifying that the shaman (or sorcerer) is historically intermediate between the original figure of the animal master itself and the fully humanized "doctor" or priest or singer who, if we forget our debt to the first singers, represents our amnesia and our arrogance.

Shamanic rites, like tribal dances, often occur in the presence of fire. The prehistory of music is linked to fire as both environment and coperformer. Dispersed in their individual or small-group routines, the early bands of foraging people were, like all terrestrial primates, reunited intimately at evening. Some 700,000 years ago humans were keeping and making fire, which probably played a part in the genesis of human music. Not only was it a "dancer" and "speaker" but it provided a protective, warm focus of reverie and the occasion for that most elemental function of all music: the display of coherence in the arts of performance and the reintegration of the group. The descendents of those ancient hunter/gatherers who still live in small groups collect at fires to warm, cook, narrate, gossip, sing, play, and dance.

In Eastern Europe archaeologists have recovered thousands of shattered clay animal figurines, part of the Gravettian complex more than 26,000 years old, 14,000 years older than the first pots. These were not broken by accident or merely smashed. Making and unmaking were together part of some lost dance or ceremony. Making involved shaping and adding limbs and heads to a body of wet loess soil; unmaking created the music of their sizzle and pop when thrown into the fire as votive offerings.[8] Fire itself was a dancer and the human dancers moved around it while the exploding figurines reminded them of the story that their rhythms and sounds were borrowed from the animals who were the models. The thirty-five or so "sorcerers" depicted in the Franco-Cantabrian caves are apparently dancers, and a fine elk steak still gives us reason to dance, having sizzled its song one last time.

The crackling fire, around which the humans dance, is itself a resonant, contrapuntal flow with a percussive accompaniment. The icon fascinates, for in its protean complexity it catches, as it turns, the light of myriad fires, warming, while seductively casting a series of cool glances. The danced image is irresistible. "What god is not tempted, then, to dance," asks Steven Lonsdale, "as his own image is portrayed with such compliments as only the sculptor of dance can pay? Through his ecstatic prayer the dancer himself moves closer to immortality: in dancing the god he becomes him."[9]

The universal entertainment of shadow figures, like the natural flicker of shadows around a fire, or cast as silhouettes upon the wall by the skilled hands at the fireside or as back-illuminated tableaux in village festivals, reminds us that the purpose of such drama and its music is to establish the paradoxical identity of human art and the natural world. Dance, as "catching" as laughter, draws people and is believed in tribal traditions to be irresistible to sacred beings who join it. Perhaps this was simply a reversal of the actual history, in which we, observing animals for a very long time, were profoundly moved and enlightened by their rhythmic, uncanny, inspiring, mesmerizing, inviting movements. Just as animals first exemplified other aspects of our lives—the lion in courage, the bear in generation, the snake in renewal—their movements were irresistibly mimicked and stylized. Our personal impulse to dance may arise in our physical biorhythms, such as infant dandling songs which combine a rhythmic motion with song. Socially, however, dance connects individual life with the mythic beginnings of the world and the stories that contain all the model roles of the animals and obligatory duties of people. The dancer's quest, says Lonsdale, "is ultimately a spiritual one for instinctive origins. . . . The dance returns the performer to his earliest rhythm" via "the infinite wisdom that is the dumb beast."[10] In many cultures dancing is said to have been taught to humans by animals, the Dogon learning from Yurugu, the fox, the Kalahari from the tortoise.

The popping of those votive offerings in the leaping flames suggests something about the crackle of fire itself, as though it were giving voice as part of its creative activity. The oven as a womb would come later, long after birthing by chant and dance was associated with the idea of a world whose main feature was its cadences. The mysteries of birth and death are danced in Hindu mythology, just as Siva's rhythms set the universe in motion. As the dancer Nataraja, Siva is the destroying/creating cycle, transforming cosmic vibration or sound to visible light and light into the physical being of all things.

Dancing is profoundly intuitive; it suggests jumping between domains. The jerboa of Africa, the kangaroo in Australia, and the hare (jackrabbit in North America) are preeminent in enacting the leap as metaphor.[11] In burning grassland the hare waits until the fire, the transforming principle, is upon him. He then seems to plunge into it but is not burned. Apart from the flame, in the cool of moonlight, the hare is like a shadow of thought in the dark unconscious of the mind, a bringer of insight, treading the boundary between lunacy and sublime inspiration. In moonlit nights in March in the northern hemisphere, the hare dances wildly in mating, a lovemaker of ecstatic, brief, and repeated frenzy. Around human campfires in the Great Plains, the jackrabbit cavorts in acrobatic delight. Buddha was said to have taken the form of a hare—to have leaped into the fire "in a fervor of renunciation," emerging unscathed, and was then translated to the moon as Bodhisattva. The hare, unlike his friends in Buddhist tradition, the monkey, jackal, and otter, does not trick and steal to escape or comply: when asked by a god for food he gives himself. His spirit always defeats brute strength, as in the tales of Br'er Rabbit or even Bugs Bunny.

Crossing the boundary in dance between human and animal seems to open the way to be born, grow, marry, heal, denote elderhood, get on in the stages of human life, and move to the next world. Singers and dancers of the world make these passages in the company of animal guides. There is music of birth, initiation, courtship, hunting, and renewal, music for protection and purification, music to send the dead on their way. This music is played, sung, and danced.

SEVEN PRIMARY CATEGORIES invoke transformation in the course of life, or in the cycle of the human seasons, all conjoining ourselves to the Others who understood and mastered the passages of life before we arrived.

Birth music includes dandling songs and lullabies that involve holding, swinging, or rocking the baby. Many of them refer to the dance of the fetus in the womb in which the mediator is the snake, the shark or other fish, or sometimes birds who, having learned to dance from the sea creatures, then taught humans. The Dogon and the Pygmies of Africa dance the fetus in the womb, as the fetus itself dances to the rhythms of its mother's heart, blood, voice, and the other sounds heard through an aqueous medium.

Initiation and puberty invoke the totem animals of the initiates' clan and convey secret or sacred information. Each animal guide is known for a special wisdom in the advent of sexuality or adult status: the red kangaroo in Australia, the bat in Melanesia, male and female gemsbok in South Africa, spiders and turtles among the Andamanese.

Hunting addresses the hunted animals. When Blackfoot Indians dance a restitution for the killing of the buffalo, the figure of the animal is present, as the seal is for the Eskimos, the deer for the Yaqui of Mexico and Pygmies of West Africa. The modern mind has problems accepting that the animal is present in its effigy, participating in dances that bring equilibrium back to a world unbalanced by death. To outsiders the celebrants seem merely to submit to a pacifying illusion which glosses over an intractable and brutal reality. Yet such peoples feel at home with wild animals and thrive as coinhabitants with them. Foraging was the first quest: the hunter learned to mime and disguise himself as the prey; the woman, to dance the animal with communion as its goal. They went on to hunt the mystery of the animal itself and deepen the meaning of the harvest in the miracle of ecstatic transformation in that rhythmic mimicry. The drum made from animal skins gave convivial form to the dancer's own pulse.

Courtship was played and danced until the twentieth century by the French and English as the pavanne, named after the peafowl with its "formal" display. In Italy it was the capriole, a bawdy goat dance. Birds are often the mythic preceptors, as they themselves dance their courtship. Among cultures who dance the birds in marriage, the cranes and grouse are especially prominent, the cranes by the Watusi of Africa, the mountain cock dance, the *Nachsteigen*, in Bavaria. Cattle-oriented people dance cows or bulls, the Bushmen the baboon. A widespread greeting is, "Who do you dance?"

Healer, guardian, helper—all are renewing and supporting specialists within the context of local myths. The eland, ichneumon, and mantis are danced to these ends among the Bushmen, using rattle and drum made from the skin of the eland to invite its powerful spirit. The tarantula bite in Neapolitan custom is cured by dancing the *tarantella*, inspired by the flailing victim of the spider's bite.

Fertility—conception, gestation, and birth—is framed within the larger fecundity of the earth. European mummers, costumed street performers in animal guise, danced the hobbyhorse in the Middle Ages as a contest between winter and spring. The reviving horse (probably based on ancient horse sacrifices) renewed the year for the satisfaction and fertilizing of Mother Earth. The mummers then danced with the local maidens to spread the mood and power of regeneration. In Mali they dance the marriage of antelopes, representing the sun and the earth, as homage to the spirit who brought millet to the people. The Bobo of Upper Volta, masked as butterflies, dance to purify the agricultural violation of the earth and to bring rain. The Hopi of New Mexico dance as antelopes and with snakes to end winter and bring the spring rain, the Antelope Youth and Snake Maiden personifying the tension between opposites—hot and cold, fire and water, frigid and thaw—resolved in the music that brings them together.

Death or mortuary music frequently calls upon underworld guides: nocturnal and shadowy beings such as snake, owl, caterpillar, and turtle. Some dancers play the role of the totem animal of the dead person to soothe their soul with hospitality. Others are intended to drive the souls away or to include the dead in the dance. Occasionally masks are worn that trap an alien spirit which can then be burned. The most elaborate funerary dances may be the snake and fox opposition danced by the Dogon of Africa.

AGAIN AND AGAIN the old myths say that "the animals taught us this." The dancer's experience is a return, not to another time, but to places in one's self. Dancing the animal transports the dancer toward what animals have not forgotten. There is a loss of the ego and recovery of an "essential biological kinship," a quest which is "ultimately a spiritual one for instinctive origins."[12] In our world, what an odd pair: spirit and instinct. In his book on animals and the origin of dance, Steven Lonsdale speaks further of access to "the infinite mystery of the dumb beast." Did we not mistakenly conclude centuries ago that infinite possibilities were our reward for escaping our beasthood?

BY OBSERVING THE musical metaphor and animal dances of other cultures we see how widespread they are in linking people to each other and to the natural and invisible worlds. Animals are understood to have been the first musicians, halfway figures, mediators, like us and yet themselves. In their music we acknowledge that the wisdom of nature is active spiritually in diverse forms and is accessible to humankind through art. The similarity and difference of culture and nature comprise a puzzle given its most articulate expression in the personhood of animals, whose most profound message to us is music. Just as the flicker of the fire enlivens the shadows on the rock shelter walls, the player, singer, and dancer unite with the animal spirit momentarily. Music is performed as successive spirits along the path of life in birth, mating, healing, planting, hunting, reproducing, maturing, or dying as episodes coded by animal signals that correspond to the passage of the seasons, the daily renewal of the sun, the flowering of plants, and the migrations of herds.

In this sense "folk" music is not a museum collection of exotic relics, arcane superstitions, occasions of beautiful but obsolete bits of a lost innocence, or anthropological specimens. The images of the animal body as musical instrument singing its own songs, the dances in which the swimming shark teaches us the movements of the fetus, the turtle going from land to water portraying the passage from childhood to maturity, the coyote demonstrating the stealthy moves of the hunter, the crane showing the elegant encounter between male and female, the bear giving the healing aster plant to the humans, the owl leading the way into the final darkness—all are lessons in palingenesis. The idea that such animal paradigms are irrelevant belongs to our own posture of alienation and arrogance.

Aristotle and Constantine mark points in the departure from a world in which music seemed to capture its essential quality. To Aristotle, as a pagan, appearance seemed central to identity and truth. He noted that the masked performance was a convention for elaborating classificatory changes and lived in a time that still believed in the rituals of dance and mimesis. But Aristotle's objectivity foreshadowed our own culture. When the streams of Greek and medieval Christian thought were synthesized by Thomas Aquinas in the thirteenth century, the modern world began to formalize its belief in the soul as a unique, transcendent possession of humans who had no need of the music of animal powers. The true spirit was unchanged by outward form. Indeed, metamorphosis was seen as the work of the devil. The Church had, from the fourth century, prohibited dancing as part of its liturgy, despising the erotic element and the concept of shape-shifting. In Christian miracle plays, Satan was a dancer. Whatever dance may have remained in the Christian rituals of its first two centuries has long since vanished. Everywhere "pagan," "heathen," and "blasphemy" were identified with the sensuous drumbeat and the transporting force of the dance.

But neither the Greek philosophers nor the Christian theologians could have foreseen the corporate industrial world which followed in their wake. Though its roots may be traced to Plato or Augustine, they are not to blame for the elimination of music as access to religious participation. We master movers of the world, progressing by expanding our power, seem outside the choreographed flow of things. Daily life is less ordered by customs associated with the sun or moon, the seasons are not celebrated, anniversaries are minimized, festival occasions are commercialized, and the holding strength of tradition is more tenuous.[13] There are fewer confirmations of kinship in terms of either social or ecological ties. This discordance of personal life is highly stressful. Events do not memorialize or reanimate familiar patterns. In seeking change we perpetuate the problem in more change and more fragmentation. Modern culture not only magnifies this dislocation but perceives it as the human condition. The traditional role of participatory religion, making the cosmos whole, has eroded. Art, always the instrument of religious language, continues as a healing medium—not religious but therapeutic—in a visually disintegrating world. The loss of traditional dance in our own time cuts the individual off from the accumulated wisdom of dance as it was derived from the nonhuman Others. The loss of this connection is reflected in modern popular lyrics that search endlessly through the meanings of love, given and lost, and the convulsive dances of elemental rhythm. When it is too loud the music becomes increasingly tactile and less kinesthetic, as though the physical effects were changed. It becomes a massage rather than an ambient sign. The deafening music deprives us of hearing itself, as though we must shut out the sound world we live in and seek a therapy at the more regressed level of touch. Rock music abjures the nuances of sound in the search for an autonomous self—solidity grasped as though from the "rock" beneath our feet. But the rock is more like the sway of the cradle and regression to a tactile environment with its beat.

The immense popularity of modern music indicates a widespread yearning. It is as though we realize that the harmony and lyrics help glue the razzled scraps of daily life together. In the 1960s young people in Europe and America recoiled from the culture that spawned them. Among the characteristics of the counterculture was its own music—not only listened to but performed, played, sung, and danced. Basic rhythm, essential percussion, words that assert a collective independence, music that moved them: all this seemed like a great tonic for the sanity of thousands of protesting, drifting individuals.

The "naturism" of the hippies of the 1960s was similar to that of antinomians of all times—a return to searching for one's roots not only in small-group dynamics and music, but in the larger rhythms best experienced away from built environments, returns to the land, to "companion animal" relations, astrological and dietary changes related to "natural" seasons, and so on. Mickey Hart, a drummer for the The Grateful Dead, who is sensitive and literate beyond many of his contemporary musicians, believes that the private search for the way to personal maturity leads deep into the past and into precivilized traditions.[14]

Folk and gospel music, tribal dance and song, spontaneous small-group and individual minstrelsy, the participatory modus of groups like that of Paul Winter and his consort, the singsong of religious chant and traditional anthem—all testify to the resilience of our own animal music in an era of noise, tunesmiths, electronic simulation, and the concert. How strangely distant are animal sounds from the modern symphony hall and the drawing room, the choral recital, and whole gestalt of "classical" music. Formal composition and performance, allied to mathematical, architectural, and astronomical monumentality, is, like the formal gardens of the seventeenth century, a microcosm of an utterly detached observer in a rational universe. The power of classical music has to do with esthetic exaltation and the genius of its artistry. Its audiences are deeply moved by the orchestration of the worldview of a society grown dense and layered with pyramidal power, distanced to the point of vanishment from spontaneous participation. Only the maestro and soloist are free to add their virtuosity to that of the composer, to interpret the script like a pope or a prophet. No matter how "inspiring" Bach or Mozart may be, or how elegant the opera, their performance is a rite of inhibited, solitary discipline among ranks of instrumentalists directed by a conductor before schizoid audiences sitting in rigid decorum, drifting in head-centered labyrinths of imagery and repressed emotions. Frozen in isolation and silence, the spectators are permitted to respond by clapping at approved intervals or, at the end, shouting "bravo!" in a kind of stereotyped anarchy. Symphonic music is indeed great in its skills, technique, and esthetics; but it is also the supreme articulation of the dissociated state of our species and our personal lives. It co-opts the melody and rhythm so essential to our health by subordinating them to execution and complexity, denying them ecological, egalitarian, and participatory function. In its amazing scoring the music disintegrates our connection to nature by making elaborate musicality an end in itself.

The music of animals, and our music as animals, do not exploit the rational/romantic, lonely, literary self. It is alien to the isolationism of both rock and classical music. In modern rock music we regress to the sway of demented captives or the sheer noise of a lunging, screaming mass. Before the grandiose orchestrations of the symphony we are intellectualized out of our bodies. In both cases we have been seduced by celebrity. We seek to preserve the genius of composers and the perpetuation of "immortal" works, as though the very meaning of civilization were the eternalizing of Great Art and High Culture. No doubt, in the history of small-scale societies of the past, genius came and went unrecorded. Even among crickets there may be occasional, relative geniuses. But their music is so much an inner part of things, of passage itself, that no individual cricket (or the leaping which it inspires metaphorically among humans) could conceivably be "immortalized," as the trilling songster achieves its supreme effort only a moment before it is snapped up by a hungry shrew.

OUR ATTENTION TO animal sounds (as well as those of all of nature) is not only about discovering the music within ourselves and around us but the importance of listening. We seem less in danger of losing our rhythmic impulse than of attending to the nonhuman voices. Ecologist Paul Sears, who was an accomplished violist and a master at reading the landscape, cautioned his family and students, "Never ignore a sound." Overwhelmed with noise in the modern city, it is not surprising that we seek to escape it and yet fear silence. Taking students into the backcountry of Baja, Mexico, my fellow teacher and I were appalled in the evenings, at the campfire, to be always encased in the sound of guitars. And so we lost opportunities for that perfect absence of the sounds of machines and domestic animals, which was of course not a silence at all, for the night was peopled with faint footfalls and calls, a delicate web of whispers and acoustic tracery which seemed to be joined even by the stars.

Music communicates confidence to the whole body when meaning is beyond pictorial and verbal sense. The effect is to return to the continuum of the forest, shore, and savanna. Instrumental music is a recovery of the calls and songs which gave form to the Paleocene world from which we came. It is odd how coalescent the whole can seem, a sort of logic of the euphonic fen. Music renews, integrates, and solaces amid the discord of modern life by returning us to the universe at the base of the brain and the stem of human time where unseen musicians created islands of order, however dark the night.

We do not need to reenact Orpheus, to instruct the animals by tootling meaning into the dark, but to hear that nocturne already present, as it was when the first person ever thought of shoogling to its rhythms.

| # Ontogeny Revisited: Teddy, Pooh, Paddington, Yogi, and Smokey

Bear ritual points to integration of the personality following a period of dissociation.

J. L. HENDERSON

THE BEAR IS the most significant animal in the history of metaphysics in the northern hemisphere. The great circumpolar "bear cult" is the salient religious and ritual association of people and a wild animal. Among tribal peoples, the spirit of the bear prowled the three levels of the cosmos—sky, earth, and underworld—controlling the seasons, the coming and going of game animals, and the destiny of the dead, all forms of palingenesis, the Great Round. In the night sky the bear is the constellation Ursa Major (later known as the Big Dipper or the Wain). Its seven stars and others nearby, Ursa Minor, and Boötes, make up hunter and hunted in the traditional lore of the northern world, narrated as a perpetual chase whose energy turns the celestial wheel of the sky itself.[1] The Navajos marked skins with the Great Bear constellation, which was "keeping watch" in the winter sky. The ancient Greeks celebrated the same seven stars as Callisto, an expriestess of Artemis, the bear goddess. To the Chinese Taoists the Great Bear was the throne of Shang, the supreme deity, and was represented on jade *kuei*, the tablets offered by Chinese emperors at the Altar of the East, on the morning of the spring equinox.

When we moved into cities we forgot the traditional stories of the bear. But as an archetypal figure underlying the forms of culture, it persists in our dreams and imagination as though some tracks were pressed into the human nervous system during the ice ages, leaving in our innermost natures a kind of preconscious expectation that Ursus' shaggy presence will give insight into human problems. In a world which no longer perceives the bear as a religious figure, a sign of the calendric year, or as the energy driving the universe, that archaic impulse emerges in yet other ways.

Chief among them, like an ancient thread by which we are connected to the past, is the bear's transformational power, so that in different aspects it can represent the stages of our lives. The bear is a mother almost as great as the earth in whose womb it renews and perpetuates itself by bearing its young. It is the nurturer without peer, who leads, demonstrates, punishes, and rewards its offspring until they are almost mature, taking them along with her into the den a second and third winter. Because of this mastery of the life cycle, bringing forth new bears and guiding them through the stages of their lives, the bear is, according to J. L. Henderson, the model for the elder who presides at initiation ceremonies. The bear is an elder among the animals, as solitary as a shaman, knowing the healing herbs and baths, ranging across the landscape as if space were time, keyed to the season, wise in the stages of becoming.

In its humanlike qualities—upright-standing, taillessness, whole-foot-walking, omnivorousness, curiosity—it has been at the center of our attention from the beginning. In our imagination we have lately reversed the imagery, so that the bear is no longer represented as a shaman or mediator of social movement but is itself projected as the phases of life. Its resemblance to us is fractured into human age groups. Thus the bear in the popular culture of Europe and America in the past century has taken shape as steps in growing up or as a serial of five siblings, each five years apart: infant, five-year-old, juvenile of ten, adolescent, and young adult.

Youngest is the baby bear, whom humans sometimes yearn to cuddle (not knowing about its irascible nature and sharp claws). The teddy bear is the newborn's prosaic form, the companion of thousands of children since it was invented by a Brooklyn candymaker. It is a doll-like animal, a fuzzy, comfort-giving object which can be ignored, forgotten, dragged about by an arm, flung down, hugged, sucked on, slept with, and talked to. Attached and detached according to whim and need, the teddy bear is the ultimate transitional or comforting object needed in moments of stress, fuzzily identified with the self, and equally fuzzily with the not-self. The delight in its soft coat reminds us that we are primates with a much longer memory of clinging to a mother's fur than being held against a naked breast, primates who may have begun to wear skins when our genus was still young. The teddy's nonhuman face triggers that other deep desire of the early speech learner for identifiable Others; other humans are vaguely look-alikes, with personal names but without the simple categorical identity of "bird," "dog," or "bear." The stuffed bear is not only the silent and ever-loving presence. It is also we ourselves as infants, an object lesson in babies for adults as well as children.

How did we get along without teddy bears for so long? No doubt other objects filled the need more or less: dolls, pets, the little blanket, the stuffed stocking. It is possible, too, that we did not need the teddy so much, that there is something about infancy in the modern world that cries out for just such a puppet.

At first merely a dependable presence and plush comfort, the teddy doll gradually becomes a companion for games of pretend. It is given parts in little dramas. The plump teddy bear anticipates its fictional expression in Pooh Bear, A. A. Milne's chubby little friend of Christopher Robin, along with Tigger, Kanga, and Eyore. We have arrived at a child's garden of the human community in the guise of species differences.[2] Pooh is clearly the animated teddy, with his same tubby preoccupation with honey and other food, set amid the dawning realization that there are others with their own ends and occasions which can be distracting. Christopher Robin and the human world are available for the child to identify with, but not yet so interesting or so entirely self-serving. The creatures have not come to the human child but he to them, as he transforms them into stand-ins for a basic range of social circumstances. The stories give an episodic life to the bear—now stuffing itself with food rather than stuffed with straw—in which a pattern of personalities and situations has begun to shape a small cosmos. All plans have a short range and relatively uncomplicated interval; they can be abandoned without second thought and new ideas pursued until the protagonists are again diverted. It is the play world of the small child given the oral enticement of the adult's telling or reading (who is rehearsing his or her own Pooh memories).

If Pooh is still too callow for human society, this is where he is succeeded by Paddington Bear, who strives to join the modern human world. Underneath he is still the same chubby child, but now we find him in stores, banks, medical clinics, taxis, houses, markets, and the whole panoply of the civilized community. His innocence and naivete do not make him an object of derision, and he is constantly learning how to get along in an adult world where he is accepted for what he is. If Pooh is the scion of a genteel country house, Paddington is the city cousin, grown up by comparison, but still the little bear underneath, living in a literal rather than symbolic world. If the music in the teddy's life is the lullaby and in Pooh's the nursery rhyme, in Paddington's it is a simple song. Paddington accepts this issue of money and telephones, though they can puzzle him; he is out to conform and become part of it all. Paddington is the civilized little bear of latency, at the main table, while Pooh is still thinking mostly of his stomach and presumably sitting in a high chair. Pooh has few if any clothes, while Paddington is seen in proper attire. Paddington has stirred from the garden and the yard—not to the secret places and adventures of Pooh but to the shops. He is the consumer and conformist in us, whose bearishness will gradually recede beneath a civil exterior. Teddy had no teeth, Pooh his baby teeth, while Paddington is growing his adult set.

Beyond the juvenile, Paddington, comes the adolescent, who is seen in the cartoon of Yogi Bear. Yogi's presence in this sequence is much more tenuous and recent, as though we ourselves had only just begun to appreciate the definitive characteristics of nubile youth and to find a bear to match it. He is an inhabitant of television and his name is a pun on Yogi Berra, an American baseball player and manager, known for his malapropisms. Yogi Bear lives in "Jellystone" Park where his companions include a small bear, named Boo Boo, the park ranger, and various tourists. As a park bear, he still dwells outside the "real" world: he is protected and yet free. The kindergarten from which he has graduated has been left behind for a preserve where he can play at adult strategies without paying the penalties of adult life.

Frequently boasting that he is "smarter than the average bear," it is Yogi's destiny to outsmart himself. His schemes usually run counter to the rules and to reasonable caution. Unlike the compliant Paddington, Yogi is not out to fit into the world but to test himself against rules, authority, and custom in the form of the park ranger. In his breezy restlessness and optimism, tie and hat only, he is reminiscent of the schoolboy or college freshman of the movies. Bubu, like a little brother, still lives in the shadow of adult omniscience. Against his own inclinations he is carried along with Yogi's enthusiasms, a near victim of the inevitable collapse of pranks. Unlike the urban adults in Paddington's experience, the ranger is the mix of bureaucrat and policeman, the modern descendent of the medieval sheriff, partly the perplexed buffoon and partly the stern parent. Yogi's transgressions are trivial and heedless rather than criminal. Since he usually loses in the end, the lesson is that excitement is possible in bending the rules. Even if he is unlikely to succeed, he will follow his ill-considered gambits, which forever take in his little follower and keep Yogi in a friendly adversarial relationship rather than isolated from the protective figure of the ranger.

Presumably, when he matures, Yogi will emerge from his feckless adolescence to become Smokey, an invention of an advertising agency for the U.S. Forest Service. He has graduated from freedom-loving pubescence into the bureaucracy itself, personifying the superego. Although inspired by an actual bear who was burned in a forest fire in 1944 and spent its life in a zoo, Smokey leaped from the zoo onto a thousand posters, leaflets, and billboards, acquiring the blue jeans of the firefighter and a campaign or ranger hat.

Two generations of Americans have grown up under his accusing gaze. The rebellious phase of life is past. He tells us that forest fires are started by our carelessness. He has become the arbiter of our behavior as campers, hikers, and tourists, the picky and humorless elder whose sole interest is to ensure our ecological rectitude by complying with his rules. He needs only his ranger hat and no-nonsense work pants to remind us of his police duties and his practical blue-collar associations with security and productivity. Smokey is the ultimate administrative mouthpiece—not an actual servant of the public who own the forest but the embodied threat of punishment and governmental disapproval of those who would endanger what will become logs and paper. It is clear that he is no grizzly, whose survival is threatened by the building and extension of roads necessary to take the saws to the trees and return the logs to the mills. He is an ordinary black bear whom we may suppose was at times a Teddy, Pooh, Paddington, and Yogi, and who has become a man in a false face. If Pooh and Paddington are also disguises, they are at least helpful fictions as we grow into our human condition. But the Smokey mask conceals a deeper, authoritarian mischief, corporate demagoguery, the cunning of expropriation and the politics of deception by appeals to our sense of security in the guise of the bear.

Some final figure of the bear is lacking in this cycle. We leap from adolescence into conspiracies of greed and power by those who are actually the former teddy bear owners. The tender, surrogate bears of our youth vanish into the figure of a man who has assumed the lineaments of a bear, whose real interest is cutting down trees, whose exhortations, cautionary admonitions, rules, and objectives are the euphemisms of "forestry" and progress.

Perhaps we must return to Goldilocks: the little girl who comes upon the house of the bears in the forest, enters into a puzzling play of roles, and ends by breaking things. She too is caught in the drama of ontogeny. She tries the food, belongings, and furniture that signify the bear's domain, from baby's to mama's and papa's. She flees in fear as the bears return in dismay. Perhaps the missing figure of the bear in the sequence from Teddy to Smokey is not an ultimate amalgam of human and bear at all but simply the return of the bear itself. Smokey is not the last act but a misstep, a caricature. The true elder is perhaps the real bear, an otherness to which the humanized bears of childhood are preparatory. We graduate into the world of otherness, not away from it.

We are brought back, finally, to watching and pondering the extraordinary life of the bear. The world of real bears is not always safe or amusing, nor is it exploitive and despotic. It requires many generations of human attention, the pooled accumulation of knowledge, to grasp the full outline of the bear's life, the efforts of poets, mythologists, metaphysicians, and artists. It is not only its likeness to us but its difference from us, working in tandem, that shape the watershed of mature reflection. That dyad is a talisman for pondering analogy, kinship, and otherness. It was perhaps just such reflection that got the Great Bear into the sky, revealed the healing arrowroot, presented the underworld as a place and the religious possibilities of burial as a future, made parenthood self-conscious, gave us the verb and model for birth, navigation, and carrying along what must be transmitted, and provided an overview of each life as many lives. Even so, unless we are satisfied with the id-world of Teddy and Pooh, Paddington's conformity, Yogi's hedonism, and Smokey's authoritarianism, we may seek to recover the final bear, not part of a modern burlesque, not a man in a furry overcoat, but the wild Other who is bound to us by our membership in the natural world rather than parody.

PART V

The Cosmos

*Longer than memory we have known that each animal has
its power and place, each a skill, virtue, wisdom, innocence —
a special access to the structure and flow of the world. Each
surpasses ourselves in some way. Together, sacred, they help
hold the cosmos together, making it a joy and beauty to behold,
but above all a challenge to understand as story, drama, and
sacred play.*

14 | The Meaning of Dragons and Why the Gods Ride on Animals

I see the tigron in tears

In the androgynous dark,

His striped and noon maned tribe striding to holocaust,

The she mules bear their minotaurs,

The duck-billed platypus broody in a milk of birds.

DYLAN THOMAS

CULTURE TEEMS WITH animals who have no exact equivalent in nature. A huge, make-believe fauna of monsters, prodigies, and wonders slithers and swarms and storms through all of the arts, as though the natural world were somehow deficient. It is necessary to ask: To what end are these creatures of the imagination? Are they really substitutes for ordinary animals or do they have their own purposes? What are they, where do they come from, and what are they doing here? What is their truth? If fantastic animals have a special role in human thought, what is it? In the long view, what begins with the symbolism of natural animals progresses through the mixing of parts, toward the combination of animals and humans, culminates in a breakaway of purely human deities, preceded by their becoming jockeys.

Bestiaries, or books of animals, existed in the past and included many fabulous forms, but during the era of exploration and empire confusion arose about the physical reality of griffins, sphinxes, mermaids, and basilisks. Many strange creatures were being introduced to Europeans: rhinos, giraffes, and sloths, along with those now called "mythical." About the time of Marco Polo in the thirteenth century, the *Physiologus*, or dictionary of animals, was expanded to include many new beasts regarded as actually living somewhere. For about four centuries it was unclear which of the strange animals in the bestiaries, and those described by sailors or half-mythic travelers like Prester John, were "real." It was not until the eighteenth century that science began to sort it all out, excluding mythical animals as merely the inventions of primitive peoples, as superstitious mistakes, or frauds and dreams. They were relinquished to folklore or to a lexicon of curiosities in fiction and art. Modern editions of medieval bestiaries entertain the reader with reproductions of the old woodcuts of such "imaginary animals." In part debunking and in part amusing, they look back to a world in which people supposedly believed that the basilisk, phoenix, dragon, roc, and griffin were "real," or, in the case of the Aesopian fable, to the quaint morals which they represented.

Despite their uncanny appearance, the physical reality of such chimeras is due to the correspondence of their parts to those of natural animals. Because of it the geographic search for bizarre forms dies hard. Disenchanted, modern city dwellers may look upon pictures of dragons and unicorns as fantasy, but are ready to "believe in" Bigfoot or yeti, the Loch Ness Monster, celestial extraterrestrials, and "aliens" from other planets. Somehow dragons meet heartfelt needs, like food requirements for certain nutritional compounds, indeed, as compound animals.

DRAGONS, PERHAPS THE best known of compound monsters, are composed of bits and pieces of birds, snakes, and quadrupeds. Just as anatomical parts are combined in dragons, so too are the ideas which those parts traditionally represent. In the Orient the dragon is a renewing and unifying participant in seasonal celebrations—the winged snake linking birth and death, fire and water, male and female, death and life. In popular imagery in the modern West it is simply a terrible demon to be killed by heroes or a silly cartoon figure. But behind the modern trivia are images of the masculine and feminine. In medieval Europe the dragon was a fire-breathing old grump guarding his gold beneath a mountain. Since gold was a patriarchal symbol of virginity, the dragon was therefore associated with the violent, virile control of women that dominated Europe for three thousand years. Its tenacity and ferocity against masculine challengers and its ownership of the cave in the earth are suggestive of that association.

More fundamentally, the dragon's ancient life-fostering association with water was feminine. In ancient Egypt, the animals of the sky mother and warrior sun god were the cow, gazelle, falcon, eagle, lion, serpent, fish, and crocodile. But the importance of water itself meant that none of these creatures alone could signify its manifold forces. Chris Knight, an anthropologist, observes: "Symbols of the life-giving and the life-destroying powers of water . . . were invented by combining parts of them to express the different manifestations of the vital powers of water . . . immense, brilliant, carrying riders between fire and blood, sun and rain, marriage and kinship . . . a single worldwide mythic *thing*," a male inversion of female power, a "logic of menstrual synchrony inverted into a logic of male pseudo-procreation."[1] Dragons are men's fear of women's procreative power, focused and incarnate in the scaly-she-monster, the fear of female periodicity with its lunar connection in which "men attempt to cut women's ropes—their Serpents and Dragons." Menstrual synchrony is the tendency of the menstrual periods of women living together to become clustered, and therefore for the women to ovulate at the same time, become pregnant, and give birth in batches. Since it is periods of light that set the menstrual clock, the addition of moonlight to daylight in societies without artificial light locked this cycle to the moon's phases. Men were then confronted not only with women's mysterious fertility and lactation but the celestial flourish of these powers.

The serpent-killers of early historical times represent masculine initiative, heroes mastering procreative magic by defeating dragons: Marduk slays Tiamat, Yahweh throws down Leviathan, Zeus fights Typhon, Indra opposes Uritan, Perseus slays Medusa, Herakles defeats Ladon, and Saint George lances the dragon. Dragonlike monsters in the Bible also represented political enemies of the Hebrews, characterized as figures of chaos in their amalgamated construction, a hideous mélange, the farrago of a disordered world. The pastoral Hebrews were obsessed with purity of type, derived from Genesis and expressed in the biblical injunctions against animal hybrids and human miscegenation. It was the awful amalgam of the apocalypse which "came up from the sea. . . . The first was like a lion and had eagle's wings . . . and made stand upon the feet of a man, and a man's heart was given to it. . . . Another like a leopard which had upon the back of it four wings; the beast had also four heads and dominion was given to it. . . . A fourth beast . . . had great iron teeth and ten horns, and among them another little horn . . . [and] in this horn were eyes like the eyes of man. . . . The same horn made war with the saints and prevailed. . . . The fourth beast shall be the fourth kingdom upon earth . . . [and] shall devour the whole earth. . . . And the ten horns out of this kingdom are ten kings that shall arise, and another shall arise and subdue three kings." That old serpent, called the Devil and Satan, is like a leopard with the feet of a bear and mouth of a lion, ridden by a woman, "mystery, Babylon, the Great, the Mother of Harlots and Abomination of Earth." The "beast with seven heads" is Rome or "the Prostitute," death itself.[2]

These apocalyptic dragons combined what was most terrible: mixed breeding in any form, the woman as harlot and prostitute, and enemy kingdoms whose cities, like Babylon, were mixtures of tongues and cultures, all signifying corruption. Such amalgams of horns, heads, wings, and claws, symbolizing foreign kings, nations, the goddess, wilderness, transformation, earthliness, the sea, death, and cities, are the combined evils of this world. Even if these horrors were the hallucinations of individual fanatics, we can sense the fear of disintegration they convey. The injunctions against hybridity faintly echo the childhood modeling of primal cognitive categories on anatomy and animal types, during the acquisition of speech, raised to a zealous idée fixe.

APART FROM DRAGONS, who combine parts from as many as six animals, most synthetic beasts in art and mythology unite only two beings. Many of these are human-headed animals or animal-headed humans. The human-headed animal, such as Pan the goat god, links instinctual appetites or animal intuition to an individual personality, making Pan a degraded or special human. Alternatively, the animal-headed, human-bodied creature such as the minotaur is a humanized animal.[3]

Angels, the 3,000-year-old spirits in human/bird form, have been rudely debunked as physically impossible by literalists who argue that they lack a keel-like breastbone and the huge pectoral muscles necessary for flight. But angels were not the product of ignorance. They were powerful spirits, interlocutors in the layers of the empyrean between heaven and earth. Like dragons since the sixteenth century, which declined into whimsical illusions, gargoyles, and childish make-believe, angels degenerated into ornaments in art and architecture, fossils of their original purposes.

Not far removed from the angels, the best-known animal-headed humans in Christian tradition were the four evangelists. They descend from an old tradition that Yahweh, the Hebrew god, rides animals with faces of lion, bull, eagle, and man.[4] These were related in turn to the tutelary genii of the Near East, where they represented the four seasons and four points of the compass, each becoming a star in pre-Hebrew cosmology. They reemerged as Matthew, Mark, Luke, and John with the faces of human, lion, calf, and eagle, each with six wings. At first guardians of God's throne they declined into mere similes in the books of Ezekiel and Revelation: Mark with his "gospel of majesty" as the lion, Luke who emphasizes sacrifice as the ox, Matthew in the flight of his prose as the eagle, John with the dignity of (the godlike) human form. Attempts to despiritualize and yet perpetuate the animal powers show how Judaism and then Christianity assimilated the old theriomorphic deities by first reducing them to minions and homilies. In the Middle Ages these hybrid images faded from sacred art as the Church progressively dispensed with animal imagery. In long perspective, the four sacred powers, the bull, lion, eagle, and earthly human, in post-Neolithic cultures of the Mediterranean and southern Eurasia slowly yielded their place to humanized deities over perhaps five thousand years. During that time, various combinations of the three animals and humans represented the transition. A history of the myth and the iconography of bull, eagle, and lion demonstrates the fluctuating fortunes of many of the old animal deities. The complexity of that history with its geography of intercultural exchanges is indicated by looking at the lion.

PERHAPS THE MOST difficult and lively hybrid image in the history of the Western imagination is the lion-headed man-snake. It begins in the Egyptian deserts, where the sky was seen as partitioned into sections, each dominated by spirits in the form of a great zoo controlling life on earth. From there the lionesque figures are taken up in Greece and Rome, where they illuminate the nature of the human soul and are incorporated into biblical and other Near Eastern cosmologies. Finally the ancient Egyptian deities are reshaped in Christian Gnosticism, denoting not reverence but hatred for the ancient animal spirits, a ready source of the images of demons for medieval Christianity.

Both snakes and lions were immensely powerful beings in ancient Egypt, spiritually and physically. Among the many cat-headed deities in Egypt the original was Sekhmet, a late bronze age mistress with breath of fire. She was goddess of the desert and its storms which, as punishments, were "messengers" of retribution. Such deities in half-animal form were protectors and punishers in a world where the weather was controlled by them or by the chief spirits they represented. Storms were likened to the tawny beauty of lions on the one hand and to their voracious maw on the other. The passion of weather was like that of lions, a reflection of human passion. Sekhmet was connected with sex via Bes, a phallic combination of lion head and human glans penis, an allusion to the vigorous sexuality of the lion and the generative power of the sun, of which the golden lion was an analogue. Lion images dominated the sacred statuary of many temples. Leontopolis was the city of the lion cult, where lions were worshiped even in a Jewish temple of Yahweh in 162 B.C.

The pharaoh, or "lion king," who mauled his foes, was represented as the Sphinx, the binary spirit-creature with the body of the lion and head of the divine sun king. The agents of the Sphinx, messengers or punishers, could take many forms because the sun god, Ra, had many animal guises tied to the sun's "change of faces" as it passed through different sectors of the sky in the course of the day and the seasons. The most crucial seasonal event was the flood of the Nile, which was controlled by the rise of Regulus, the brightest star of the ecliptic, in the constellation Leo. Leo was also the astral sign of the pharaoh's horoscope, hence a nativity star, the lore of which combined observations of earthly lions with astrological theory.

These realms of lion power—the punishing and fructifying solar disk, certain constellations and their influences, the storms, the great river, fire, the cycle of the seasons, sexual passion—were undoubtedly represented in their early forms as the animals themselves. Then, by commingling the lion with the snake, Wadjet, a vast scope of earthly, cyclic, incandescent, and underworld powers were fused with the efficacy of the lion spirit.[5] The flow of cultural influences from Egypt brought the snake spirit, Wadjet, and lion deity, Sekhmet, into Greek religion, where Horus (Apollo) and Helios are represented on gems with serpent staff, orb, rays, riding a lion, or sitting on lion-supported thrones. The lion's fury and eroticism made it the fitting id of Dionysos, depicted riding the lion. This combined figure, Chnoumis, is known from carvings on gems which were worn as amulets against personal misfortune, especially illness. In the sky, Chnoumis was the first spirit or decan—each decan representing a planet—of the constellation Leo. In the human body, Chnoumis ruled the stomach and heart by "decan therapeutics." (The other thirty-five astrological decans ruled other parts of the body and sections of the zodiac, each supreme in succession as the sun crossed the sky.)

The combination of the snake and lion modified its old Egyptian fieriness in Greek Orphism, a mystery religion of the sixth century B.C. It told of the triumphs of the gods and goddesses over old earth-spirits, men with snake-tail and wings, who were defeated and "exiled" to Egypt.[6] In Orphic cosmogony a sidereal bestiary was the culminating astrological principle, in which the snake, Chronos, signified celestial cyclicity. Time was essentially bestial because it is "measured by the revolution of heavens swarming with animal powers."[7] Phanes, the sun, passes across the sky through the zodiacal signs, represented as bull-, lion-, and snake-bodied. The Orphic hymns conceived of these animal phases as "swallowed" by Zeus, whose head is in the sky so "he might as well have all the heads of the different beasts that dwell there growing from his shoulders."[8] The ability of Zeus to take any form is derived from the sun's passage. Egyptian Chnoumis, Greek Chronos-Herakles, Roman Saturn, and Gnostic Yaldabaoth were "experiments in polycephalism," having the heads of different animals of the zodiac, each evoked as the sun passes through it, the passage governed by the cyclicity of time as the snake. Each of these spirits could also influence souls at birth, as they came from the planets, and so govern the human personality.

In the first century, the Greeks, like their Egyptian antecedents, saw souls transiting to and from the Milky Way through Cancer and Capricorn, coming down via the gate Leo. The soul itself was imagined by the Greeks to have three tiers: one was human, associated with the head; another was the lion, equated with the heart and nobler aspects of courage and magnanimity; and the third was the "many-headed beast" connected to the belly. The tripartite soul was the means of "the pacification of wild beasts, both those within us and those without us as prophesied by Isaiah."[9] The "lion" heart, tamed and trained, would help control Typhon, earth-beastliness. But as the later Greek philosophers ranked rationality over passion, their confidence in the lion's honor declined. The Stoics denied that passion was noble. They and the Cynics regarded Herakles as a model of heroic virtue and his animal opponents as evil, although only three of his twelve labors were against beasts (hydra, lion, and boar). The twelve labors of Herakles correspond to the twelve zodiacal signs through which the sun passes in a year. "Anger" was the lion's weakness, its fiery temper producing conflagrations like love and vengeance. Plato described the soul as a many-headed beast, a lion, surrounded by a man—they controlling him or he them—and he joined the lion to the snake to typify degenerative emotion.

Heroes like Zeus and Herakles, defeating the monsters of older ideologies, set the stage for the perception of demons as combination animals for the next two thousand years. As feathers, scales, and fur dropped away from the features of the gods, the animals themselves were increasingly debased. The sun's daughter, Echidna, was the half-human/half-snake mother of Hydra and Chimaera, wife of Typhon, the dragon with wings and many heads, whom Zeus would overpower.

Egyptian ideas and images made their way north not only into Greece. Mithras, the Persian deity adopted by the Romans as a solar god, was also envisioned as lion-headed. But the background of Mithraic ideas was more deeply rooted in the concept of a cosmic conflict between good and evil. The fused image was seen as representing human misfortune, as the leonine face of the sun struggled to free the soul from the earth-snake body.

Among the Romans, lionesque shoulder ornaments were worn on the toga, representing the star Regulus, their design corresponding to the hair whorl on the shoulder of young male lions. *Chnoumis* became known throughout the Roman world in the late second and early third centuries as the cosmogonic fire, creator and controller of the cyclical eternity of the heavens, as Khnum had been master of the Nile in Egypt. As we have seen, many of the Mediterranean gods and heroes—Chronos, Herakles, Phanes, and the Orphic and Mithraic lion-headed, snake-bodied deities—were all similar responses to the idea of an "astrological pressure," the power of the skies over human destiny.

THE LATER GREEK puritans influenced the Gnostics, a heretical Christian sect that also had roots in Persian duality. Their ideas, which would disproportionately affect all subsequent Christianity, were devoted to the transcendence of spirit over matter. They adopted the Greek triple-parted soul in their astrology in the second century, drawing upon the animals of the Egyptian zodiac as imagery for physical passion. The Gnostics saw the rational soul as encrusted with animals "who assimilate the lusts of the soul to the likeness of themselves—wolf, ape, lion and goat."[10] This infection imperiled the soul's transmigration to heaven. The evil, leonine maker of the material world and his bestial adjutants pervaded the cosmos, the "lair of the lion and snake."[11] The animal aspects of man correspond to the creator's bestial self and the body's ravenous appetites, especially to the lion's passionate sexuality. This leonine demon in the self could be mollified, however, by coaxing it into willing servitude or starving it into submission.

The Gnostic tradition of a leonine creator is a story of the struggle to rescue divinity from "the cosmos and its rulers." In this nature-rejecting philosophy, the leonine/serpent demiurge was the creator of the material world and man's bodily aspect. Humans "devour" and are devoured by the part of that creator lent to physical existence, and a struggle occurs in each person in which matter and spirit try to consume the other. The soul is composed of elements of the world-ruling powers and their celestial counterparts, the archons. This association of the animal lusts of sky and earth is like an illicit mating of the primal archon, Yaldabaoth, and human female, Arrogance, with human passions as their bastard offspring. Man, trapped in the archons' physical world, was ordealed by water and fire, the water from the serpentine, a seven-headed demon of desire, defiling its host. Earth and sky demons, combined as the snake-bodied dragon with the lion's face, must be "fought" with fasting and virginity. This association of lion, fire, and sex was lodged in the lion's bone marrow.

That the sky should be filled with animals in the traditions of almost every people on earth is one of the most surprising and inexplicable characteristics of our species. Each sign of the zodiac (from *zoion* or "animal") is a part of the sky through which pass the principal planets. The whole is divided by astrology into twelve sections, each thirty degrees wide, with a planet and sign expressing the planet's character, named for the constellation to which it belongs. In Gnosticism, archons guard the gates of the seven planets, endowing the immortal soul with animal passions as it descends into birth; and at the end of each human life the archons, still in the form of monstrous animals, guard the way to heaven. Rising up, the soul encounters these bestial spirits in sequence through the seven celestial spheres, Yaldabaoth last. Each gate requires a password or else the soul returns to earth in the form of that animal guarding the gate.

In the Old Testament, Yahweh was described as a lion in strength, boldness, ferocity, and protection, similar to the Egyptian Mios, a "fiery-eyed" god of storms and lord of heaven. The Gnostic haters of the Jews went the Mithraists one better: they made a vice of Yahweh's leonine power as well as his snake spirit, identifying him with the chief archon, lion-headed Yaldabaoth, combining them as Yahweh/Yaldabaoth, royal perhaps but demonic and voracious. The Psalms provided the Gnostics with the source of this portrait of the creator as lionesque, enabling them to identify him as an erotic, lion-headed leader of the animal-like archons. The "lions" of the Psalms merely ate men and sheep, but for the Gnostics they were world rulers and soul destroyers. Adopting the Egyptian imagery of Leo, the Gnostics equated the midsummer sun with Yahweh's evil and regarded his angelic horses with fire-breathing lion's faces as metaphors of oppression.[12] The serpent part from the Egyptian Chnoumis provided this complex lionesque head of Yahweh/Yaldabaoth/Sekhmet with Wadjet's snake body. So the Gnostics described the god of the Jews as a lion-headed serpent, a harshly vindictive creator of the material world.

The seven Gnostic theriomorphic archons are bizarre forms of angels, descended from the Egyptian astrological punishers and put to use by the Gnostics to revile the earth and the god of the Hebrews. When the Gnostics translated the Old Testament's seven wrathful angels of retribution into the seven planetary archons, guardians of the seven heavenly spheres through which the soul must pass at death, they redefined angels as demons. The angels, originally Mesopotamian, had counterparts in the Egyptian sacred seven: serpent, raptor, ram, sheep, wolf, lion, and panther. All these became Yahweh's vengeful ministers.

The sun's passage evokes the many-headed Yahweh/Yaldabaoth, sometimes as the snake-lion Chnoumis, but also as a series of animal heads, showing what power is exercised at a given time. From animal-headed to animal-masked to the animal demon, the chief of the seven archons was envisaged as a snakelike, humanoid body with heads of lion, bull, snake and eagle, bear, dog, and so on. According to the religious historian Howard M. Jackson, "only Egypt furnishes . . . a setting in which Yahweh . . . in lion-headed human form may be clearly explained."[13] Deriving their terms from the writings of Paul, who pictured Yahweh simply as an inferior angel, the Gnostics changed him to an alien spirit, hostile to the human soul, a saturnine, repressive, planetary power, a baleful, retributive killer, the star of Nemesis. The profound astrological mysticism of the Chaldeans and Persians, adopted and transformed by the Gnostics, made Yahweh/Yaldabaoth into the chief of the constellations of the zodiac, with its "bestial inhabitants of the Egyptian sky."

In the gospel of Thomas, promulgated by the Gnostics, the tamed lion was salvation being "devoured" or controlled by humans. To be eaten by the lion was to be dominated by animal instincts. Hence the paradoxical Gnostic saying, "Blessed is the lion devoured by man, foul the man the lion eats." The appetites threatening to enslave man, led by a lion-headed chief, amounted to a battle for possession of the soul, pitting restraint against passion and blessedness against defilement.

Gnosticism's worldview, perhaps the most virulently dualistic metaphysics in Western history, infected Christian asceticism with "the prevalent Graeco-Oriental Dualism, which dominated all thought. . . . If a man is to escape the natural world he must renounce his physical nature."[14] The lion's role in this was its association with royalty and divinity—as sacred power shifted from earth to sky. Its round face and great mane, the "heat" of its fiery nature, its nocturnal, predatory passages, its creative and destructive energy, its habitation at the desert edge, that most otherworldly of places—all contributed to its role in the iconography of ascetic, otherworldly religious ideas.

The transition in which the natural world was to be condemned as evil was marked by the collapse and reversal of older traditions in which the sustaining of earth in its cycles and its timeless unity were epitomized by the snake. To those for whom the hybrid figure of snake-bodied lion was an amalgam of reconciliation of sky and earth, the hybrid image was an integration and resolution. But to others, for whom not only the earth but even the heavens were contaminated, the lion-snake-man combined celestial and terrestrial evil spirits. The Gnostic's empyrean was a set of nested spheres, a great zodiac of bestial powers who blocked the soul's ascent to heaven, just as the soul itself was a battleground as the Greeks had said. The debased earth was linked with the animal side of humans, presided over by a lion-headed, snake-bodied master of demons. The immensely complex history of the composite forms of lion, snake, and human shows how arcane and esoteric the meaning of such monsters can be. It also reveals a background to the Christian intolerance of Jews. As modern, educated people who repudiate religious bigotry, we can see that such hostility, which appears to be arbitrary and individual, can be reinforced by a long, tenacious infection of whole segments of society, culturally born with old fears and images that tend to focus frustrations and anxieties toward Others, human or animal.

HYBRID FORMS CONTINUED to embody ideas about the world throughout the Middle Ages. In Europe the Wild Man was a hairy creature, half-animal/half-man, an apostate from Christianity living outside civilized norms. He could be seen in stained glass and other ecclesiastical art as a costumed participant in festivals throughout medieval Europe. In these enactments he threatened women particularly, was captured, and then was paraded. His animal body and immortal spirit were balanced between heaven and hell. The wild woman was said to exist, too, but the hostility toward women was too great for the docile apelike figure and was more often projected as the witch.

While this recluse in the woods no longer reminds us of our soul's vulnerability, the Wild Man survives in anthropological terms. Monkeys and apes are a constant reminder of our primate heritage, and the myth of a "missing link" excites the imagination. A modern form of the Wild Man is the yeti or Bigfoot, a creature reported from the Himalayas and the American Northwest forests. The ape is the animal shadow of the yeti's image, as though we were haunted not so much by our fallen state as by our primate ancestry.

Public sympathy for the yeti shows that attitudes toward apes and the wilderness have changed. Although researchers talk about collecting a specimen of Bigfoot, the popular mind imagines a creature "like us" yet rare, perhaps an endangered species, one that we might even like to visit or interview. The wilderness no longer threatens as it did in medieval Europe; indeed, it has itself become endangered and in short supply. We camp and hike "in nature," as though emulating the fallen, asocial, sinful Wild Man of the past, an aspect of ourselves we cannot give up. Instead of fearful, unnatural beings, the friendly but grotesque forms are now sad and pathetic as we become wistful and nostalgic.

Two thousand years of Western demythologizing brought an end to most traditional monsters, making them merely pathetic. Those old hybrid figures seem bound to us by feelings that are both strange and playful. There was little fun in trinities, in the devil as a goat-legged, bat-winged anthropoid, or in apocalyptic dragons and the rest of monotheism's devils. A different feeling appears in Shakespeare's circle of half-human/half-animal figures, for which we feel a bemused sympathy which was new in the West. In *A Midsummer Night's Dream*, Bottom is enchanted as an ass, typically enacted as an animal-headed human. Indeed, Jeanne Addison Roberts observes that "the merging and blurring of figures is characteristic of the whole play [with its] yoking of surprising opposites. . . . Miranda is both a worm and a goddess. . . . Iris is linked with sheep; Juno with peacocks, Cupid with wasps and sparrows. Ariel . . . flies, swims, dives, rides, and is linked with winds and flying things: owls, bees, bats, chicks. Puck . . . appears rather inexplicably as a water-nymph, and as a harpy—both hybrid forms. . . . Caliban appears from the start as a man-animal hybrid . . . 'fish-like' and 'puppy-headed.' . . . The Rigid Order of *A Midsummer Night's Dream* has been turned on its side, and living beings must seek new relationships."[15]

An Eastern equivalent of Shakespeare's enchanted beings is the Hindu Ganesha, a lively, eccentric personality, an elephant-headed, roly-poly "human" who wears a snake for a belt and has a rat companion. The myth of the elephant god's incarnation as Ganesha, the son of Parvati and Siva, is widely known in India. A divine boy, born from his mother's sweat and unknown to an absent father, stands guard over her bedroom and is decapitated by the returning, impatient Siva. Ganesha is restored by application of the nearest head to be found, which had belonged to Gajendra, the lord of elephants. Ganesha personalizes the traditional gatekeeping role of the sacred elephant. Associated with boundaries, passages, doorways, he is the master of undertakings, beginnings, and openings and the remover of obstacles along the way. His image is saluted by truck drivers about to start through a mountain pass and is venerated by those beginning any venture—a trip, a new business, a marriage, even opening the shop for the day.[16] Demonic obstructors are therefore Ganesha's enemies. His celebratory sway over commencements makes him a dancer. Failure to honor Ganesha can result in calamity.

There is ambiguity and ambidexterity about Ganesha, who is both obstacle and remover of obstacles. Generally his demeanor is benign, and he is a beloved, pot-bellied, one-tusked, devoted son, very much of the everyday world. In the festivals celebrating him, his clay figure is afterward dissolved, a mime of ends and beginnings. Even in its briefest outlines, the Ganesha story shows how complex such a hybrid figure can be, how many historical and mythic threads wind into it. There is about Ganesha shrines everywhere a genial and amused familiarity not unlike the tolerant affection expressed toward street cows and temple monkeys, which has no equivalent in the Western world.

Ganesha is a descendent of the sacred figure of the elephant during the thousands of years prior to Hinduism. The yoked elephant is wonderfully gentle, but the bull is dangerous when in musth, at mating time. It is difficult now to appreciate how important the elephant has been to the human imagination. It was a major figure in Paleolithic art and thought; but as the last ice age faded and mammoths and mastodons vanished from the north, it receded in the consciousness of northern peoples. In Asia the elephant was tamed and became familiar in a way that its ancient ancestors and its African cousins did not. In India the rain clouds that precede the annual monsoon are "elephants." The elephant's trunk is associated with sacred snakes. Elephants were the mounts on which the gods defended the world in all four directions. Siva was said to have once produced an army of elephant-headed men, Vinayakas, to fight demons. Live elephants and their sculpted images are abundant in processions and temples today, performing ablutions and giving blessings.

Scientific study of the elephant's sacredness imposes an excessive seriousness. The first human laughter may have been at its incongruity: Such ears! Such a nose! Such teeth! Like the fat man who is always lighter on his feet than we expect, the elephant is full of contrasts—tenderness and fury, the mountain who is said to fear the mouse, the greatest of beasts who can move through the foliage like a whisper, the guardian son of gods with the grotesque and benign visage, behind which is power over the fate of all beginnings. But how is one to bring elephantine reality into the human context with its need of explanations and narratives of origin? There is about it the necessity of a grafted significance, given concrete form in the elephant-headed, human-bodied son of the god and goddess. The tale of Ganesha is one of decapitation and restoration into what became, perhaps, better than the original. In a world of disintegrating connections, the image of Ganesha is a perpetual joy in India.

ALTHOUGH IT LOOKED back to the old Christian idea of a degraded human as part animal, a new meaning of the amalgamated form became prevalent in eighteenth-century newspapers and books—as a lampoon of prominent individuals. In secular forms such hybrids continue to occupy the human imagination and express disjunct ideas. For example, the art of caricature stitched an identifiable human face to an animal body. It depended on a conventionalized public recognition of the appearance of politicians and others and a stereotype of each species. Caricature attacks and ridicules the target by identifying it in one dimension and exaggerating some supposed quality or characteristic: the arrogance of the lion, the balkiness of the mule, the deceits of the fox. Caricaturing is a strange descendent of our ancient human proclivity for identifying by dissecting and reconstructing. Indeed, the process has become an instrument for the butchering of reputation.

The practice of individual assassination by caricature was extended in the nineteenth century to social groups by such artists as Grandville, whose parodies of class, bureaucracy, professions, races, political groups, and social systems was accomplished by attaching human bodies to animal heads. Grandville's little burlesques are widely reproduced as drolleries in modern literary journals, much the way marginal illustrations enlivened the pages of medieval manuscripts or grotesque sculptures appeared in eighteenth-century Italian gardens.

THE PERCEPTUAL BASIS for the conglomerate animal is the primal experience of butchering: taking animals apart along the natural joints, thinking of them as made of parts, and recognizing the delectable differences in organs, muscles, and bones. Early in human existence we became psychic disassemblers, able to perceive and to conceptualize things as assemblies. As butchering became a gesture, the parts became words and dismemberment became analysis. The integrity of parts allowed each to bear not only its own taste and function but its own associations, to denote mobility, habitat, virtue, or other qualities. Wings, for example, abstracted flight, ascent, speed, aerial existence, seasonality, ethereality, or overview. As signs of powers, the anatomical parts became a lexicon of religious discourse, the connectors to the cosmos of abstractions, of which it remained only to reassemble parts from various animals in order to meld a seeming disjunction of concepts and construct and communicate philosophical ideas. The dovetailing of pieces of animal anatomy to make fabulous beasts is part of our intellectual heritage: significant images for making seamless what was a synod of ideas to start with.

The animal world is like a workshop, with all those bits available for dovetailing ideas into an organic whole in the form of novel animals. The process occurs in all cultures. It drives the creation of precolonial South American monsters, for example, before which the modern museum-goer is stunned. The Llama People, who vanished from the west coast of Peru two thousand years ago, compounded beasts from seven beings: snake, bat, cat, toad, tortoise, bird, and human. Their function, says one scholar, was to emphasize the multiple nature of Ancestral Beings, for which there was no single image and no one myth.[17] Of the process in general William James says: "Elements are withdrawn from their usual settings and combined with one another in a totally unique configuration, the monster or dragon. Monsters startle neophytes into thinking about objects, persons, relationships, and features of their environment they have hitherto taken for granted."[18] Such dis- and reintegration is fractal; the face is a part of the head, but the elements of the face can be isolated in turn and recombined to make other faces, the guises of ancestors, ghosts, or demons, the masks of which "teach neophytes to distinguish clearly between the different factors of reality, as it is conceived in their culture."

Before books such monsters were the ideograms of complex ideas. Such beings gave concrete expression to linked ideas, especially in religious instruction. As the individual within a cultus seeks spiritual orientation, composites are especially appropriate because the novices, as young adults, often perceive themselves (like the parts of animals) as socially dissociated. The task of their latent maturity to be engaged in "the resolution of things into their constituents, isolated, made into objects of reflection by componential exaggeration," followed by recombination as uncanny beings to give them concrete imagery.[19] The lineages of those ideas were like roots, each with its own animal sign. As a storied eschatology, they were cast in the web of cosmology, an evolving tale of movement, marriages, disarticulations, and recombinations.

If we were still hunters and gatherers, great horned bipeds might embody our transcendent ideals. In a world of pastoral ideology, our political power, esthetic ideals, and libidinal impulses would center on centaurs, flying horses, and unicorns. If we still lived in an urban-agrarian city-state, religious imagery might celebrate the minotaur. These successive figures are all part of the Western heritage.

There is no end to the story of the composite animal. Linked to commerce in the twentieth century, it has become abundant in advertising, especially as animal-headed humans. If Marshall McLuhan was right—that an electronic tribalism is replacing national chauvinism—the appearance of composite figures in commodity huckstering may be a way of giving perennial concepts their due in modern dress.

GRAFTING BODIES TO configure or juxtapose metaphysical ideas has been explored here as an evolution of sacred animals into combinations, eventually to the human/animal dyad, and finally to the end of the animal part to leave humanlike deities.

Another track of this history, parallel to the one described, is not the fusion of animal forms with the human figure but a special association. It begins with animals as holy or numinous and culminates in a deity or pantheon of gods who appear in human form. The intermediate stages are represented successively by the combined human/animal figure, the human deity riding on the animal or drawn in a carriage by it, and finally by the human deity accompanied by the animal which is sacred to the god or goddess. The middle stages of this sequence—in which the deity stands on the animal, or rides on it, or is pulled by it in a vehicle—are widely found all over the earth.

The thirty-two forms of Sakti, a feminine form of the great god Siva, are also mounted. Such an animal vehicle is the sacred being's *vahana*. Typical of those ridden, the peafowl is a prominent example. In India's traditional south it is the steed of Lord Maruga, the supreme god of the Dravidians. The peacock is the syllable *"Om,"* the sound of the universe, and is often depicted with a deity on its back. Various spirits, presiding over learning, especially Saraswati, the goddess of learning, ride peacocks. The peacock is the companion of Kartikeya, as well, the son of Siva and Parvati. There too the peacock as Mayura is the killer of serpents and keeps in check *ahankara*, the coiled serpent of the ego. And the peafowl is, of course, an eater of snakes and thus an appropriate symbol of a shift from such primal forces.[20]

In the wonderful biosociety of Hindu India, all these stages exist simultaneously: some animals are merely themselves (the gecko lizard); others are deified (Nandi the bull); many deities are mixes of human and animal anatomy (such as Ganesha); another group comprises the mounts or draft animals of the vehicles of the gods and goddesses (the horse); some animals (the dog) accompany specific humanized deities and are sacred to them; and there are finally those sacred beings who appear solely as human. In India, as in classical Greece and to some extent most cultures of the world, it may be possible to discern the outlines of a history and a psychology of this immensely diverse subject. The problem and its summation, according to the author of a book on the peafowl, P. Thankappan Nair, is that "the coming of Buddhism in Asia and Christianity in Europe confirms the historic fact that all civilizations, sooner or later, replaced symbolic animal forms with human figures."[21]

Eric Neumann says: "We know how often the goddess appears as an animal, as cow and swamp bird, as ewe and lioness. She also 'is' a fish. She bears the fish's tail of the mermaids; in her character of Artemis, Aphaia, Dictynna, Britomartis, and Atargatis, she is associated with sacred fishpools, and the lower part of the Boeotian Great Goddess consists of a fish and waves. Later, however, she ceases to 'be' the goose itself, but rides on it or wears its symbol on her cloak; and she ceases to 'be' the lioness, but stands on it. She no longer 'is' the serpent but is accompanied by it. At this higher stage, she becomes a goddess in human form, ruling over the animal kingdom; and in alchemy she was still represented as Earth Mother with upraised arms, feeding the beasts."[22]

If the earliest image of deity was a "Great Goddess," why was she envisioned as a "fish" or a "swamp bird"? Neumann may be wrong. More likely, her powers were originally conceived as aspects of (and under the aegis of) a variety of animals whose specific characteristics suggested their respective talents or domains. In ordinary experience it made no sense to collect those diverse qualities into a single figure, although some special animals, like the bear, could themselves represent the collectivity, as though modeling the human necessity of conceiving plurality as a unity. In time, the animal appears in company with a human deity. But to see the animals as subordinate to a ruling figure in human form required a radical shift in metaphysics that came with coalitions of chiefdoms and the early agricultural empires with their sacred queens and kings. The drastic alteration in the embodiment of cosmic energies and process by clumping them in a superhuman effigy had to do with a new sense of human control—at first over certain plants and animals, perhaps in driving and corralling half-wild goats and sheep. At the same time, class structures representing the domination of some humans over others made use of the images of drovers as social metaphors.

Direct control of animals almost certainly lent itself to modeling authority as parents, who were embedded in the infantile, unconscious mind as omniscient, archetypal figures. The dream of being carried by animals may be part of this infantilized psyche, just as we are each carried during our early life. Wanting this to continue is a regressive fantasy. The creation of "bearing" mythologies, in which people are carried about and protected by metaphorical sacred powers, suggests the failure of separation within the family as a social dysfunction of society.

During the emergence of human cultures, in the two million years of the Pleistocene, there was a dynamic tension between our slowed development or infantilization and the need for culmination in a mature sense of objectivity and independence; between neoteny, which made culture and learning possible, and psychological growing up; between the slowing of individual development to extend the acculturation process and the mitigation of infantile behavior. Maturity required that the individual separate from parents and others in whose care the necessary immaturity was protected. Neutralizing the regressive effects of extended immaturity was the task of an ecological elderhood and a calendar of development that makes up a substantial part of the customs of primal societies. If those practices of ritual or conceptual integration were crippled or abandoned, neotenized cosmologies (as the extension of infantile psychology) would have produced the ideological myths of the sacred family, of being carried and tended, and all the other egocentric behavior and id activities associated with immaturity. Animal carriers would replace a sacred, zoological comity in a wild, nonhumanized terrain.

The depiction of divinity in animal form represents not only the primordial but an ecologically efficacious religious attitude. Cosmologies with hybrid beasts, whose bodies are part human, indicate an early step in the progression from primitive to civilized state in which the anthropomorphic form is emerging, like a butterfly from its cocoon. The human hero standing over the slain monster, wearing its skin or other parts, signifies an emergence from this combinational state in which the animal part of the body is detachable as clothing, the power having been forcibly taken rather than shared or inherited. When symbolic human figures ride or stand on animals, a further distancing and a more explicit domination has occurred. Sigmund Freud once observed that the reason why animals in dreams are the "bearers of the soul" is their motility.[23] At this stage the humanized deity can still "become" the animal from which it descended, for the animal is one of the deity's avatars. Toward the end of the series, the humanized deity is accompanied by its particular animal, which, though still sacred, seems only a shadow of its original and becomes just an emblem of the humanized deity. Finally there is the "civilized" state in which only the human figure remains, in whose stories there may or may not have been an animal form or companion. But the arrogance of humanism assumes that the deity is really "like" a man or woman. The animals do not disappear in such societies; they become slaves and pets and their iconography becomes a subject matter in art.

All primal societies assert that they are "natural," confirming values and practices by reference to organic, geomorphic, or celestial models. As this assumption was modified by peoples with domestic animals (that is, whose relationship to animals became social as well as ecological), the natural forms were replaced by animal figures with human heads. In highly centralized states and urban societies, the animal images became idiographic and alphabetical; the reference had shifted to words and speech itself, spoken from a human mouth, or that of a distant, unseen god.

The record is one not only of replacement but of the transfer of powers wrested from nature. The myths of the anthropomorphized gods, the mainstay of modern "world religions," assert that a supreme deity authorized human viceroys (Moses, Jesus, Buddha, Mohammed) to reveal the truth. They succeeded in the course of history by replacing a series of numinous animals with combinational figures wearing horns and skins, then gods and goddesses whose animal guides and patrons run alongside, then deities in vehicles being pulled by animals, riding or standing on them, as the saints stand on beasts in a thousand cathedral naves. Finally, the power is completely transferred and the historical sequence exhausted: the deity without the animal.

By the time the Greeks were making written records, the birds, who had in previous centuries portended something, were reduced to messengers: the eagle of Zeus, the raven of Apollo, the woodpecker of Mars, the owl of Athena. Roman parallels followed. The Christianized Romans kept the animal messengers already in the Jewish Bible, demonized the composite monsters made from animal bits and the witches' animal familiars, and put Christ on a white horse. The city-state and centralized, "pastoral" empires, committed to aggrandizement and defense, invented a symbolic panoply of men mounted on horses, llamas, and elephants and their ritualized riding (continuing today in the rodeo). To ride with a saddle and bridle, or to drive with harness or halter, enhances and makes explicit the human domination—first to head the animal in a chosen direction and finally, in the end, to make it pull the whole paraphernalia of our human circumstance. Mounted or standing upon an animal, some kind of human or superhuman either directs or absorbs its special power.

Some such progression leads to ourselves, urban people, more distanced from animals in their daily lives and more anxious about their identity and themselves as humans. However humble, our stance before a divine humanoid presence signifies our loss of ecological humility and the narrowing of our spiritual perspective. The series that began with the dispersed energies of the cosmos, represented concretely in the different species of animals, ended with deities in human form who co-opted the animal powers by stages, through intergraded form and riding, signifying the human control of nature.

15 | Augury and Holograms

Take heed what time thou hearest the voice of the crane
  Who, year by year, from out the clouds on high
  Clangs shrilly, for her voice bringeth the sign
  For plowing and the time of winter's rain,
  And bites the heart of him that hath no ox.

HESIOD

WHEN STUDENTS IN my class on animals and culture consider oracles, augury, omens, or divination, the effect at first is silence. Even in a world of daily weather forecasts, predictions of freeway traffic, imminent volcanic eruptions, and the newspaper horoscope, nothing seems further from modern taste than "superstitions," wherein the call of a frog foretells the future, or seems more repulsive than the close examination of warm liver or small intestines, taken still steaming from the body of a sheep. With the exception of popular astrology, no rite today seems more suspect or remote than trying to assess the future by scrutinizing the flights of birds, as did the ancient Greeks and as do modern Navajo Indians, the colors of cows, as did the Egyptians and as do the modern Nuer, or the livers of sheep, as did the Mesopotamians and as do modern pathologists.

Both the old and the modern practices necessitate watching animals closely, studying the details of organic form and structure as they relate to some larger whole, for there are subtle connections between the behavior of animals and earth events, between the color of an organ or the texture of a mammal's pelage and the universe. The natural world is rich in signs—as when dogs and other animals foretell earthquakes by sensing the first pulses of earth movement, or a zoologist, reading the chemical signatures in the viscera of a hare or a sea lion, can anticipate the imminent death of large numbers of mammals and a windfall for the scavengers of their bodies, followed by hard times for the predators, including humans.

There are many synonyms for such foretelling. The abundance of terms is a measure of the importance of such ideas in history. "Oracle" refers to *dike*, the Greek notion of judged fate, a deserved destiny. Oracles traditionally inhabit sacred places, especially caves, and have a perennial, almost timeless, existence. "Prophets"—individually unique, often chosen, and itinerant—are part of biblical tradition.[1] "Augury" depends on natural signs, or omens, which need interpretation. It is a form of divination that signifies rather than pronounces. It portends through a net of relationships. It lays a matter open by disclosure rather than depending on inspiration, as among oracles and prophets. Augury is less pretentious than prophecy and often depends on reference to animals.

Among animals the bird is seen as part of a greater system through which it moves, rather like time itself, a fleeting signature taken in at a glance. The cow, to those who see it as a microcosm, does not flash across the earth but remains to be studied out, not as a fugitive bit of the whole, but a contraction of it. While the whole animal represents the cosmos, the sheep's stomach, liver, or entrails are the tacit parts, as pregnant with meaning as the womb is with new life.

As the old, preclassical, Mediterranean world declined, the holy connections of animals went with it. Priests began to look beyond nature for truth, and the spiritual apostles and deities of Western religions became human in form and otherwordly in orientation. Our modern perception of animals as automatons denies the significance of augury as a marvel fraught with sacred presence.

THE BIRDS WERE among the *numina* longest to survive. Many qualities touch them with such powerful magic—flight, migration, song, eggs and nests, seasonality—that even the ideologies of transcendence retained them as angels. Bird life is a highly visible, poetic analogy to human life. Individuals in many species share food as a social bond in courtship and care for nestlings. Many dance and sing in formal, almost ritualistic, fashion. Their plumage is like costume: put off and changed in different seasons, repaired and replaced when tattered, the feathers together and individually beautiful to behold. Living or moving together, cooperating, roosting in great murmuring flocks, they are among the most social of animals. In flight they wheel in synchrony, as though keyed to some central mind. Yet they are kin to the earthbound, cold-blooded reptiles, egg-layers with scaly legs and toothless mouths. Some hibernate. Finally, they are masters of travel, mysterious appearance and disappearance, astonishing migrations.

"They come and go where man and beast cannot go," writes Jane Ellen Harrison, "up to the sun, high among the rain clouds; their flight is swift, their cries are strange and ominous, yet they are near to man; they perch on trees, yet they feed on earth-worms; they are creatures half of Gaia, half of Ouranos."[2] The Etruscan word *templum* is a space marked out, as in "template," "contemplate," "temporal," that section of the sky across which the augur was expected to fly. The first temples were probably oriented shrines without roofs. Unlike the medieval cathedral or "city of god," they were open to the natural world rather than an escape from it.

Birds are auspices, a term derived from *avis*, which appear and disappear and are thus like the past and future. They are not only omens but "knowing" in the year's unfolding in the sense of expertise, bringers rather than passive signs. In Old World traditions, the crane's arrival told when to plow and the swallow's when to prune. Calls of the cuckoo forecast rain because the cuckoo was part of the rainstorm. Notions that birds signified oncoming weather was wedded to the older idea that the bird made the weather, brought the rain, the thunder, the sunshine, and the spring. "Birds are not, never were gods . . . but there are an infinite number of bird-sanctities."[3]

Ordinary folk, who did not expect signs from the great gods, watched for such omens as the magpies in a flock:

> One for sorrow,
> Two for mirth,
> Three for a wedding,
> And four for a birth.

Bird lore was so rich that mastery was regarded as the "mantic art," from *mantikes*, "prophetic." When there were neither clocks nor calendars, the birds signaled the season's activities with precision. The foretelling by animals was a matter of practical things. As the early Greek poet Hesiod put it:

> Lucky and blessed is he, who, knowing all these things,
> Toils in the fields, blameless before the Immortals,
> Knowing in birds and not overstepping tabus.[4]

Among farmers who had no printed calendars or agricultural colleges, the key to the season's round lay in traditional signs. Ignoring the signs, even by a few days, could put a crop at risk. "The small Boeotian farmer is not a sceptic," observes Harrison, "but a man hard pressed by practical necessities. What really concerns him is the weather and the crops and the season; how he must till the earth and when. . . . It was all you could do to keep body and soul together by ceaseless industry and thrift, by endless watching out, by tireless observance of the signs of earth and heaven. . . . But first and foremost you should watch the birds who are so near the heavenly signs, and who must know more than man. This watching of the birds we are accustomed to call the 'science of augury.' . . . In its origin it is pure magic, pure doing; the magical birds make the weather before they portend it."

Our loss of sensitivity to birds is most evident in the spiritual destitution of the woodpecker, *Picus*. Zeus in a Greek myth stole the scepter of the green woodpecker, Keeleos, the old rain and medicine king of Eleusis. *Picus* was a numen who "enshrines a beautiful lost faith, the faith that birds and beasts had *mana* other and sometimes stronger than the *mana* of man . . . in their silent, aloof goings, in the perfection of their limited songs, are mysterious still and wonderful." One could divine by them, but more important was their presence in "making" or "bringing."

In many traditional societies, the woodpecker's drumming signaled the onset of the rainy season, making *Picus* the "rain bird" of Old Europe. The human significance was that ceremonies to start the rainy season required drumming, and so the drummer was a rainmaker. The bird did not accompany the human drummer but was said in the beginning to have "taught" the skill of bringing on the spring rains, drumming as the woodpecker does in making nest holes, eating, and signaling. Like the woodpecker's nest cavity, the first drums were made from sections of hollow logs. Associated with the tree—in myth, the great tree of life—which it entered and left (bringing its young forth), the woodpecker had a special place in the human sense of creation, moving between the ordinary world and the womblike spaces within the living trunk. It is not surprising that, as a prototype of all founders, the woodpecker had a close association with holy trees—which, according to folklore, were saved by it—and with the wolf, as the two of them saved and nourished the abandoned twins, Romulus and Remus, the founders of Rome. The white eggs inside the tree seemed to be the model of potency and becoming.

As the forests went down, the woodpecker receded from everyday view. Another rainmaker, the cuckoo, likewise diminished, although both persist at the most conservative fringes of rural society. In rural America the cuckoo is the "rain crow," for no apparent reason, although it takes no special knowledge to observe that it characteristically calls as clouds cross before the sun.

As KEYS TO foreknowledge, the birds were available to everyone. The butchery of animals as a regular household activity meant that most people in the past were also familiar with entrails. But their reading required special training. Extispicy is prediction from the examination of organs. The word is related to the term for "priest" in Latin, "diviner" in Etruscan, and "liver" in Chaldean. A prayer was often recited, the animal sacrificed, and dissection begun. It was rather like an expanded medical diagnosis based on rules of interpretation. The lungs, gallbladder, heart, stomach, large and small intestines, some bones, and especially the liver were carefully examined for deviations in color, atrophy, hypertrophy, displacement, unusual shape, or marks according to clay models of the organs in health and disease.

The reading of entrails was a science and an art. In the high civilizations of Mesopotamia, special training and sophisticated analysis involved a technical terminology recorded in the cuneiform manuals of ancient Persia, directions to accompany the potters' replicas of sheep guts. Typical was a set of twenty-four Babylonian clay tablets on birth deformities in lambs, ewes, mares, sows, and bitches. A library of these commentaries or interpretations, the *Summa Alu*, was written in arcane euphemisms for anatomy, such as "door to the palace" for the esophagus. Among the Akkadians and Mesopotamians these models and manuals were frequently mentioned in the stories of the kings, for extispicy was regarded as an essential rite prior to all important undertakings.

At its most elemental level such reading of the organs answered a single question with a yes or no, but in its more advanced form the diviner could read the organ like a landscape, translating its details into coming events. As a result, decisions were made and actions considered according to the oracular organ. Implicit in the practice was the assumption that cosmic powers dictated future events and that animals sacred to those powers mediated knowledge of them. In our era people may see in this only delusion, but it should be noted that a pathologist looking at a dead liver can give a history, and a physician testing a sick liver can predict the longevity of its owner.

Very old stories of the organs of the viscera described them as if they were species—the liver, heart, and lungs as kinds of beings—and imagined them as having lived outside of bodies at the beginning of the world. Having moved inside the animals, they became beings within other beings, the organ acting as historian and oracle about the life of the whole animal. In somewhat the same perspective, animals themselves are inside nature, interlocutors about that larger realm and closer to its sources of cosmic power. Perhaps the worms in the livers of sheep in the ancient world spoke with authority of bad times somewhat like the guts of reindeers in modern Lapland tell eloquently of radioactive fallout from the disaster of the Chernobyl nuclear power plant in Russia, both forewarning of human misery.

The reading of organs is done within a field of relationships of other organs, the whole organism and its circumstances, through a series of levels extending away from the animal. A modern parallel to the construction of a universe by the scrutiny of a small part of an animal can be seen in the paleontologist's reconstruction of whole animals from a few bones or teeth. This is done by knowing the relationship of muscle size and mass to bone topography, the structure of jaws necessary to support different kinds of teeth, an understanding of the rules linking tooth form to food habits, tooth and body proportions, awareness of similar teeth and bones of animals living and fossil, the musculature as indicated by the details and shape of bones, general principles involving the relationship of overall size to parts, and knowledge of the climate and habitat of other animals or plants with corresponding anatomy. Taken one step further, the environment of an organism can be reconstructed, or a whole habitat mapped out, with some fossil pollen and the remains of a few plants and animals: ecosystems provisionally recreated, in the mind's eye, from some small bits. Likewise, in the past, unusual behavior was considered significant and thus was provoked, as in sprinkling water on a bull to irritate it to action. There was also the study of *exta*, that is, scats or droppings.

Among the ancient Akkadians fine distinctions were made between situations and consequences, between details of the organs, prediction, and degrees of avoidability. Some events, even though portended by an omen, were not in fact inevitable and could yet be deflected. In such circumstances a ritual was performed "to turn aside."

THOSE TODAY WHO are inclined to peer into the future prefer computerized probability systems, recreational palmistry, or the astrological columns in the daily newspapers to the reading of entrails. Yet the observational basis of extispicy and prognosis continues in our time in the guise of autopsy and forensic medicine, as retrospective foretelling. We "look into" the living entity by using sound and radiation, as we do surveillance at a distance. The monitoring of wild animal movements is widely predicted by the use of radio/satellite combinations.

Both the archetypal psychologist, Carl Jung, and anthropologist, Claude Lévi-Strauss, speak of "synchronicity" as linked events that are perceived in terms of prediction. The recognition of linked events that makes prediction possible depends on the fervor of observation. In his essay on modern extispicy, "The Hare and the Haruspex," biologist Edward Deevey reviews the mystery of the rhythmic rise and collapse of lemming and rabbit populations in the Arctic: "Driven by the same Scandinavian impulses that drove the Goths—depression, irritability, berserking"—the tundra rodents erupted every fourth year in Norway, at which time they were eaten by reindeer, fish, gulls, owls, and all other big predators. Flooding into lowland cities like a plague, spreading tularemia, the neurotic and aggressive lemmings then fell over dead. With less aggression, snowshoe hare populations followed a similar pattern of numerical success and demise. For centuries the cause was sought in the climate, predators, sunspots, disease. It was not until the ancient practice of liver examination was revived in a more clinical setting that the real story emerged, as if the victim of a murder had left an incriminating note. Operating through the brain's adrenopituitary system, the stresses of crowding, poor food, increased fighting, and sex overtax the capacity of an interlocking system of brain, endocrine glands, and liver. The animals' livers, in which blood sugar is stored as glycogen, atrophy. Hypoglycemia or blood sugar deprivation follows, leading to hemorrhage of the endocrine glands such as the pancreas, convulsive seizures, and death.[5]

The ancient Babylonians knew that the liver contains about one-sixth of the body's blood and understood that it is the center of processes connecting the organism to its environmental circumstances. They would have appreciated this analysis of the crazed lemmings, as well as other clinical diagnoses, showing that the liver, a blood purifier by concentrating chemical poisons like insecticides, can itself become a poisoned meal instead of the traditional pièce de résistance. Given Dr. Deevey's data, a good diviner might have predicted the disruption of human society and the eruption of Gothic warriors that changed the history of Europe in the eighth and ninth centuries.

A quarter century after Jane Ellen Harrison published her lament on the lost reverence for the green woodpecker, *Picus*, American robins began dying on city lawns from pesticide food-chain accumulation, and mercury dumped into the sea in Japan was ingested at fish dinners in Seattle. On the coasts of Oregon sea lions began killing their pups, driven mad by the accumulated insecticides in their body fat. Radioactive selenium from the testing of atomic bombs entered the soil, grass, cows, milk, and the bodies of children, where it replaced calcium in the structure of new bone, its radiation causing leukemia and other cancers. These relationships were understood by dissecting experimental animals. Such portentous doings seem close to the world of the ancient Greek farmer, a dangerous and fabulous "conjunction" in which things participate in unexpected ways.

"ELECTRONIC MAN," OBSERVES Marshall McLuhan, is like the sleuth or the Paleolithic shaman—making the leap from here to there in imagination with the power to show the wholeness of that which appears disparate to others. We live in a fragmented world, but we assume, or hope, that it is coherent. The magician and sleuth remind us what the paleontologist, physiologist, and computer model-maker know: that we learn or surmise sequences from scattered clues. In magic the audience sees only beginning and end, while the skill of the magician hides the connection. The process is an extension of our old Paleolithic talent of tracking. Divination and scientific hypothesis are like a mystery in which a master detective experiences a flash of intuition, a leaping ahead from the evidence. He is then able to prevent further mayhem, as the hunter is able to ambush the moving prey and the oracle to warn. All the great fictional (and perhaps real) detectives have the ability to hold disarray in a kind of perceptual space, to single out and rearrange critical pieces like a paleontologist with a pile of fossil bones on the table before him—that is, to ask the right questions and finally put them together like a puzzle that sends its ramifications far, wide, and ahead.

The principle that relates all this inductive thought is its emphasis on events. The things are important—the wand, magnifying glass, deformed liver—but what they portend is a set of happenings. Claude Lévi-Strauss proposes that the distinction between nature and culture is fundamental to human consciousness. He says that nature and culture can be taken as metaphors of each other, in which the relationships are the substance of the analogy.[6] These relationships are defined by events, as in the "process philosophy" of Alfred North Whitehead. Events rather than things are the true concreteness of the world. To José Ortega y Gassett these events were "circumstances." Likewise, ecology is focused on dynamic symbioses among the organisms or between them and their environment. Animals stand forth as actors or the symbols of action, perceived as verbs and bearing consequences in their tissues.

If all of the genetic material, the DNA, which instructs the genesis and activity of organisms were read out according to the prior conditions to which the DNA is adaptive, a world—or part of a world—could be recreated from that information. Every animal contains a model of the world, from the perspective of its own niche, in its genes and chromosomes. From a bit of viscera a universe can be glimpsed. A liver, like the rabbit itself, is a kind of hologram, a photographic plate created by using laser beams to record diffraction patterns of light based on the interference from two different sources, one reflected from an object. When light in the same frequency strikes any part of the plate, the whole three-dimensional image is recreated. As in the genetics of an organism, there is redundancy. A small fraction reproduces the whole image. Human memory may work in somewhat the same way—as patterns produced by waves of impulses traveling along networks of brain cell connections.[7]

Every animal is a holographic memory of a constellation of circumstances: the remembrance of its own evolutionary past and the energy patterns that impinged on it over many generations. Every detail of an organism, its color and form, even its behavior, is an expression of such a mnemonic, a memorial of interaction with a world. Some species, as well as their parts, cows as well as their livers, encode wholes that can be translated. Even fossil remains enable a paleontologist to construct a complete animal from a few teeth and bones, and a painter to create a landscape, a diorama, from the animal.

The idea is like a mosaic of otherness hidden in the single being or part. Once we understand some of the consequences among details, we are able to reconstruct a story from linked bits. The detective reconstructs the past; the oracle reads the evidence forward in time.

"MONSTRUM" AND "OMEN" come from a common root word and have to do with presentiments of disaster. When I ask my students what group of people these days are most closely associated with the image of "doomsayers," a few will mention hell-and-brimstone preachers, especially as we approach the third millennium, when there will surely be those on the rooftops waiting to be snatched up from Armageddon exactly as they are, wearing the same sweaty underwear and with whatever small change happens to be in their pockets. But there are usually a few students who give the professor the answer he wants: the ecologists.

One or two members of the class might even know Rachel Carson's 1964 augury of the birds, *Silent Spring*, though none can be expected to have seen Fairfield Osborn's *Our Plundered Planet* or William Vogt's *Road to Survival* from the late 1940s. There are abundant later titles, although Paul and Ann Ehrlich's *The Population Explosion* is perhaps the best known. But why, I ask my students rhetorically, should this one minor branch of biology have produced two generations of catastrophe-minded writers widely scorned by the champions of progress as "pessimists"?

It was not, I think, because they came from gloomy households or had melancholy genes. It was because there is something intrinsic to the study of symbioses, populations, and the flow patterns of elements and energy that predisposes observers to prediction. As the biblical prophets also knew, being a Cassandra is no fun, and I know of no ecologist in the past half-century who has enjoyed being a forecaster of debacle. What has made ecology the modern form of extispicy? Ecological concepts are embedded in process and time. Fundamental among them is the notion of succession, whereby communities of organisms pass through predictable and recognizable stages from a loose assembly of a few "pioneer" species on bare rock, lava, sand, or mud flats to relatively dense, species-rich syndicates with deep soils or stable pond environments (or, conversely, are degraded and diminished). The criteria by which the state of the community is assessed—as collapsing, arrested, or moving forward—are certain "indicator" organisms, the naturalist's augurs.

Other naturalists, such as Lorus and Marjorie Milne, who wrote several dozen books about "nature," did not normally trade in cautionary tales. But their books are full of wonderful House-That-Jack-Built stories. Ecological Just-So tales usually turn out to have unexpected causal connections and endings opposite from what our unecological, shortsighted, cause-and-effect assumptions lead us to expect. We kill hawks in the expectation of benefiting quail, only to discover that the rodents, normally controlled by the hawks, eat more quail eggs. Crocodiles, thought to be the enemies of people who live along African rivers, turn out to be their friends because they eat baby hippos, thereby keeping the population of hippos down, with the result that the vegetation along the rivers is adequate for the hippos, who do not wander into the village precincts seeking food and killing people. Obviously a good ecologist, knowing something about crocodile populations, could predict whether more people or fewer would be attacked by hippos.

It is equally clear that no ecological augur will have a perfect record, since the same characteristic of nature that makes the initial assumption wrong—its interactive complexity—can overturn a prediction. Even so, such a "conjunction" of events was probably important for divination in the ancient world and makes sense to my students in the present.

16 | Bovine Epiphanies: Fecundity and Power

The bull is sacrificed to the Olympian Zeus who stand there. . . . On his head are horns; he is bull-faced. It is usual to say that the god assumes the shape of a bull, or is incarnate in the form of a bull. The reverse is true. The sanctified animal becomes a god, sheds his animal form, or keeps it as an attribute. Any animal may rise to be a god, but he must first become sacred.

JANE ELLEN HARRISON

PERHAPS NO ANIMALS are as closely identified with the spirit of agriculture as cattle. Yet in the first five thousand years of agriculture—the early Neolithic—it was not so. Most cultures living by combinations of gardening and foraging conceived life itself in the image of a great goddess, a lady of the beasts, sister, wife, daughter, crone, and mother. But about six thousand years ago, as agriculture settled on single crops and surplus production in the interest of the state, her image as the fertile mother eclipsed the others and the cow became her avatar. The idealized cow, and then, thirty-five centuries later, with patriarchy, the etherealized bull, drove the philosophies of early civilizations. This cattle fetish was rewarded by the release of the stored fertility of the world's tropics, especially the subequatorial dry belt-lands and savannas, but in time it produced great tragedies on a scale unknown to tribal peoples.

Enthusiasm for the bovine image was inherited by the first planters from an earlier time when the splendid beauty of the great aurochs, or wild cattle, was beheld in ecstasy by hunter/gatherers, whose metaphysics was conceived as a pantheon of animal spirits. Some of the oldest evidence of this concept is the twenty thousand years of paintings in Eurasian caves, in which the images of aurochs are prominent, along with drawings of horses, antelopes, elephants, lions, and bears. In a very early temple at Catal Huyuk in what is now eastern Turkey, carved, horned heads adorn plastered walls on each side of a large figure, referred to by archaeologists as a human goddess "with teddy bear ears." That it was sacred and feminine is likely; that it was "human" is the wishful thinking of scholars eager to see the Great Mother instead of a bear. The figure is a human/bear hybrid, sitting in the posture of a bear, representing the transition from sacred bear to cow goddess, the surrender of the adjudicating bear spirit to that of the tamed milk-giver. Bears would meld, not into cows, but into the keepers of the cows—the human figure. The horned heads accompanying the bear/woman figure at Catal Huyuk are not "bulls" as the archaeologists claim but cows, since both sexes of the wild and early domesticated cattle were horned. There is little evidence of a sacred bull at this time, while there is much of the cow. About the same time as the Catal Huyuk temple was in use, small, half-human/half-bear figurines were made in large numbers in the early villages in Eastern Europe in the form of a bear-headed woman, sometimes holding a bear-headed infant in her arms.[1] She is very close in spirit to the madonna on the wall at Catal Huyuk.

Cows and bulls were derived from the huge wild aurochs, *Bos primagenius*, six feet tall at the shoulder, by domestication begun more than eight thousand years ago, based on evidence from the Euphrates in Iraq, the Nile valley in Egypt, and the Indus in Pakistan. Drawings show early cattle in Sumerian barnyards. The barn itself was part of a temple complex, a wattled structure or sacred "byre," whose gate of reed bundles represented stylized female genitalia, signifying fertility and motherhood, for which the cow was an epiphany. The great religions of the archaic states began in such bucolic huts, and ceremonies of fertility and sacrifice were conjoined with those of sowing and reaping.

At the core of this shift to holy cow milk is the nature of the cow as mild and bland. Domestication changed the animals toward blunt, nonselective sociality, diminished intelligence, and increased lactation. The use of cow's milk may have begun with wet-nursing in early farming settlements: human fecundity had doubled; epidemic disease made more orphans; and nutritional and genetic inability to breast-feed increased among women. The first substitute for the overburdened or dry mother, if no other lactating woman was available, was almost certainly the goat or sheep rather than the cow. But the quantity of milk was greater in the cow, and it was quantity that impressed the agricultural mind, bent on productivity and the amassing of surpluses. Goat milk was better for people, but cow milk's high butterfat content made human fat, which became the badge of affluence.[2]

Milk was the alter ego of the goddess herself; her bull calf was the animal of sacrifice, removed in order that she might bear again and come fresh. Used as an adult for breeding and pulling, the bull, in its seasonal roles, was not so much a snorting god-king as a benign son who would service its mother's sexuality, die as a sacrifice, or till and drudge as an obedient ox. The death of the bull calf provided meat at no cost. With a little agrarian imagination, its sacrifice could be turned into a negotiation, a "feast" for the spirit of the life cycle, an arbitrary deity who controlled all productivity. Like the moon and the bull, plant (and human) life were subordinated to a feminine principle, which itself was given symbolic form as a solar eagle or lion-bird who nibbled away at the waning moon-bull. As a conventional rite of the season, the bull's sacrificial death could be seen as essential to the renewal of crops in the cycle of life.

Later inverted as the letter "A," the ideogram of the mother-goddess was a triangle with horns that held the recumbent new moon. From the front, the cow's skull with its horns also resembles the human female reproductive system: the cranium as the uterus, the horns as its twin oviducts. The passageways of new life, the horns were like a gate to renewal, hence a "way" to consecration.

Everything about cattle suited the preoccupations of the farmers and their city cohorts: the overt and perpetual reproductive cycle of the animals, the cows bawling for servicing or gestating or lactating, the tireless bulls, the rivers of milk and semen, the urine and dung like brown gold, as fertilizer, fuel, medicinal substance, the flesh enough to share with the gods, and the skins for making garments, sandals, and drums—in it all the blatant oversize performance of the agrarian dream of plenty, of greed and desire.

IN RETROSPECT, IT is hard to miss the shared symbolism of the human primate's genderized sociality, obsession with sex, and milky mother love, given frantic expression in the early agricultural emphasis on producing and reproducing. The feminine principle, earlier signified in dozens of ways in plants, animals, water and earth forms, begins to appear as a cow goddess all over the ancient world. At the mouth of the Indus, the goggle-eyed Zhob was incarnate as a cow. In Egypt the cow was Hathor, the celestial deity with the moon between her horns. In ancient Sumer, at Lagash, where Ishtar was the goddess of flocks, she was represented as the horned Ninhursag. Her temples were the home of sacred herds, and milking scenes appear in the ancient drawings. At Ur on the Euphrates she was associated with the mountain from which the rains in season came to the upper watersheds, swelling the river like a pregnant belly or a full breast. The heifers became cows which made milk, new cattle, and, when they got old, hides and meat.

Regular human access to the cows meant locking the bulls out except when the cows came "in season." Containment of the bull in a ring appears more than six thousand years ago. Only a small number were needed for breeding, because cattle are a "harem-keeping" species. As these became "the bull of their mothers," their human keepers discovered that inbreeding was the quick way to the selection of traits that differentiated the tame from the wild, increased milk production, and made the cow ever more bland and sedate.

Only the odd bull was kept, either to fertilize cows or, castrated, as oxen, to pull and carry. The bull god was both son and lover, a consort whose principal function was to fecundate and die in sacrifice. His segue into the underworld marked the central rite of the dead prince. As the spirit of vegetation he would come again in the spring, brought forth by her as a new groom, a divinity who would flourish and then die again. In Babylon the artificial mountain or ziggurat was both tomb and symbol of the goddesses' fecundity, the grave of the end-of-summer dying god, Geshar or Tammuz. In Egypt, Sekhmet, the lion goddess, was an aspect of Hathor the cosmic cow. The monthly drama of the moon's loss abbreviated the cycle of the year and symbolized the dying and resurrected deities, the seasonal vegetation gods, who faded with summer's end and the cow's diminished udder. In Egypt the dying prince was Osiris, who descended into the underworld and was released by Ishtar-Aphrodite, whose animal was the lion. The bull god, as a king, died in a ritual regicide dedicated to the earth and underworld goddess, "paced by the beat of a bull-skin drum, the rhythm of the universe, a wisdom beyond death and time."

In Crete, the "horns of consecration" were the major multifold symbol, best known from the terrace of the palace at Knossos, but also appearing as paintings on temple walls, some with a middle piece so that they resemble a figure with arms upraised, or accompanied by two birds.[3] The same figure is known from an Egyptian custom of burying a bull except for its horns. The Cretan horns also correspond to a cleft or double hill, and there are objects connecting horns with snakes. It is all one web: the imagery of the mountainous womb, the cleft mountain as vagina, breasts, or horned gate, the pillars demarking the sanctuary, the goddess as virgin and horned mother, the celestial cow who was sometimes a lion, the holy milk and sap and sacred herd, the dying, renewing divine sons, Osiris or Horus, the death-sleep, like the sun's disappearance beneath the earth, and morning awakening, and the season's renewal, all comprised in the sacred cow.

Representation of the fertility goddess as a tree was linked to the similar appearance of the milky sap of the sycamore and the fig, evoking the old, archetypal reference to the tree bole as womb, cradle, and coffin, part of the vegetative capacity of the genetrix independent of a male counterpart. Indeed, the passionate concern for fecundity had flourished with the yield of cultivated spices, fruits, and nuts. Holy trees and plant substances notwithstanding, even though people depended much more on plants for their food than on animals, the symbols of fecundity were mostly animals. In those ancient, Near Eastern subequatorial regions, the land open to soil leaching and desiccation, peoples hungry for crops and greater human numbers, there the cow represented the acme of renewal and patroness of women in travail—a birthing probably experienced a dozen times in the lives of women whose babies were removed for wet-nursing or weaning so that she could become pregnant again.

In the eyes of the West, the "sacred cow" of India is an appealing and gentle figure, an exotic mélange of peaceful naivete, ethical innocence, kindness, and enduring wisdom. Esthetically, it is not very different from the moo-call of bucolic tranquility. The effect of great limpid eyes and creamy prodigality, the divine but subordinate vigor of bulls and heavy reassurance of oxen, the playful exuberance of calves, the kine of Theocritus and the bovids in Dutch landscape paintings, advertising by the Borden Company depicting Elsie the Cow, the dairy lobbies: all were a sensibility developed over seven thousand years of cow tending and the adulation of fat, grafted onto an older love, generated by a hundred thousand years of watching wild cattle and stalking. But as people took control of the lives of cattle, an earlier gender equilibrium was first tipped toward the feminine by the cow-keepers, in the form of this bovine epiphany, celebrating creativity, earth, place, and the cycle of life and death.

The cow in India is associated with an indigenous, cattle/Great Mother cult among Dravidian small farmers who could afford to sacrifice goats but not cows and whose reverence for cows was probably very similar to that of the Sumerians.[4] But the cow's spiritual status in India was also influenced by Aryan invaders who entered India from the west in the second millennium B.C., bringing milk oblations, cattle sacrifice, dung and urine purification rites. No other animals were so important among the Vedas, in whose sacred books the cow is earth, mother, and the cosmic waters. At Harappa, in the Indus valley, the cow was associated with sacrosanct enclosures and pipal (fig) trees, connoting maternity, its early metaphors being clouds, the sacred bringers of rain. Thunder was like the lowing of the herd, and the cloud-cows were regarded as the mother of Agni, god of lightning and fire, twin of the major Aryan deity, Indra. In one story Indra shatters the dragon-demon, Vrtra, whose captivity of the cows caused drought—thereby releasing the cow-clouds with their heavy udders and rain. The waters were to the earth as milk to people whose cow-dung-fueled fires created edible food from inedible, uncooked rice.

Inheriting the primacy of the cow from the Vedic books, the sedentary Hindus, influenced by Gautama Buddha, emphasized the cow's sanctity by ceasing to ritually kill and eat it. The Brahman priests' search for a primal force led them to food in the figure of a great spotted cow, Ida, the "butter-handed" and "butter-footed" one. In myth she becomes Aditi, goddess of motherhood, ancestress of the world and the gods, a sky presence. In this way, the veneration of cattle outlived the old core of Aryan cow sacrifice.

To the west, however, beef eating spread and the bull prospered. In the Indus valley, Rudra was a destroyer deity, incarnate as a bull, whose image, drawn on rock, reveals that the bull had already been bred to exaggerated proportions. The Aryans venerated Dyaus, the Vedic great father/bull, who fertilized Prithvi, the earth-cow, producing all the gods—including the bull gods Parjanya, Soma, Agni, and Rudra, from whom the Hindus obtained Siva the destroyer, he of the lingam, and the black bull Vahana, or "Nandi," Siva's vehicle and consort, invoking the mystery of the transcendent god who is also immanent. All Siva temples have a Nandi statue facing the entrance.

The Aryans were beef-eaters and milk-drinkers, sky and sun worshipers, who killed livestock in dedication to their gods, whose attention to the annual rains was linked to a sky god in conjunction with the female earth. Renewal involved a sacred "marriage" and the continuity of the male line. In the West, the cow's importance as bearer and nurturer diminished in importance, compared to the bull's fertility and his immolation.

THE ARYANS TOOK the bull idiom to the valleys of the Tigris and Euphrates rivers, where the obsession with cows diminished with the rise of conflict and struggles for power. There, with the conflict among empires, the focus would slowly shift from the cow to the bull. Above the reed gates, which were themselves ideograms of the goddess in the Mesopotamian temple compounds, was the crescent of the Ur-moon-bull-god. But the full authority of the bull arrived with pastoral nomads. It is one thing to be a tender of milch cows and another to roam the wilderness with a lean larder on the hoof, expanding the herds by keeping some castrated bulls as steers, stealing cattle, and fending off bandits. For pastoralists the herd was not merely an adjunct of grain but capital itself. The metaphors of the dominance of bulls over steers would lend itself to struggles among men and such oaths as the biblical "swearing by my father's loins"—which is to say, the presence of testicles. Nomadic, pastoral cultures became muscular and chauvinistic, emphasizing masculine concerns about power, control, virginity, honor, and shame.

As the Near Eastern lands of the agricultural city-state filled up with people, as cavalry and metallurgy appeared, the cow mother was subordinated to the bull father. Priestesses may have tended and milked the cows, but as the cities grew, the cattle had to be grazed farther out, necessitating male herders and generating a culture of cattle apart from the precincts of temple and home. In expanded trade, cattle were used as currency and subordinated to breeding records dominated by pedigreed bulls. Like the bovine herd, the cities themselves and the great fields of grain gradually became colored by the culture of nomadic pastoralism, the most trenchant patriarchy the world has known.

Nomadic pastoralism gave the West its shift from hieratic to dynastic society, from the timeless reality of the cyclic cosmos toward history. In the new mythology, the goddess made male gods and was fecundated by them, but a warrior bull-god made the world—at first from a female body and eventually without her help at all. From ancient Sumer, 4,500 years ago, there is a terra-cotta plaque of a bull with its foot on the holy mountain—the horns marking the point of contact between earth and heaven, from which comes rain and dew and the river. The Hittite pastoralist invaders from the north influenced the history of Egypt in these matters, leaving a heritage also in Palestine, Jordan, Syria, and Lebanon of the bull god Baal. Among the Hebrews there was the Bull of Jacob, or of Joseph, and the sacred image of the golden calf. Yahweh emerges from Baal and other weather and storm gods, givers of rain. Gods such as Teshub and Ramman were represented by the sign of the bull. Baal was later forbidden as an idol by the Jews as their old mountain and weather deities ceased to be given natural form and coalesced into a single, celestial humanistic god. Inveighing against the worship of such bull idols, Moses had three thousand idolators killed.[5]

In Egypt, also, the bull's prominence rose with military unification and empire, in the form of the Apis bull, the god Hap, whose sign was the crescent moon. The bull was associated with kingship as elsewhere—and as Apis, the king's soul joined Osiris as the "Bull of the Underworld," who dies, passes below, and is reborn annually or in special twenty-five-year celebrations. Osiris, the deity associated with everlasting renewal, was represented on coffins as a bull, and oxhide covers were placed over the dead. Moreover, Ra, the sun, was the "bull of his mother," reproducing himself daily. Ptah, the forerunner of the transcendent creator, was incarnate as a black bull with the wings of a falcon. The bull king was stud of a lineage, his alter ego the sacrificial bullock or the ox who energized the passive fields. The cow, mother and feminine, was to become a mere vehicle. As the "nourishment of kings," milk became "the symbolic focus for a body of metaphysical thought," over which fell the shadow of the all-powerful bull.[6]

The shift of sacred gender in cattle—from the emphasis in Sumerian cow byres to Assyrian, Hebrew, and Egyptian bull kings—reflects the power of empire, the demise of the planter's deities, and patriarchal control over women and a diminished Mother Earth. Fecundity was newly interpreted, not as resident in the earth, but in the weather, just as purity and punishment were not in the soil but in the sun. As sacrificial rites became increasingly like bargains between men and gods who thought like men, the death of the royal beast took on the character of negotiation or conciliation with an arbitrary self-interested ego.

Thus as agriculture and cities came into existence and slowly encroached upon the foragers' world, between ten and five thousand years ago, the bear and lion, slow to reproduce, not being herd-followers or placid munchers of hay, unsuited for domestication, could not minister to the new ethos. At times they were seen as its enemy. The yield by captive animals of work, milk, manure, urine, bones, skins, and flesh on demand had to be consecrated in new religious forms. New questions had to be asked of the ideological meaning of the human/animal relationship. To whom would the ceremonial butchering of a bullock be addressed symbolically? In what sense would the animals continue to be emissaries, and between whom? Would these domestic forms—objects of human creation—be consumed as a sacrament or sacred communion as had the wild forms? How would animal powers be perceived and represented?

AMONG THE ANSWERS was the rite of sacrifice, the primary religious act of agrarian society, which would become the central act of Christianity and also the bullfight. It was a revised form of the prehistoric idea that the hunted animal was considered a gift, in a sense self-given or self-sacrificed. In some of the old ceremonial feasts of the Pleistocene, a portion of the killed animal was set before the head of the prey itself—a paradox of the animal's eating itself, resolved in the belief that its living spirit survived beyond death and was present. This separation of the spirit and body, first envisaged by hunters, was enlarged and altered by farmers and herdsmen. Cattle, holy as they were, became the principal currency in transactions with sacred powers who seemed more like humans than the old animal spirits. The domestic slaughter became a tithe or deposit in contractual ceremonies with a deity who, "receiving with one hand, is obliged to give with the other."[7] Sacrifice shifted the moment of truth from an unforeseen, improvised hunt to a scheduled death at an altar in which privileges were bought with meat.

The shift can be seen in the altered meaning of Ursa Major, the constellation of the great bear, said by hunters of the northern hemisphere to cause the sky to revolve over the earth. By ancient Roman times the same constellation would become the cowlick on the shoulder of a young bull, held in hand by the sun god, Mithras. This bull "moves and turns the heavens around."[8] Between the sacred bear and bull there was an interval of several thousand years during which the blood, milk, and semen of sacred cattle became the supreme epiphanies.

In the death and resurrection of the bull as the symbol of the agricultural year could be heard faint echoes of the Paleolithic celebration of the hibernating-renewing bear, slain by hunters, not in the name of a god but the bear itself, as Jane Ellen Harrison says, "a high sanctity preceding and begetting gods," killed not to bribe or please but "in order that he may be eaten . . . because he is holy." To the old classicists this seemed like "the uncouth predicament that you must eat your personal god," a dilemma eventually resolved by separating the sacred animal from the deity, to whom it was "given" at the beginning of the agricultural year, the Twelfth of Artemision (precursor of Easter), which marked both the blossoming of flowers and the rising of the dead—the resurrection. The bull's life "begins with the sowing, is cherished through the winter . . . and in the early summer dies to live again in the people through the medium of the sacrificial banquet. . . . In him is concentrated as it were the life of the year."[9]

Sacrifice began as a rite using domestic animals or plants to impel the rain cycle of the year—a death that oiled the wheels of the seasons. In time it became a bargain, atonement, or appeasement, calculated in advance to deflect evil or bring victory in war as well as rain in season. In ceremonies of transition the victim's death could represent the end of an illness or the end of childhood, anticipating health or maturity. Plutarch, the Greek historian, wrote of the god Dionysus, responsible for new births and growth, that "he suffers unto winds and waters and the earth and stars," regenerating by first, enigmatically, "rending asunder and tearing limb from limb" whatever is to be sacrificed. Immortals, breathing the smoke from such a sacrifice, became benefactors.

Like the Hindus, the Greeks knew that something must die for something to be reborn. The blood of the victim could be seen as food for the god in authority. The rite is a fresh enactment of the martyrdom of a hero who was a voluntary sacrifice, who reincarnates the spirit and identifies the human participants with the sacred figures of a myth. An iron-age ceremony, the Bouphonia, a communal feast with barley cakes in a cave, celebrated rebirth. Several bulls were allowed access to an altar baited with grain: the first to approach was accepted as self-selected immolation, a logic reminiscent of the widespread hunters' theory that the quarry was voluntary. Feasting at this shrine was followed by a mimic death and resurrection of initiates in the presence of a stuffed bull to which a plow was yoked. Wooden statues of the Olympian gods were carried as horned guests: Dionysos Tauromorphos, Horned Iacchos, and Zeus Olbios. A sculpted sarcophagus of 1500 B.C. shows the bull's blood carried to a sacred olive or cypress tree, linking the bull to earth fruitfulness. Elsewhere the slain bull was buried and altars were added to the shrine. Dying ceremonially, the bull renounced mortal life to redeem humanity in the context of the seasonal cycle—an outgrowth of the bull-of-his-mother who must die applied to the rites of passage of young candidates for manhood. To some it still seemed like murder, so the executioner in the Bouphonia "flees" as a slayer, is caught, and is exonerated in a "trial" (again echoing the ancient custom of hunters who protested their innocence to the spirit of the animal they had killed).

To the early Greeks the sacrificed animal was, says Harrison, a "year-daimon," showing the assimilation of the annual cycle to a cosmic cycle. But the classical Greek had a different idea about immortality: "He shakes himself loose of the year and the produce of the year. He demands a new honor, a service done to himself as a personality." Such bull-deaths exemplify a revolutionary shift from the idea of a year-spirit to one of immortality. The cycle of periodic reincarnation was broken in favor of deathless immutability, "which is really the denial of life, for life is change. . . . Such is the very

nature of life that only through the ceaseless movement and rhythm of palingenesis is immortality possible." In refusing the function of the totemic
animal, discarding the daimon of either earth or sky, abandoning the year-
daimon in search of no-death or re-rising, the Periclean Greeks attained a
"barren immortality."[10]

Heroes took over bull power in the pursuit of eternity. Sacred bovids, beginning with the fecund cow and advancing to the virile bull, were finally degraded to mere substance in the hands of humanized deities. Throughout
the Old World men interceded in rites associated with the goddess and her
"corrupt" tyranny of the animal demon. This masculine triumph was foreshadowed in the myth of Enkidu, a Sumerian hero of the first bullfight,
marking the transition of bull power to men themselves.

Powers over which humans had felt no control had been concentrated in
theriomorphic deities such as the cow and then the bull, ritually enacted in
the "giving" of the offering or the self in sacrifice. But as men became more
confident of their superiority in the world, they conceived the cosmos in
terms of ever-new history instead of cyclicity, altering the old myths from
Enkidu to Mithras. Among the Greeks, one of the labors of Herakles was the
killing of Geryon, a bull god of the river Tartessos. Although the slain bull
continued to be offered as a sacrifice, it was to themselves that men dedicated
the bull's power.

The cancelling of the contract of subordination to the bull can be seen in
many myths. Zeus, as a bull, seduced Europa, a daughter of the king of Phoe-
nicia, then swam with her on his back to Crete, where she gave birth to (king)
Minos. The king's wife, Pasiphae, copulated with a sacred bull and bore a
monster, the Minotaur. It killed Minos' son, Androgeos, and to it Minos was
required to feed seven boys and seven girls every nine years, taken from the
defeated king of Athens. Theseus, son of an Athenian king, ended this human sacrifice by capturing the Marathon bull and sacrificing it to the gods,
and with the help of Androgeos' sister, Ariadne, killed the Minotaur. And so
the Greeks freed themselves from the domination of the bull god.

But the breakaway took many centuries. Wrestling with the bull by the
old Mesopotamian mythic heroes was emulated throughout the Mediterranean world. Along the Indus, in Mohenjo-daro and Harappa, bull games
were played in 1500 B.C., depicted in the rock art of North Africa and paintings from Crete, where the bull is shown in art being sacrificed, played with,
and garlanded.[11] The Cretans played with bulls by vaulting over them or
onto their backs in beautiful leaps, apparently a test or display of adolescent
acrobatics which also seemed to mock the bull's eminence.

Defeat of the sacred sun-bull by men was the theme of Mithraism, the official Roman religion prior to Christianity. Its ideas of sacrifice, sacramental feast, and bloodbath would not be lost, however, only transformed by Christians. An "oriental," heavenly bull, created by the sun god Ormazd, was captured by the young sky god Mithras, who sacrificed it at Ormazd's order, transforming its blood and body into the source of life. In the liturgy of Mithras, bull gods turn the "axle of the wheel of heaven."[12] Ritual slaying usually meant cutting the throat; its blood allowed participants to be born anew. Out of such motifs dedicated to Cybele and Attis (Persian and Phrygian), the Mithraists developed a regenerative rite which in turn became "the nearest approach to the religion of the Cross," although detested by Christians as a travesty on the "sacrifice of Calvary."[13]

Ceremonial reenactment of the mythic slaying was the Taurobolium, "the most impressive sacrament" of the so-called mystery religions, in first-century Rome. The candidates for initiation entered a pit under the altar and were bathed in the dying bull's blood, drinking it, as it poured down upon them in a baptismal shower. For them it was the sun-bull's blessing, first obtained by Mithras himself in combat with the animal, achieving forever its blessing for the devotees. Novices were said to be "born again into eternity."[14] Great care was given to the head, the blood, and the bones. Severed heads became sacred bucrania on a shrine. Sacrificial altars, abstracted in form from the shape of the womb, were designed to channel the blood to save for ritual uses in drink and other sin-cleansing. This "rich, philosophic religious mysticism of closing paganism, in sublimated form, has become the perennial possession of Christianity."[15] It does not stretch the imagination to see the similarity between Christianity and the Mithraic celebration of the sun's birth on December 25, its ideas of good and evil as heavenly and earthly, and the power of the blood of the sacrificed one.

On the hill where the Mithraic bulls were sacrificed the Vatican was built. In Christian iconography the tender, loving bull survived as the apostle Luke, who was represented into the Middle Ages as a bull-headed man. He is celebrated in an annual festival in Andalusia in which a drunk bull is paraded in the street and caressed by the people. Substitute the even meeker lamb in the ceremonial death, and it is not difficult to see how Constantine and his pagans could be converted by the Christians. Mithraism, with its sacred bull death, widely practiced from Britain to Africa and the Atlantic to the Euphrates, was declared unlawful by Theodosius in the fourth century. A century later the Church, in the Council of Toledo, described the devil as a black bull.

Still the bull's special power persisted. The Celts celebrated a similarly mighty Druidic bull, and the Norsemen put horns on their hats as the warrior's emblem, made horn cups from which they pledged their fealty in drink, and blew on horn battle trumpets. Even the image and power of the wild bull persisted. Special hunts of the aurochs continued into Greek and Roman times, transformed into virile demonstrations of the confrontation between powerful, heroic individuals and isolated animals. Caesar used wild aurochs in bullfights in the Colosseum in which the venators used cape, shield, and horse. Combat with the bull was one way to obtain its blessings, which were shed on other useful animals and plants, such as the grape and wheat. In ancient Thessaly bulls were speared from horseback with the help of dogs.

Thus was the wild defeated by the domestic and the domestic controlled by men. The constellation that for circumpolar humankind was the means by which a great goddess turned the universe in its nightly cycle was now identified with the shoulder of a bull in the hand of a bull-headed young divinity. From Mesopotamian bull-grappling to Cretan bull-vaulting and garlanding to the Spanish bullfight appears to be a complex historical sequence.

In Christian societies the sorting out of pagan rites from orthodox practice did not end the appearance of bulls in sacred precincts, as the bullfight became associated with Christian festivals and was sponsored by the Church. The bullfight in primitive form was already in Spain by the time of the Moorish invasions in the seventh century, with formal processions, music, and dedications. By the eleventh century Christian knights competed with the Moors. Pope Alex VI, Rodrigo Borgia, brought the Spanish bullfight to Rome, took the bull as his crest, and had his Vatican apartments decorated with paintings of the ancient Apis procession. His son became a famous matador. Bull sticking was so tightly linked with Christian religious celebrations that two hundred bulls were killed at the canonization of Saint Teresa in 1622. After 1700 the sword was used in a contest with bulls by professionals on foot. Vincente Manero sees the modern bullfight as Christian catharsis: the bull is the beast in man—the beast with the seven horns of sin—overcome by the angel in him. In due course the contest was rededicated to the admiration of virility, to the victory of disciplined skill and personal virtuosity of the matador over brute force, and to the hero of the reconquest, El Cid, Rodrigo Díaz de Vivar, a famous bullfighter.

Bullfights may also have been related to a display of heroics in combat with natural forces in the form of the bull by the Visigoths from the north. The bullfight went to North America with the Conquistadores. In any case, the Roman arena, the Latin Fiesta de Toros, and the bullbaiting in Europe and America seem to be secularized, masculine fantasies of combat whose psychological functions have outlasted liturgical needs. The bull ring, says Victor Turner, is a setting in which "paradigms are transformed into metaphors and symbols." Mediating between the living and the dead, the bull ring, like the playing field or the law court, is a modern "arena of ritual sacrifice and ultimate truth."[16]

The corrida—the running of the bulls—and fights between men and bulls or between dogs and bulls became celebrated spectacles of the bull's power in terms of its defeat rather than the blessing of the animal. The running and the bullfight may, in some way, perpetuate a feeling that the killing of the animal tests the state of human grace. Like bull riding and bulldogging in the rodeo in the American West, it is a reenactment of the struggle of men to wrest power from animals as a continuing measure of civilization. Otherwise it is emptied of meaning if the animal is slaughtered without formality and its spirituality subordinated to the numb brutality of a mob.

The bullfight may not now involve a formal quest of fertility or magic power so much as the vicarious expression of repressed hostility, displaced aggression, or the release of the frustration that results from the clash of defiant individualism and social restrictions of family, church, poverty, class, laws, and so on. One wonders if this could not always have been part of the sadistic process of public butchery or the venting of discomfort by young men in pastoral societies, who are uneasy in the control of their fathers. Love them as they might, pastoralists often beat their cattle, and kill their fathers.[17]

CATTLE ARE A kind of universe, not only because they create pastures and meadows, the architecture of pens, lanes, and barns, the symbols of milk and horns, the social prototype of the herd, and so on, but as a microcosm echoing a larger metabolism. James Fernandez lived in the upper rooms of a peasant household in northern Spain, below which the livestock were housed. "The stable was a part of family life," he says, "who are attentive to the signs rising from the dark below: the shifting of weight, the rubbing of a flank against a post, the signs of the laborious bedding down, an unexpected bovine cough or eructation, and even the sound of a new calf suckling. One senses that the family is about as attentive to these nether regions as they are to their own visceral processes—quite without considering that their own visceral processes eventuate in the stable. A house without a ruminating, rumbling, wheezing, shifting stable is a moribund house." This consciousness is an example of "that dark arena of our interest where thought emerges from non-thought—where ideas arise out of that embedded condition." And, he adds, "symbolic inquiry recognizes how embedded ideas are in images and objects—among these the body—and how poorly understood are the procedures by which ideas are appropriately squeezed from their embedded condition."[18] Thus does the perception of the body sounds of cows precede the sense of self and place for these peasants, an "ordering of experience a stage earlier . . . a presentational symbolism" revealing a "profound organismic organization" that keeps ideas grounded in reality. It is a physiological commons that bridges human inner states and the perception of similar states in animals: repletion, evacuation, nurturance, satiety, sleep, waking, and other analogies, precursors to metaphors whose reference is far wider than the household.

The rumbling of entrails of cattle as intimations of life has further extension in the scrutiny of cattle by pastoralists who see the world writ small in them. The Masai of Africa, for example, also reason in terms of cattle. Westerners who think of the Masai as engaged in an "economy" of cows may confuse this kind of thought with irrationality, but it is strongly practical. For them cattle are not a commodity in a modern market system. The object is not to produce beef but to increase the herd, to engage in exchanges, gifts, and theft as traditional social practices. Their personal relationship to the animals is strongly marked: the Masai herder may recognize as many as one thousand individual cows. Cattle names are derived from a combination of the name of an ancestral cow and a descriptive term—"Nkeyi" means speckled daughter of a certain female. Further descriptive terms indicate whether they were acquired by theft, exchange, gift, debt, or bridewealth, tell their sex, maturity, and wholeness, and identify color, form, and patterns. In this way the group's social ties are embedded in the configuration of the herd. Bulls are not highly thought of, being wiry, scrawny, and mean, so that few are kept whole. Castrated, as oxen, they are admired, and the strong, beautiful, and sociable ox and cow are the bovid pair which symbolically and linguistically represent the people themselves. Cattle are a composite array. All of life's meaning is, by some miracle, prefigured in the animal. Not only does it mirror what is already seen to be unified, as in the human clan or lineage, but it is the key to all disjunction, as though its very configuration held the gift of transcendent insight. Doting on cattle seems to nourish the intuition. It is not that the problem at hand is like the array of color patches in the herd, in their shapes, tones, location in the body, distribution among the individuals. Rather, the patches are patterns that inspire imagination about all other patterns in life.

Little boys from the age of six spend their days tending the grazing cattle, so that by the time they are twenty or so, owning their own herds, they have spent thousands of hours watching and comparing and abstracting. Typically the owner is present at the kraal as the little boys drive in the cattle at evening, counting them, watching for missing animals, noting newborns, marking the ill or injured, spotting individuals not one's own, observing the position of each in the herd, and in general accounting for a herd that may number several hundred. These skills of recognition and accounting are done by "chunking" information in descent lines and grouping their ownership among the observer's different wives or by other means. In one herd of 54 there were 13 representing matrilineages of which 80 percent were females, 16 of which had calves, along with 27 heifers—all named. The cattleman also counts the animals in his head while going to sleep and then dreams the herd as well.[19]

AND SO WE come full circle, or at least to a cognitive obsession that has replaced the old metaphysical one. At first wild cattle were drawn and danced with great exuberance in caves and around fires. In history, the cow herself became the focus of idolatry. Then, as the fecundity of the earth was seen as a result of rain from heaven (perhaps as irrigation replaced hoe farming), the cow's impregnation by the bull was seen as the crucial event. As human populations increased and grain became the mainstay of life, milk became the food and bath of the privileged, the cow receded, and the bull advanced. Just as the bull dominated his harem, women were dominated by men and the earth by humankind. The sire generated genealogies, as did kings, and the bull (male) prospered on the cow and his queen. These embodiments of power were enacted first as the incidental sacrifice of the bullock, then as the main symbolic rite of renewal in a sacred regicide, and finally as martyrdom in a willing atonement.

Cow and goddesses in the Latin world were subordinated to bulls from whom heroes in turn wrested their power. The old epiphanies of the snake and bear—wild, androgynous, creator spirits—were replaced by the domestic pair of bovids with their milk and semen, the fertility of the earth by manure, and its cultivation by plow-pulling. The change of exemplary myths from the wild creatures of the earth to domestic cattle carried with it a brief interval of the celebration of the feminine and its symbolic motifs, undermined almost from its beginning by the motifs of the bull idiom: centralized power, discursive elucidation, the skills of record keeping, the logic of causation and of strategy, the lapse into dualities, the loss of confidence in the "genius of place," and the emergence of the animal as symbol instead of embodiment. Mounted pastoralists created adventitious, competitive, political organization, as the combination of cattle-keeping with horses released pastoralism's potential as the instrument of political empire and aggression. Although a minor economy, pastorality is the stylistic precursor of modern world cultures, perhaps in part because its mobility anticipates the traffic of industrialized internationalism.

The way of life among mobile herdsmen is essentially a culture of hierarchy, theft, rebellious sons, and competitive use of the earth. The North American ranchers, the gauchos of the Pampas, the Somali and Mongol cattlemen, the Australian cowboy, and other bearers of the bovine idiom continue the mindset of ancient, mounted cattle-keepers, their ideology and ecological effects, be they Incan or Aryan.

We need not regret the passing of the old milk worship with its paean to a lost, goddess-centered celebration of udders and butter. Cow's milk has turned out to be one of the great poisons of the world. It may have prolonged many children's lives, but it has surely (like domestic beef) shortened the lives of an equal number at the other end of the span.

More than axe or fire, cattle-keeping is the means by which people have broken natural climaxes, converted forest into coarse herbage, denuded the slopes, and turned grasslands into sand. The overuse of beef and milk as human food, leather as an interface with the world, cattle as intermediate hosts of human diseases, the hoofed causes and inheritors of erosion, are the effects of eight millennia of a misdirected human ecology. If the auroch was the most magnificent animal in the lives of our Pleistocene ancestors, in captivity it became the most destructive creature of all.

17 | Lying Down with Lambs and Lions in the Christian Zoo

The wolf also shall dwell with the lamb, and the leopard shall lie down with the kid; and the calf and the young lion and the fatling together; and a little child shall lead them. And the cow and the bear shall feed; their young ones shall lie down together; and the lion shall eat straw like the ox, and the weaned child shall thrust his hand into the den of the basilisk. . . . Dust shall be the serpent's food.

ISAIAH 11:6–8

ALTHOUGH IT IS difficult to characterize Christianity's attitude toward animals as anything but a vast agglomeration, certain themes and biblical chapters mark its mainstream. If Judaism is a "diaspora," scattering tiny colonies worldwide, and Islam a rolling stone, Christianity may be thought of as a sponge which sucks up everything with its converts, and re-gurgitates creatures transformed, from toads who incarnate the devil and demons that look like bats, to lambs with whom lions may lie down. The rec-ord is ambiguous because of its catholic heterogeneity—from Francis' compassion for birds, Jerome's benign menagerie and demonic harassers, and Anthony's anathema for snakes to the horses and dragons of the Apocalypse. It is helpful first to review the source of biblical natural history and parable with some glimpses of recent spokesmen and then to examine three major Christian concepts of animals: their behavior in the garden; their salvation in the ark; and their association in the cave. Then it may be possible to say something positive about the effect of the sponge on the animals.

Western Christianity took a critical turn away from the natural world in the fifth century by separating the divine from the physical in Jesus and, by implication, in all "incarnation." The most vivid survival of its earlier sense of cosmic unity may be seen in the living fossil of the Coptic Church in Ethiopia. Excommunicated and isolated from the rest of Christianity fifteen-hundred years ago because of their objections to that split of the divine from the natural, the Coptic monks today still see divinity in animals and plants. In some cases their temple and monastic grounds contain the last bits of forest in lands otherwise reduced to sand by the abuses and rapacity of Muslims, and later Christians and others. The sacredness of nature in Coptic theology is in harmony with and partly a relict of the culture of the pharaohs, with its profound concept of the holiness of all life. As the best evidence of that indigenous philosophy of old Egypt, the Copts may be more informative than modern treatises on "the history of Egyptian religion" by scholars who seem blinded to nature by their training in Western scholarship and the sterile effects of academic life on their own experience. Other examples of the amalgams of "pagan" beliefs occur everywhere that Christian evangelizing has altered the religious landscape, usually with harsher effects in later centuries as the Church became increasingly anti-natural.

Western religious ideas of animals, called "theological zoology" by Florence Murdoch, comprise two thousand years of "grotesque parodies of natural history" adapted from the work of a pre-Christian Samian slave named Aesop and the Greek *Physiologus* of the second century B.C.[1] Origen in the third century, Basil and Ambrose in the fourth, and Eusthathius in the fifth were also tireless writers of animal parables. After the ninth century much of the earlier animal symbolism, such as saints and evangelists with animal heads, was prohibited. What remained "in both the Old and the New Testaments was seized on with some relief" by new converts—no matter how improbable the stories—who would otherwise have been deprived of a mythical zoology.[2] Apart from biblical parables, Christian animals mean either demons to be trampled or adoring faces in the manger scene. The tradition of the virgin birth, from which Christianity took the nativity, was originally set in a cave with many animals, as depicted in late Byzantine art.[3] But it bore too many pagan traces for the Roman Church, and the cave was traded for a barn with its cattle.

With the Christianizing of the Romans in the fifth century, the way was open for evangelizing northwestern Europe. Officially Christianity had opposed learning from animals from the fourth century on. The meaning of animals was no longer searched out as a lifelong vocation in the totemic sense; nor was contact made with elemental powers through shamanic trance, or seen as coded divinatory messages, as had been its history. Tropes and parables did not require one to look at animals at all, but the missionaries had neither the power nor the means to plow paganism and its animals under. It took centuries to recruit enough converts to go out and cut the sacred trees, smash the effigies, and burn the drums. Meantime the Church, the great catholic sponge, incorporated ancient holy days into its calendar, reinterpreted Germanic lore in biblical terms, redirected archaic rites into Christian ceremonies. Churches were built on the sacred sites of wrecked pagan temples, as though to replace them, but in fact absorbed ritual customs from them. Animal deities did not disappear but were transformed into secondary spirits, masking the beasts in Christian homily.[4]

Along with Scandinavia, Britain was the last stronghold in northwestern Europe of indigenous religion. The Irish, Welsh, Scots, and English were a trial to the nature-scorning Church for fifteen hundred years. The signs of the seasons for agriculture and hunting—and the compass of the sky for navigation, hence the celestial realm and astrology—were of great practical importance to coastal peoples. Gildas, a cleric, was sent in A.D. 540 to Glastonbury to attack astronomy in the Celtic church as witchcraft. He and others uprooted the Roman Mithraic rituals and concepts of the universe and opposed the Arthurian myth of a stoic earthly paradise and Arthur's ideal court with its many astronomers. When William the Conqueror arrived in Britain in 1066 he inveighed against the Anglo-Saxon bishops, whose scriptures, *The Chronicle of Nennius*, depicted the universe as a circle in which the sun was a great horse, following Heraclitus' use of Hippoi (horses) as the symbol of God's power and a vast mathematical system embracing sacred numbers, time, and space.[5]

The twelfth and thirteenth centuries were diverse and contradictory. There was Pope Innocent III's "contempt of the world," Saint Anselm's equation of sin and the use of the senses, and Thomas Aquinas' reconciliation of Greek logical and Christian moral thought, the philosophical bases of human dominion over nature that would characterize the next eight hundred years. In some ways these centuries were the acme of European paganism in Christian guise. Gothic cathedrals were brilliant integrations of the two. In stone, Christ stands on the defeated beasts. He is not riding them, as did so many of the old Greek deities and Hindu gods, signifying shared powers—at least not overtly. But the very presence of so many animals was astonishing in the temples of a religion that, along with its sister traditions, scorned all images. A new vision in stone and glass, born from pagan/Christian France, simulated a world ripe with undiscovered allegories in nature.

Prelates, designers, and builders of the Gothic cathedrals worked plant and animal motifs into every space, as though the building were itself a forest. Saint Bernard of Clairvaux railed against them, indignant that "Christian mysteries should be clothed in Pagan allegory. . . . What mean these ridiculous monstrosities in the courts of cloister, these filthy apes, fierce lions, monstrous centaurs, here a serpent's tail attached to a quadruped, there a quadruped's head on a fish?"[6]

Bernard's contemporary, Saint Dominic, left an indelible episode, as described by a thirteenth-century nun, Saint Cecilia, and recounted seven hundred years later by Aldous Huxley: "In the Crusader's elegant little church a Maronite service was in progress. The words of an incomprehensible liturgy reverberated under the vaults, and above the heads of the congregation a family of sparrows was going unconcernedly about its business. Nobody paid any attention to their noisy impudence. The little creatures were taken for granted. Along with the ancient stones, the altar, the intoning priest, they were an accepted feature of the Sunday landscape, an element in the sacred situation. In this part of the world, birds seem to be perfectly compatible with monotheism. . . . Ancient traditions die hard, and perhaps the birds we had seen owed their immunity, in mosque and church, to the buried memories of a religion far older than either Islam or Christianity."[7]

Saint Dominic preached from behind the grille to the sisters: "His theme was devils; and hardly had he begun his sermon, when 'the enemy of mankind came on the scene in the shape of a sparrow and began to fly through the air, hopping even on the Sisters' heads, so that they could have handled him had they been so minded, and all this to hinder the preaching. St. Dominic, observing this, called Sister Miximilla and said: 'Get up and catch him, and bring him here to me.' She got up and, putting out her hand, had no difficulty in seizing hold of him, and handed him out through the window to St. Dominic. St. Dominic held him fast in one hand, and commenced plucking off his feathers with the other, saying the while: 'You wretch, you rogue!' When he had plucked him clean of all his feathers, amidst much laughter from the Brothers and Sisters, and awful shrieks of the sparrow, he pitched him out, saying: 'Fly now if you can, enemy of mankind! You can cry out and trouble us, but you cannot hurt us.' "

"What an ugly little picture it is!" observes Huxley, "An intelligent and highly educated man wallowing in the voluntary ignorance of the lowest kind of superstition; a saint indulging his paranoid fancies to the point where they justified him in behaving like a sadist; a group of devout monks and nuns laughing full-throatedly at the shrieks and writhings of a tortured bird."

Yet it was the same century as Saint Francis of Assisi, to whom the birds were indeed fellow beings. Although historian Lynn White says that society is "tinctured with orthodox Christian arrogance toward nature," he chooses Francis as the saint of ecology.[8] Noting that Francis was neither a pantheist nor a believer in the Hindu concept of the transmigration of souls, White sees him as truly Western, his ideas a "god-glorifying panpsychism." Francis urged the birds to praise God, and when the wolf of Gubbio "ravaged" the countryside Francis persuaded it to desist, and it died sanctified. He revived a Greek idea that creatures have a *gnosis*, the capacity to recognize divinity, and thus witness and celebrate creation.

Perhaps only a modern historian in desperate times could have proposed this solitary wanderer as the spiritual father of ecology. Francis' ideas of the animal soul may have been inspired by twelfth-century Cathars, who in turn were influenced by East Indian theology. It is likely that Francis was more pagan than tradition allows, glorifying animals while not explicitly denying doctrine, a heresy thinly veiled in inspiration. Little is known of his private life. As for the Church's response, the Franciscan order retains almost nothing of Francis.

Franciscans today keep such places as the Church of San Remedio in the Italian Tirol, whose founder, five centuries before Francis, shared his cave with a bear. One might suppose the bear to have been a "little brother" of Remedio. But it killed Remedio's ass. Let the bear replace the ass as Remedio's beast of burden and mount, said the bishop. Keeping a slave bear seems very unlike Francis, but the episode falls within the convention of the animals' recognition of his sanctity and tempers or subverts a Franciscan love of animals with the conventional legends of the saints and their devoted beasts.

Albert Schweitzer is often put forth as a more recent model of the Christian love of animals. His "reverence for life" along with Francis' friendship with the birds and mammals have been a motif for "animal liberation" and "friends of animals." Ecologically, however, Schweitzer remained the swineherd he was as a boy, who did not hesitate to butcher his pet pig at Lambarene when it intruded on Sunday services, like Francis, who was rebuked by Brother Juniper for damaging private property by cutting off a live pig's feet. Schweitzer retailed misinformation about the ferocity of gorillas and kept his private little zoo of tightly penned antelopes.[9] Insofar as the ideal animal protectionist today is vegetarian, neither Schweitzer nor Francis fits the definition. The latter was so uninfluential that the mountain where he wandered near his home town of Assisi has long since been stripped of its bird life, paralleling the modern exploitation of the upper Congo.

Rounding out the thirteenth century's diversity was Frederick II, pope and emperor, who disdained the bestiaries for the study of real birds. There were shifts in the Church toward literality, the physical elevation of the "host" for adoration, emphasis on the incarnation and the dogma of transubstantiation, the feast of Corpus Christi, the collection of saints' relicts, and increased emphasis on the madonna. From about 1196 to 1229 devotion increased toward the physical substance of deity or incarnation, hence the deifying of matter. Nature became increasingly valued in its physical reality.

Between the thirteenth and nineteenth centuries Western civilization achieved little on behalf of wild animals. Bad weather, wars, and plagues hammered Europe in the fourteenth and fifteenth centuries. Feminine earth spirits and animals were believed to consort with such evil forces. The persecution of witches with their animal familiars, the public torment of animals, and the prosecution of scapegoats greatly increased. Evil was incarnate as demonic animals, the "beasts" of the Psalms: death, devil, and anti-Christ.[10] This imagery overwhelmed the lingering traditions of nature love from the thirteenth century and reaffirmed the Christian attitude toward animals as envoys of the dark side of a dualistic world.

OF THE BIBLICAL and hermetic foundations three are paramount: the garden, the ark, and the saintly rapprochement to nature. The first, the Garden of Eden, the earthly paradise, is the landscape expression of the concept of peace and brotherhood, misapplied as the ideal interspecies relationship of all time—the way it was at first, the way it should be now, and the way it will be in eternity. One need not be a Christian, Jew, or Muslim to be affected by its dreamy perfection that has so saturated the yearning of half a world for so long. But when we examine this image closely we find an imaginary world at odds with the natural one. The Eden idea is not one of a beautiful earth, but one flawed along with humans. It expresses the fear and hatred of both animals and other peoples, using sheep—or lambs—as the measure of the perfect religious subordination.

Friendship of all the animals and people is described in Genesis and the book of Isaiah in the Old Testament. Yet Isaiah rails against the powerful Assyrian whose "land is filled with horses" and who has "idols . . . which he has made for himself to adore, moles and bats . . . who seeks for pythons and of diviners who mutter in their enchantments." After the Assyrian has been humbled with the rod of Jesse "and the thickets of the forest shall be cut down with iron, the wolf shall dwell with the lamb," and so on. Isaiah's vision is not the original biblical Eden, which was perfect in the first place—no wild horse-lovers, no friends of bats living in a terrible forest. But these two paradises converge: one from the original creation without pain, time, bloodshed, or carnivory; the other to be created by men who will subdue the pythons and Assyrians, drawing on a ferocious, maudlin, infantile concept, a land of counterpane.

Isaiah's seductive dream, with its images of sweet warmth and the reconciliation of worldly antitheses, has pacified the hearts of generations, yet its tender notions conceal one of the most hostile statements to nature in Western thought. The hated Assyrians are like the "viciousness" of nature. Many Bible animals are metaphors of different nations. Of the crushing of Babylonians, Isaiah says: "The wild beasts shall rest there and their houses shall be filled with serpents; and ostriches shall dwell there, and the hairy ones shall dance there. And owls shall answer one another there."[11] Owls, ostriches, and serpents signify the post-apocalyptic desolation. In Isaiah the intent of the dream is clear—a future harmony in a world of discord and a more or less secular counterpart, the making of a defanged nature in a new garden, strongly suggestive of the lost paradise.

To the biblical fallen world our culture has added Greek teleology and modern progress, while a world filled with conflict (resulting from Adam's sin) continues not only as dogma but as Marxist and Spencerian nightmares emphasizing competition and combat, misunderstood "Darwinian" evolution, and obsession with life as a struggle. In the biblical Eden every creature was provided for and none attacked the others. The Lord said, "Wherein there is life, I have given every green herb for meat." So here they are, the asp, the lion, the lamb, and the man, all quietly together in perfection. No more biting each other's throats or dying, all adoring the Divine Being and submitting to his will (in other words: his human agents). But what is the preunderstanding of this animal garden? It is a friendly anarchy, each species chewing grass, tending to his own needs, without treading on the others. Who makes the decisions, gives orders; who runs and fetches? Does the grass care?

And what is that about a little child among the beasts? It is the first kindergarten. In the Pseudo-Matthew's Gospel of Infancy, dragons, lions, leopards, and wolves come to admire the sacred babe, and, like the messiah predicted by Isaiah, he grows up playing with the lion's cubs.[12] A Hebrew book, the *Apocalypse of Baruch*, contemporary with the Gospels, says: "And wild beasts shall come from the forests and minister unto men, and asps and dragons shall come forth from their holes to submit themselves to a little child."[13] From a biological perspective, this emphasis on the child reminds us of our primatehood. Most giant primates don't have much to fear from large predators, but they do fear for their babies. We and our primate cousins no doubt have a long genetic memory of infants snatched up by hyenas and leopards. Predators are reticent about attacking adults of any large species if the young, the most vulnerable, are available. Hence the sparing of infants would be the final test.

"Paradise" and "park" are closely related both in history and in the popular mind. A nationally known American radio commentator, Paul Harvey, once made the point to his audience that the trouble with the preservation of nature in the national parks was, what with wolves, lions, and other predators, that there was no place for "mercy." Mercy is that forgiving posture of God in the presence of sinners. The newsman's comment expressed the biblical view that all nature fell with Adam and Eve, prior to which there was no death in Eden, hence no eating of one animal by another. If this is what nature should be, or was, it follows that the celebration of nature along the lines of the bloody, fallen reality simply perpetuates the error. Ecology, a system relating species through food chains, is itself a result of the Fall and should be replaced with a family-like, intraspecies model. Two big steps in this direction could be taken by simply eliminating all the animals that eat meat and keeping those that eat grass. The lions that lie down with the lamb, the asp with whom the child may play, can be simulated in some form of virtual reality.

In light of the biblical assurance that "God so loved the world," we suppose there must be a divine love of animals. It is held to be demonstrated in the incarnation. But a love of animals does not require that they themselves be sacred or spiritual presences. They may be dear to their creator and therefore justify man's respect without being numinous. Kindness to animals and the exercise of a "caring" dominion—these notions refer to our own souls, not theirs. Historically the animals fade except as parables or as the objects of which we are God's stewards.

As the modern world becomes more crowded, poorer, and increasingly violent, swept by epidemic and starvation, we can expect that vacations from reality will hold greater appeal. One of these vacations is the psychedelic dream of the garden, be it fetal or final. The failure to see that ecology is only a metaphor of human society and not an extension of it can plunge us into desperate efforts to bring animals into that garden. Our success will end their wildness or even their existence.

IF THE GARDEN of Eden is the first foundation, the second foundation is the ark. The design of Noah's ark was based on the Babylonian mathematical conception of the universe. Instructions for the original ark's construction first appeared in the Epic of Gilgamesh, some three thousand years before the Old Testament. It was to be a square of 120 cubits, six stories high, subdivided within each story. It was, says Alun Llewellyn, an allegory, a graph of time and space, protecting the world against chaos, based on astronomy. Its uses of multiples of 6 and 10 were those from which all modern measurements are derived.[14] That the original model was mathematical was mostly lost on the Hebrews, whose notions of the earth had, in the meantime, been dominated by disasters, especially floods in the valleys of the Tigris and Euphrates, probably known to Abraham and his kin or perhaps common knowledge among the captive Israelites in Babylon. The Flood commemorated in the Bible was probably a composite of many floods, the magnitude of which were multiplied by upstream deforestation and overgrazing by domestic cattle.

The procedure on Noah's ark was that the unclean beasts were taken on board by twos, male and female, the clean by fourteens, males and females. The unclean were defined in Genesis as those with ambiguous characteristics, such as cud-chewers with cloven hooves. One may wonder in this rash cleansing of a corrupt world why the unclean beasts were taken at all. Generosity of spirit does not seem to be evident in the whole episode, in which Noah is simply doing as told.

The rain of forty days probably reflected the monsoon climate of ancient Mesopotamia, overlaying a half-forgotten memory of Paleolithic traditions based on the forty days of the bear's hibernation in the Mediterranean region. All sorts of transformational intervals of forty days or forty years settled bearless into Western myth and ritual. The Flood signified a recalcitrant nature, unruly and out of season, unlike the gentle swelling of the Nile. As the waters went down, Noah got the sign of land from a pigeon, having first tried a raven who went his own way. Once on land, Noah "took every clean beast and every clean fowl and offered burnt offerings on the altar" — which is to say, he sacrificed a lot of animals as part of a contract. Ritual sacrifice defines the culture of dominion and negotiation as distinct from that of primal peoples who encountered the cosmos modestly in terms of guesthood and the gifts of nature. The burnt offerings were, so to speak, the price the species paid for their tickets.

As for the ark itself, there are those who still seek rotting ship timbers on mountainsides somewhere in the Levant. Some believe that a bearded old man at one end of a gangplank counted heads, as pairs of animals lined up and trotted aboard, and then took them on the world sea. This literalizing suggests that once we abandon mythic thinking, it is extremely difficult to regain the metaphoric handle. And when mythic places lose their role in the imagination, they become the property of realtors and other positivists.

The ark is the prototype of all animal saving, a shorthand for the right attitude toward nature. Noah's ark is the model for the zoo and the nature preserve. It is the protected place in a chaotic earth. When God made Noah his keeper of the animals he was presumably speaking to us. Zoos, which were until recently display arenas with cells for captive animals, are now changing just as madhouses and prisons changed in the nineteenth century. They are a refuge for delinquents who cannot fend for themselves, the dangerous, indigent, and incompetent—that is, all the unlucky animals whose native habitat is appropriated for cities and crops.

The inmates of this sanatorium are incurable, however, incapable of being turned into good citizens. For centuries animals have been seen as demented persons. In the courts they now have "legal standing" equivalent to that of fetuses and psychopaths. Like idiots savants, their uncanniness is both fascinating and grotesque. Some are esthetically exquisite, like the giraffe, while others, notably the primates, driven crazy in their prisons, seem to exemplify our worst failings and to mock our attempts to do them justice. It does not take an "anthropomorphic" delusion to see their situation—not as a metaphor for our own urban existence but as an extension of it. They masturbate obscenely, scream purposelessly, torment each other brutally, kill their babies, and throw their shit at spectators. The despair and depression of captive apes is heartbreaking unless one can accept the illusion of the happy prisoner. Reptiles and fish glumly wait and punctually die. The birds furiously make the best of it if their cages are large enough. Otherwise they tear out their own feathers. Compassion demands amelioration, so the zoo becomes a park and its bears go into pits, reminiscent of the way post offices and banks have removed the bars between the teller and the patron.

But all this is secondary to the new-wave zoo-news, where a great genetic repository is being created.[15] The several pairs of chromosomes in each species, two by two, have marched aboard for the long trip to nowhere. The genetic ark, disguised as a zoocraft for leisure-time entertainment, fulfills a scientific component by finally reducing the inhabitants to their most essential constituents: their DNA. It is necessary to breed them in captivity as they cease to be found in the wild, where reproduction is complicated and life fulfilled by the nuances of choice, environmental requisites, the subtleties of courtship, and especially by the psychological health of the animals. Reproduction can be forced. Condors and cranes, unable to soar, will not reproduce in zoos—will be hence artificially inseminated, the triumph of the breeder's art, long since taught to every Future Farmer of America.

Like farms, zoos select offspring with inherently docile behavior and thereby alter the organisms to endure destitute environments, institutional diets, and human handling. Gone is the genetic diversity of large, wild populations. Zoos now advertise themselves as repositories of endangered species. These number in the thousands and are unsuitable for public display. Who wants to see a hundred species of snails or flies or scores of rare minnows? Who is not familiar already with aquarium fatigue? The history of zoos began as the hobby of monarchs, grew to public entertainment, and now emerges as a veterinary mission, but it will be shouted with ever-incisive fervor that the cost of saving rare snails would feed several homeless people.

Hard times have come upon the world, and we, like Noah in the Bible, must partition space. Boundaries were marked in ancient Greece with stones soliciting protective ancestors. These sculpted herms were in time given heads. "When the herm gets a head and gradually becomes wholly humanized, among a pastoral people," observes Jane Ellen Harrison, "he carries on his shoulders a ram, and from this Ram Carrier, the Criophorus, Christianity has taken her Good Shepherd."[16] An admirable image: we are to carry the animals, provide for them, and protect them. We will be responsible not just for the sheep but the giraffe and all the vanishing species of butterflies and freshwater mollusks and bats and on and on.

The wild relics will be housed in preserves or buildings, probably, like Illinois pigs. These sanctuaries will become cosmic vessels for the animals of creation, like Noah's ark. Nature manages to copy ideology, and ours is the era of the individual. The Noachian custodial mind is highly personal. Its vision of nature is that there are not deer but "Bambi," not dolphins but "Skipper," not elephants but "Jumbo" and "Dumbo." Unfortunately for individuality, you can only save individuals temporarily, because, like us, each has to die, and their replacements are a function of species populations. Noah's religion forbade the notion that the individual is a momentary embodiment of the species, an evanescent being in a natural continuum. And so we rescue animals by removing them in helicopters from valleys being made into reservoirs, by giving sick or "lost" whales expensive injections or help from ice-cutting ships, by protecting rhinos in Africa and lions in India from poachers by attaching a warden to each animal. Sets of twos and sevens are driven into captivity under the watchful eye of a ship captain and the media. The whole phenomenon of nature is compressed into an event recorded on film: the benign driving, capturing, transferring, treating, and fencing of animals.

The zoo presents itself as a place of education. But to what end? To give people a respect for wildness, a sense of human limitations and of biological community, a world of mutual dependency? The zoo has early analogues and antecedents. King Solomon long ago had the idea of animals as education: "What we call the book of nature was to him a vast and many-volumed book on all phases and features of human nature, in which the world of lower creatures was held up to a man as a moral mirror in order that he might see therein the reflections of his own vices and virtues."[17] The Gothic cathedrals were themselves such a universe, miniatures of "the city of God" where every creature was a word of the divine thought. In these places animals wait in their paragraphs and woodcuts, stone cubbyholes and stained-glass stalls, adoring and submitting, abstracted from habitat. In spite of Saint Bernard's objection to their presence, the message is about captives. Like the stony placidity of spirit-broken gorillas and elephants, their sculptured and printed counterparts suit their moral function. Novel as the great cathedrals were in the twelfth century, the animals did not arrive there by chance but came from other barns—Romanesque and Cluniac copies from Byzantine carpets and tapestries, from brooches, miniatures, arabesques, and manuscript illuminations, from old bestiaries, astrologies, and mystical medical dictionaries. The cathedral gargoyles and their fantastic kin are like rhinoceroses and hippopotamuses, once dangerous and numinous, now ridiculous. They have the same message as the essayist who says, "A World Safe for Rhinos Is Not Best for Men."[18]

Enclaves "protect" the defeated, as they did American Indians, by assigning them to reservations and then eliminating them outside the sanctuary. Who or what is nature to be saved from—our atavistic impulses? Our atavistic "nature" gets its exercise by walking through the zoo. Zoos provide us with saturated opportunities to feast the eyes on animal diversity, a deep hunger that resonates with their strangeness and our essential humanity. Western moral, ethical, and spiritual dimensions of the matter are outside of nature. Sacred groves did not exist when all trees were sacred. Places only become holy by being set off from nature (the cathedral) or because of historical association with a divine outsider (Mount Sinai, Mecca, Mount Calvary). Compassion, responsibility, and stewardship are all predicated on the superiority of the caretaker over that which is controlled.

Noah's ark, the ecological form of the enclave, takes true counterplayers out of our lives. From the ark we move to Spaceship Earth with its unfortunate imagery of captain, crew, galley slaves, and those left behind. But Earth is neither an ark nor a spaceship; it is not "going" anywhere; nor is it drifting out of a bad time and waiting for a new landing. It is not even floating. As Margaret Mead once observed, island peoples either recognize their spatial limitations and the interdependence of all the inhabitants or they fail and take nature down with them—that is, the nonhuman inhabitants of an island whose only possible future other than catastrophe is measured by its own diversity and freedom.[19]

The myth of the ark raises lifeboat questions. How were those in each species selected (and what about all the insects)? The question is related to a similar metaphor—the notion of natural resources as a raft supporting only a limited number of humans in a world on the brink of devastation because of human overpopulation. When biologist Garrett Hardin noted that the raft will not support everybody, the mob's fury was his reward for truth. Like the bearer of bad news Hardin was blamed for the news and condemned by instant moralists who were bad ecologists.

Among the animals, chaos is already here in the extinction of species and the shunting of selected large, familiar animals into enclaves. In war, the wounded are sorted by triage: those who need help, those who can wait, and those beyond help. As the planet becomes destitute, the triple doorways of triage will open first to the rich; second, partly, to the poor; but the third class, the nonhumans, already drowning, will not be saved by our toy arks: the zoos.

AFTER THE GARDEN and the ark, the third model for Christian attitudes toward animals seems to be to go out and befriend the animals where we find them, that is, in the wilderness, bringing them kindness and brotherhood.

We could protect mice from snakes and snakes from hawks and hawks from farmers by simply putting the animals in zoos, following the ark prototype, a temporary reconstitution of the garden. Otherwise, we must *convert* the animals.

Christian saints model this concept as hermits in caves in the desert. In most places animals are dispersed and difficult to see, but the desert concentrates them because it lacks vegetative cover and because water is localized, often at the foot of cliffs near caves, so that animals come to such places from the surrounding region. In addition the desert has a special ambience because of its sensorial qualities, a model of the Other World.[20] This theme of desert holiness, where also are Satan and demons, this theme of trial and perfection, creates for the desert hermit a special relationship with animals. Saints Paul, Anthony, Columba, Godric, and Yvain are envisaged as compassionate and paternal in their zoological relations. The original model is the hermit: "In retiring to the wilderness, the hermit monk returns to a state of nature, he attempts to attain the spiritual felicity of Adam before the fall. He lives with the beasts and is recognized by them as a friendly being rather than as the born enemy of the animal kingdom . . . strives to overcome his impulses by constant self-deprivation . . . to imagine his ascetic solitude as a recovery of the happy innocence . . . in Eden."[21]

Monastic life traces its origins to seven hundred years of hermitage. This troop of spiritual seekers in the desert intended to subdue both inner and outer beasts. According to the apologists for the hermit saints, the outer beasts quietly submitted to the power of the spirit represented by the hermits. Of the inner beast, Psalm 91:13 says: "Thou shalt walk upon the asp and the basilisk: and thou shalt trample underfoot the lion and the dragon." These four beasts according to Honorius of Autum were actually sin, death, the anti-Christ, and the devil—exactly what the hermit went into the desert to defeat. The early Anglo-Saxon recluses were bred on a heroic, inherited imagery of combats with beasts and monsters not greatly different from the desert in which Anthony struggled with demons that slavered and snorted around him. The Ruthwell Cross in England is a rare record of seventh-century concepts, including a sketch of the beasts to be subdued and, eventually, their adoration of holy men. Christ stands on the heads of two beasts. Above him John the Baptist holds a lamb. Below him Saints Paul and Anthony are fed bread by a raven in the desert. The inscription says, "The beasts and dragons recognized in the desert the savior of the world," referring to Jesus' forty days and nights in the desert. But defeat of the dragons is scarcely the Christian basis for being "friends of animals."

As the hermits tamed the wild animals the latter became penitent. The statement in the Gospel of Mark that "Jesus was in the wilderness forty days and forty nights and with the wild beasts" may be interpreted quite differently than a desert in which he is surrounded by snarls and gnashing teeth. In art it is rare to see Jesus' wilderness sojourn represented with animals. An exception is a little sixteenth-century painting in the Metropolitan Museum in New York by Moretto Da Brescia, called *Jesus Among the Animals*, showing the animals in adoring humility around him, illustrating the doctrine that "the pure and sinless ones in the Kingdom of God will again dominate the animal world in peace."[22]

The critical interaction between the cave hermit and the local animals is not the hermit's intention but the beasts' recognition of the holy man: "In retiring to the wilderness, the hermit monk returns to a state of nature; he attempts to attain the spiritual felicity of Adam. He lives with the beasts and is recognized by them as a friendly being who represents a higher spiritual plane rather than as the born enemy of the animal kingdom, strives to overcome his impulses by constant self deprivation . . . to imagine his ascetic solitude as a recovery of the happy innocence . . . in Eden."[23] From being simply grateful, or adoring, the animal becomes a friendly helper.

The thesis that animals are spontaneously attracted and subject to holy beings in their wilderness sojourn is very old. Saint Columba in the fifth century and Saint Godric in the twelfth were known for their sympathy for animals. Crocodiles carried Pachomius across a river. Simon healed a blind dragon, finding "no creature beneath God's mercy." Cuthbert's wet feet were wiped by otters. A wild antelope gave its milk to the thirsty Macarius of Alexandria. Bitten by an asp, he tears it apart; but he heals and releases a little hyena brought by its mother in a sheepskin (which he keeps), making "the hyena promise not to harm the poor by eating their sheep" but only to eat carrion. Elijah was fed by a raven in the desert, as were Saints Paul and Jerome, the latter receiving a loaf each day from the ravens for sixty years: "And the ravens brought him bread and flesh in the morning, and bread and flesh in the evening."[24] The Anglo-Saxon Saint Erasmus shared the raven's bread also, as had a hundred different anchorite hermits in stories from Egypt. Nor was the tale just of Christian monks but of Muslim *welis* and Brahman and Buddhist ascetics living in caves with the animals around them.

As to the inner lion, the hermits cried: "What shall I do, for I am being driven to lust and anger? . . . It was to which David referred when he said, 'I smote the lion. I strangled the bear. I put away anger, I squeezed out lust with hardship.' "[25] And "David, when he fought the lion, held him by the throat and straightaway dispatched him. And if we too hold ourselves by the throat and by the belly, which covers the sexual organs, with the help of God we shall be victorious over the invisible lion."[26]

After Daniel and Jerome found their way into the lions' den, Gerasimus took a thorn from the lion's paw and Anglican saints Cuthbert and Guthlac of the seventh century followed suit. According to Moschus in A.D. 620, "all this was done, not because the lion had a rational soul, but because God wished to show how the first man held the beasts subject to himself, before his disobedience and his expulsion from Paradise."[27] In the retelling over the centuries, the lion gradually turns into the knight. Wild creatures spared Anthony's gardens. Paul shared his cave with a she-wolf. When he expired, his grave was dug by two lamenting lions (who sometimes bury their un-eaten leftovers). Others are kept warm and protected by lions. Lawrence S. Cunningham, in "Saintliness and the Desert," concludes that the saint's purpose had nothing to do with contacting or caring for animals.[28]

The foregoing examples are typical of the hundreds of "Lives" of the saints, neither more nor less true than Aesop's fables. Those who want to demonstrate that concern for animals is a fiber in the Christian bone can rummage through the tales of a thousand years about animals as perennial retainers and desert companions in rectitude. Even the model of Francis comes to us from the preceding seven hundred years of itinerant prophets, fugitive sheepherders, Thebaid hermits, and Irish saints. Combined into a litany, all these references to animals "document" the heartfelt community of sharing, caring, and reciprocity by which we imagine a better world.

Susan Bratton, an apologist for the ecological Christian hermit, says: "Providence is again seen as arising from or compatible with wild nature." The demons who took animal form were driven out of the cities by Christian prayers. Anthony was terrorized by them in the guise of lions, bears, leopards, bulls, serpents, asps, scorpions, wolves. This hunger to find Western tradition sympathetic to animals—and the passion to confirm that this has indeed been Christianity's true heart all along—is appealing. The flaw is revealed when Bratton says that Theon shared "his" water and Paul "his" cave with a she-wolf: the irony of the possessive pronoun speaks for itself. Legitimating the saint/beast stories as a core of human/animal bonds, Bratton links them with the "personification" of animals in Mediterranean fable, the shamanic traditions of animal helpers, Egyptian theology, the animal avatars of the Greek gods, and so on. Thus do the hermits ride on the coattails of shamanic flight and Egyptian polytheism.[29]

But the saint/beast stories do not justify the implications of the fawning lions and servile ravens, which become like faces in a congregation. It is difficult not to sympathize with Bratton's bold attempt to vindicate Christian "ecology." Who would not rather find that their familiar culture supported a true respect for biological diversity? Again and again, in essays of diligent scholarship, she shows us how the saints loved and protected animals, but we are swamped with hagiography and its one-sided perspective. The hermits "who made friends with gazelles, wild asses, and wild goats . . . that kept wolves in their cells and slept with lions" are just a little too nice to be true.

These tales have their parallels in Islamic and Hebrew traditions as well, a veneer of "documentation" to transform the myths into history, animal fantasy, eulogies in the form of chronicles. The legendary Celts and monastic and desert hermits no doubt lived in close proximity to animals; but the idea that their Christianity was exercised as a "love of nature," of animals seeking forgiveness and doing penances, is dubious at best. The big animals, "those most likely to provide certain services such a bringing a fish to a saint," were prominent; small things were ignored "because they had sinister reputations in folklore." Thus the downside can be attributed to those superstitious pagans.[30]

The strangest aspect of all the stories of saintly hermits and their slaves is not their content (myths around the world are far stranger) but the presentation of hagiography—ecclesiastic anecdote—as history. The events are recounted as if they were a literal chronology. They are a textbook demonstration of the incapacity for metaphor, the inability to accept a plural metaphysics, the absence of a sense of playfulness and irony, and blindness to true otherness when animals are subordinated to a morality. Ignoring the ecological relations and wildness of animals denies their independence and their being. The uncanniness of animals then becomes alienation and diabolism. It is this fear of otherness that turns the wheels of all monomanias— monotheism, monologue, monopoly, monotony, and monogamy—transforming the spirituality of animals into rectitude and submission. If the hermit-saints' tales are intended to be understood as truly mythic, then the meaning of the myth is the reaffirmation of the hierarchy of God, man, and nature, all controlled and valued from the top down, and a vacuum of inherent interest in the bottom layer. These stories are a form of the Western preoccupation with purity, anxiety about pollution, and the "lost" perfection, in the midst of a fearful wilderness.[31]

It is not hard to imagine the conditions of life for such ragged ascetics in North African caves and Irish bogs, demented with loneliness, ravaged by the gnawing of hunger and sexuality, feverish with infections from lashing themselves with dirty whips, wretched with the lice in their hair shirts, hearing and hallucinating lions—lions to be tamed inside and outside. Of the beasts outside, we can imagine the saints to be lying on their pallets thinking of all those lazy birds and drowsy cats—each should just bring a little something like the ravens that supported Elijah. Like the working animals on farms, the little beasts seem to owe it to us. When a bear killed Saint Remedio's ass it was put to work as his mount. Saint Gall's bear had the same contract for lifetime servitude without getting anything to eat. As James of the Bible saw it, tame them all: "For every kind of beasts, and of birds, and of serpents, and of things in the sea, is tamed, and hath been tamed, of mankind."[32] Otherwise they can keep the thorns in their feet (which are probably abundant because of the ecological degradation of the plant community due to overgrazing by domestic goats). If the hermits rightly deserve animal service in the desert, are not the missionaries in the cities equally deserving?

Apart from the conceits of the apologists, some hermits were actual persons who lived in desert caves to be "with the wild beasts." But this did not necessarily mean that beasts were at the door. Urban people always suppose that the wilderness is crowded with ferocious predators.[33] When they go there they are at first fearful, then surprised and disappointed. Most mammals, and many birds, are shy or nocturnal. Life must have been very dull for the desert hermits. More perilous than physical danger was *monotony*. Caves are highly desirable dens in arid environments. Anyone there long enough begins to notice the small creatures around them and, as they get tamer, the larger. From ants, scorpions, crickets, beetles, toads, bats, mice, foxes, swallows, flycatchers, ravens, nightjars, to jackals, hyenas, and even lions, such extraordinary company becomes attractive in one's loneliness and, for many, fascinating, not because of religion but for ordinary human reasons. The joy the recluse finds in such a wild troop is not surprising. The pseudo-historical claims that the beasts kowtowed to the hermits must allow that life in a desert cave can become an extraordinary biosocial experience and that cave-inhabiting people, especially born-again ascetics, may exaggerate the motives, behaviors, and attitudes of the animals around them. Later writers about the recluses—themselves bookish—generate tall tales which are embellished through the generations, to the benefit of the hermit's holiness.

THE SADDEST THING about the barnlike ark, the sainthood with its doting lions, and the green tunnel of the Garden of Eden is that they are so boring. The dream of animals as supplicants is desperate compensation for estrangement: self-protection in a final curl with a warm puppy. In most fundamentalist heavens there will not be much to do but join the throng praising and praying: "Every creature which is in heaven, and on the earth, and under the earth, and such as are in the sea, and all that are in them, heard I saying, 'Blessing and honor and glory and power be unto him that sitteth on the throne, and unto the Lamb for ever and ever.' "[34] One wonders whether the curse of toil— "persistent, unending, and fundamentally at odds with humankind's propensities as shaped by the hunting experience . . . the lot of all farming populations . . . a permanent enslavement to an unending rhythm of work"—would not have been better than the tedium of endless hallelujahs and adoration.[35] Only the pacified and sedated forms and their human cohorts could be so utterly banal. What the peevish Isaiah wants is to put down discord among the animals (by force if necessary): to take away their voracious natures.

Carping at the Bible, though, is like raving in an empty hall. Why punish the old desiccated words or flail against literal-mindedness? The answer is, partly, because the images continue to seduce the imagination and partly to repudiate the premise of the idiot beast. We can reconcile the Peaceable Kingdom with the realities of bloodsucking, infecting, devouring, and competing wildness by recognizing that the Peaceable Kingdom is a yearning, a literal misunderstanding of a good myth. Protecting, cooperating, and tolerating— "lying down together"—are social euphemisms for harmony. All the predation is evil only because the myths of the garden, Noah, and the saints lack acknowledgment that the world of plants and animals is a comity based on death as well as life. The Christian stories cannot accept the ambiguity of sacred beauty in the chain of predation, the paradox of the plurality of truth, which the animal counterplayers represent.

An older myth precedes hagiography: the story that, beneath the masks of feathers and fur, the animals and plants are a sacred society, rich in marriages, festivals, speech, commerce, the whole range of social intercourse. The myth was a poetic, esthetic, narrative instrument and creative translation. It honored death as a social bargain because the myth's true function was community. It honored the creatures in their difference from us and their genius. Without domestic animals, primal peoples did not confuse the fabulous confederacy of the wild with their own groups, clans, and moieties, but took them as images in a magic mirror, the ecological reflection of a social mode. The whole of natural relationships and forms constituted a discourse, a necessary truth having significance for meditations on human society. In this style of consciousness, the buffalo grazing quietly in the presence of lions is both a reality and an idea; the rabbit hunching, immobile in its crypt of grass beneath a passing eagle, a coyote eating some afterbirth, a fat tick falling onto the oak leaves from the deer's belly, all are a richness of ideas pouring into the world. Amity in this view is affirmation of the symbolic face of a universal metabolism, which is a balance of dissolution and construction. Discord in nature is superficial: behind it is a deeper concord, a reconciliation brought about not by the rod of Jesse or any other outside power but by intrinsic, mysterious agreements. The whole is bound in compliance rather than struggle, affirming order under apparent turmoil.

Thinking this way depends on keeping a certain distance. One does not bring the animals into an ark, garden, or cave, supervise or project ideals upon them, or go out to graft social justice into the biological realm. A leap is made from something given to something understood. This separation of the physical event and its life in the mind is so important that, among primal societies, close physical contact with animals was widely thought to be suffused with dangerous spiritual harm. Thus all physical approaches were undertaken with the courtesy and circumspection due to passages between realms, cautiously on customary bridges, usually through religious art.

Jung calls attention to the lack of theriomorphic elements in Christianity.[36] The Christian/Muslim/Jewish defeat of animals is the emptying out of their souls, leaving shapes of animated dust, innocent relics of a lost world. As past and future history, Peaceable Kingdoms, Spiritual Slaves, and Arks, stuffed with pacified flesh, are merely wrecked metaphors, relics of the impulse to observe and be part of the natural world, bungled projections upon that world rather than attention to the prodigality of its thought.

IN THE LAST quarter-century, theologians and others have written hundreds of books and articles defending Christianity against the charge that it is at the root of our "ecological crisis." I think, for example, of Paul Santmire's beautifully argued counterthesis.[37] Virtually all of the argument, on both sides, has focused on biblical and other doctrinal texts. What the partisans overlook are the liturgical practices: the loss of religious services that include masked and participatory dancing, the drum as a sacred vehicle, and other rites that incorporate the soil, plants, and animals in a ceremonially recreated cosmos. Only occasionally in local parishes do priests go out and bless fields, do flowers decorate the altar, are seasons acknowledged in thinly disguised, old, pagan ways recognized for what they are—baptism bringing one forth from the water, coronations as metaphors on birth, burials as spiritual as well as bodily return to earth—and seldom are animals brought to services. If the formalities approved by Rome or those accepted by Martin Luther are biologically the most barren and arrogant metaphysic, then the bending and infringement of these formalities in local practice offer the best hope for change in Christian cultures, however secular their outlook has become.

18 | Hounding Nature: The Nightmares of Domestication

There are enough mild
dull eyes of domestic brutes that we have bred
from bird and beast
to make them part alive and partly dead.
A thousand generations in a cage
makes a helpless thing.

MICHAEL MCCLURE

A 1910 ISSUE of *The Country Gentleman* contains an article entitled "Which Is Man's Best Friend, Dog or Horse?" Even allowing for millions of cat-lovers, the question would still get a response in a pub or on a television talk show. The discussion would very likely miss the baneful side of the question, however. Indeed, there is the extreme alternative that they are false friends of man. As we shall see, canine and equestrian history starts with domestication and ends with cornering nature itself, which, contracted beyond endurance, turns on its pursuers.

Horses and hounds, as destroyers of nature and humankind, can be understood more fully in the light of the history and agendas of domestication of all animals. As the presence of the dog as hunting instrument unbalanced the ancient equipoise of the hunt by Pleistocene hunters, the obsession with kine by pastoralists shaped the new regime under agriculture. It is generally believed by archaeologists that, along with dogs, the sheep and the goat were the first domesticated animals. Various theories address the means by which the first shepherds corralled wild sheep and goats.

When author John Berger calls for an end to the "marginalization of animals" by hailing "the only class who, throughout history, has remained familiar with animals and maintained the wisdom which accompanied that familiarity, the middle and small peasant," his normally brilliant intuition has been bent by the bucolic fiction—the cultural blindfold against the brutality of the herdsman and peasant who breed, castrate, brand, bob, and hobble farm animals and kill their wild cousins.[1] This modern instance of the fable of the wise peasant is a variation on the story that the manure-sunk barnyard is sweet. Perhaps the urban debacle has so overwhelmed us that the perverted animals look good by default, and we clutch the Myth of the Husbandman in desperation, ignorant of its distortion and its destruction of the wild world.

There are many definitions of domestication—captivity, tameness, utility, control, but these are mostly metaphors. Biology is relatively clear: domestic animals are genetic races derived from wild ancestors to whom they are closely related and from which they are very different. The earliest domestication of plants and animals is a boundary marker that distinguishes the era of primitive agriculture from the whole of prior human existence. While the capture and taming of individuals may have taken place thousands of years before agriculture, there is no evidence of the controlled breeding of them until the Neolithic about twelve thousand years ago. By selective breeding their owners created forms with altered tractability, size, strength, appearance, fertility, milk, meat and egg production, ferocity, and speed. The same program continues today. Desired types are kept; the others are destroyed. In the process, animals inadvertently lose the exacting combination of adaptations to the community of wild things, complex social behavior, and that poised vigilance which constitute wildness. In place of wildness are simplified reproduction, reduced resistance to disease, and loss of intelligence. Their "bred-up" characteristics are passed on to their descendents, who, blunted in mind and body, become increasingly dependent on human care.

The environmental requirements of wild animals are so complex that many will not voluntarily breed in zoos. Wild-caught ducks and geese will not mate in captivity for ten or fifteen years after capture, even under optimal conditions, whereas domestic forms breed in the second or third year of life. There is excellent evidence from the archaeological record that the earliest domestic hoofed animals—the sheep and goat—have diminished cranial capacity, hence smaller brains, than the wild mouflon and Aegean wild goats from which they were derived.[2] In general, domestic animals are fatter and heavier among birds, plumper but smaller among cattle and pigs, larger and more slender or more muscular among horses, and stupider among them all. Maturing earlier with diminished pair-bonding and mating rituals, more variation in color, infantilized and ecologically displaced, they can survive only in protected places because they can no longer compete with their wild cousins, avoid diseases and predators, or endure the physical hardships of the wilderness.

The selective breeding for a single trait breaks up a complex of genetically based and finely tuned characteristics that adapt animals to an ecological niche. As more generations are bred for "pure" type there is an accelerated disorganization of the harmonious genome, or stable form. "Improvement" in the breeder's terminology carries in its train innumerable faults of these kinds, not all of which show up at once, so that later generations are subject to an ever-widening ripple of genetic instability. The growing proportion of dysfunctional individuals among the offspring requires intensified care or more rigorous junking of undesirables. The deformities run into the dozens and then the hundreds. This means that for every "thoroughbred" racehorse at the track, others must die; for every "breeder" hen, hundreds of offspring live out their lives in small cages without breeding, their physical and behavioral deformities hidden from public view.

The domestication of plants and animals was the first genetic engineering. The questions of "biological ethics" focusing on the micromanipulation and transplanting of DNA in human fetuses, cloning, experiments toward the heritable alteration of individuals, and the "creation" of new races is actually part of a project that has been going on for some ten thousand years. Today, the four groups of domestic animals—livestock, pets, experimental, and feral—have in common a radical break from their ancestry. Livestock, grazing the ground bare around human settlements—as one can see in a thousand African villages today—denude whole regions as herds increase, becoming the "hoofed locusts" of the subtropical world. Three and a half billion cattle, sheep, goats, buffalo, and camels trample the planet. One-fourth of the global warming of the atmosphere is due to methane, 500 million tons of which is flatus from the guts of these animals.[3]

Some species and breeds are less impaired as whole animals than others. These still have gene complexes that adapt them to life somewhat similar to that of their wild cousins. The horse, house cat, camel, reindeer, and certain dogs can survive in the countryside, even wild places. Let loose or abandoned, they do not regain their biological wildness but become feral interlopers, predators on native species or grazing destroyers of habitat. Uncontrolled populations of them create their own destitution, a plight that seems to impose a moral obligation on people, at a largely invisible cost.

When I returned to western Illinois on a visit in 1985, having moved away twenty years previously, I was aware of an uncanny difference in the countryside which I could not at first identify. Finally I realized that a sound was missing—the banging of the metal lids of outdoor feed hampers. On hundreds of farms, pigs had vanished from sight and hearing. Now there were new, long, narrow buildings, into which they had been banished forever.

Eventually all domestication imposes such machines. As the domestic animals in the village were caught in modern life, it transformed their lives just as it did the lives of people. The diverse farmyard was swallowed up by corporate agriculture, the stockyard replaced the local abattoir, cattle trucks appeared on the highways, the kennel replaced the shed as production line and infirmary, the animal shelter disguised the execution of the castoff pets, and all the owners were increasingly faceless. Ruth Harrison's 1964 book on the cruelty of factory farming, with special reference to the industrial rearing of chickens, marked a new turning in the two-centuries-old English fight against the inhumane treatment of animals.[4] The techniques of assembly-line production, perfected by Henry Ford in the making of automobiles, had found its way onto the farm. Where once animals had been beaten or starved or allowed to suffer unprotected in the winter winds, now they were untouched, restrained, or confined in tight little cages in well-lighted buildings, inoculated, hormonized, and stuffed with manufactured "foods." Suddenly the humane consciousness had discovered that there were things more terrible than neglect.

No one doubts the earnest desire of agriculture to improve the lives of people, but the biological refinement and "perfection" of animal breeds approaches a grotesque final act. The rationalizations and justifications generate fantasy even among scientists. An editorial in an issue of *Science*, for example, dedicated to "species engineering," says: "Over evolutionary time the friendliest of wolves (and possibly the most intelligent) learned that wagging their tails and delivering slippers was an easier way to earn a living than hunting caribou in the wilds. . . . An original wolf might say to the dog, 'You have lost your freedom. Your obsequiousness is humiliating to the family Canidae.' The dog could reply, 'I am much less warlike, far more altruistic, and besides, it's a wonderful standard of living.' Whether society prefers to have wolves or dogs remains to be seen."[5] The writer's down-home, tail-wagging, and slippers stuff falsifies what science and even school kids today know: wolves didn't decide to become dogs and don't want to be dogs. The pipe dreams about "warlike" and "altruism" and "standard of living" are sneaky exploitations of our friendly bias toward dogs. As for subservience, the worst thing about the editorial is its breezy participation in the larger fiction that everything is getting better and better through genetic manipulation of plants and animals.

The theme of that issue of *Science* was genetic alteration: a leading article on gene management to cure human disease, followed by one on livestock, with nearly everything else on plants. For ten thousand years such "engineering" has involved recombinant DNA—which is to say, controlled breeding by dam and sire selection, castration, sterilization, insemination, and quick slaughter for the nonbreeders. The new engineering deals in direct gene alteration: duplication or mutation by radiation or other means; or gene splicing and transplants. The eight-authored article on livestock indicates a keenness to increase growth hormones in the blood of cattle and pigs. The results were pretty much the same as ordinary breeding: "improvements in daily weight gain and feed efficiency, reduction in subcutaneous fat." For centuries we have been eating meats poorer in nutrition and higher in the wrong substances because somebody wanted quicker-growing animals that would survive in barren environments. The effects on people were high levels of gastric ulcers, arthritis, cardiomegaly, dermatitis, and renal disease. And the animals themselves live out the nightmares of chaotic genetics which took their wild ancestors millions of years to adjust as a complex.

Another direction of genetic engineering is the implantation of human genes into the embryos of pigs, apes, and other animals so that certain tissues and organs develop with closer genetic similarity to the corresponding organs of people. When needed for transplanting into people, these organs are less likely to be rejected. Donor animals are therefore little more than organ gardens. Some people will complain about this "sacrifice" of animals, but it is only one more reason for the billions of animal deaths and centuries of their perversion and confinement for human use and amusement. That the animals are not "happy" in their lifetime may be difficult to prove, as they may be fat and live in protected quarters. Food markets of the twenty-first century may display "pork," "beef," "lamb," and "fish" which were never part of a whole animal but were cultured as edible tissue in a "meat" factory. Such developments represent the end of a long trajectory in which the otherness of nonhuman life is diminished, deformed, and obscured, first as captives and domestics, now as assemblies of available substances and organs.

No TRACT ON animals should omit the obligatory thousand words about Wonderful Dogs. As a dog owner I know the pleasures of canine companionship—the intelligence of the helper, the hedge against loneliness, the loyalty of the guide, the courage of the guardian, the child's friend and king's only honest subject, the medical sacrifice, even the pièce de résistance at feasts around the world.

But the dark side has been there all along and has not escaped attention in the past. It is not simply the flea and worm carrier, the mad dog, the biter of pedestrians and killer of infants, the vicious pack killing sheep, the consumer of canned kangaroo and old horses, or the nuisance on the carpet, but the black dog of hell. It is Hecate as the Black Bitch, Sarameyas, the Vedic death-dog, Odin's Geri and Freki, Cerberus, all the skulking corpse-eaters that ever haunted village streets in times of war, pestilence, and famine, the feral demons, D-O-G, the opposite of G-O-D. Centuries of perceiving human evil and sickness as canine incarnations of the devil and chaos released on earth, their mangling of dying soldiers, the auguries from their entrails and omens of their howls—all are shadows of the sinister underside.[6]

Our nearest animal friend is also the most uncanny. Dogs have long been regarded as having special powers. The howling dog was widely thought to be an omen of impending death, associated in witchcraft with the full moon. There are many rites of aversion by which animals are given the group's burden of problems and then driven away. These are "scapegoats," but goats may have been a second choice. Moreover, the gallbladders, blood, teeth, urine, and genitals of dogs have been used in healing practices—as evil against evil. The ancient Greeks, for instance, pressed a puppy to the diseased part, in the hope of transferring the disease to it, then killed it.[7] The Japanese sacrificed dogs for rain and read auguries in their viscera.

As has been pointed out in Chapter 4, anthropologists have been addressing for years the ambiguity of the dog: the loathing of it in some societies and affection for it in others. This is generally explained on the basis of its liminality: both household occupant and outsider, protector and assailant, noble and incestuous, resolute in its loyalty and filthy in its habits, intelligent and yet four-legged. These ambiguities lend a double edge to the dog which may be bent either way to suit the culture's needs. The ancient Middle Easterners rejected the dog as vile and polluting, although their Neolithic forebears had depended on it for herding livestock. When the Zoroastrians came to power they capitalized on the dog's prominence but reversed its valence. Instead of emphasizing its disgusting habit of digging up and eating the dead it was incorporated into mortuary rites. They likened the dog to fire, both protective and destructive. The hound, they said, penetrates and links this world and that, reaching across the edge of life into death. It hears and smells what man cannot, aware of those presences invisible to humans. In the Zoroastrian ceremony of the *com-i-swa* the dog was given consecrated food as an act of endearment, making it a sacred vehicle to nourish the soul of the dead. Something of this appears among the Hindus with a negative emphasis in which ritual foods are offered to Yama's four-eyed dogs of death as a kind of hedge against their hunger.

With the coming of Islam, Persian customs reversed once again, making the dog an abomination according to Shiite orthodoxy. If touched it requires a purificatory ritual. Islam expanded the Judaic tradition from Leviticus in which the dog is unclean, although all right to touch. The satanic place of the dog continues in Islam—witness a 1985 newspaper article in Tehran saying that television cartoons of dogs were a Zionist conspiracy to corrupt Muslim children, just as the Zionists had corrupted the Christians by teaching them the love of the dog.[8]

But the Arabs loved their hunting hounds so much that these breeds became an exception among the dogs: still ambivalent but central to their religious imagery of the holy conqueror. The hound and horse were companions in the myth of a messiah who vanishes into a mountain "in anticipation of a future epiphany," a royal hunter who lives beyond death and will rescue everyone. That myth tells of a quadrupedal steersman to the realms of life and death—the mortal king with his divine connections, the hunter who ensures life while dealing death, the hound who guides and protects the hunter's passage between worlds, and the horse who is the mount of both god and death. It is the final irony of the metaphysics of the hunt as a holy quest of assimilation in the cycle of existence that it should devolve into a myth of "chasing after a form of survival that no longer depends on death, hunting game not for sustenance or entertainment but for immortality itself," and therefore denying the cycle of life.[9]

The messianic zeal for a divine savior who is also king has been a Middle Eastern obsession for five thousand years. Some familiar, others little known, historical and mythic heroes were thus portrayed as mounted hunters—from the Assyrian kings to Mithras and Herakles, the Twelfth Imam, Frederick II, Frederick Barbarossa, Siegfried, and so on to every statue of every general in every plaza from Kazakhstan to Lima. How are we to understand our loyalty and love for these freaks and travesties of the wild forms? And why idealize these ecological wrecks as the best of them all?

UNLIKE THE DOG, the horse is not cognitively or metaphysically ambiguous. Its dark side lies not in its direct harm to people but in its employment in killing and enslaving people, in herding livestock in large numbers beyond the geographic and physical limits of habitat, in centralizing political power that destroys self-rule, in dismembering families and communities to an almost unimaginable extent. All of this came after many thousands of years of unmitigated admiration and attention to wild horses.

It is no wonder that we love them. Along with elephants and aurochs, horses made us human.[10] They were the object of our first thought, forethought, and hindsight. The great beasts were killed only by the most strenuous mental work, the capacity to plan and execute the hunt, and recapitulate by dance and narration.[11] According to André Leroi-Gourhan, the pairing of animal species by Aurignacian artists in the Paleolithic caves made cattle and elephants feminine, horses male. Like the respect for snakes, the love of horseflesh reaches far back into prehistory. Among the finest sculpted and painted works of the Pleistocene are those of horses. For a thousand centuries the principal meat of Europeans and Asians shifted among a half dozen large game animals—bear, reindeer, mammoth, deer, rhino, cattle, and horses—but in all that immense span only the horse bones appear in every archaeological stratum, while the others come and go. Everything we know about such hunting leads us to suspect that our ancestors had "horse" in their stomachs and in their heads in every sense: nutritional, metaphysical, esthetic, and metaphorical. Truth of the invisible and eternal were to be found in the natural world. And of all the animals the horse may well have been the most elusive and intelligent, the one deepest in human dreams and imagination, most challenging to our imagination, and therefore evoking the most eloquent responses. If there is some path by which the fervor of a hundred millennia of doting, scrutiny, love, and pursuit can find its way into the unconscious, surely the horse is among that rare fauna animating our inmost selves. In nature there may be no more thrilling figure. It is impossible to stand next to a horse even now and look at its marvelously sculpted head and face without being excited and moved.

Perhaps this prehistoric perspective can help us understand what has followed: the flowering of human and horse companionship, the convulsive dreams that haunt us still, from knighthood to the steeplechase, from centaurs and unicorns to winged horses. The tragedy in the subjugation of the horse is all the more difficult to see because of a love affair going back half a million generations.

Close up, the horse makes the heart beat faster. The sleek, clean coat has a voluptuous touch. The sensuousness of horses cannot be wholly separated from ideas about them or from analogies of their physical presence. The randiness of mares in heat, the lust of studs and stallions, these are familiar to anyone around horses. Like bulls and cows, this "laciviousness" seems exaggerated, perhaps for no other reason than their size and strength.

Horses in captivity, unlike bulls, enjoy being stroked. One rides passionately. Horseriding is a source of vestibular sensuality and genital stimulation, a dreamlike, vicarious flight, not unlike that of narcotics and intoxicants, like the rocking of a bed or a ship. No one who has ridden a horse at a gallop will have difficulty appreciating the physical stimulation, the sense of flight traditionally given empathetic reality in the idea of birds and riding the winged horse. Those who ride winged horses in stories are transported in both a geographical and psychic sense—the Pegasus effect. The energy of the rearing horse is like the rise and fall of sea waves as they break upon the shore, followed by the hooflike clatter of the backwash over the shingle. Even the swell that surges before the break suggests the momentum of the herd or the cavalry. Miniatured in the little seahorse fish it is native to the ocean. The color of individual horses is always a focus of attention. In legend these colors are often fantastic, as in the tale of Fergus, an Irish warrior, son of Ro-ech (Great Horse), in which a magic horse carries men across the sea above the reach of monsters. It has a golden body, flashing eyes, crimson mane, green legs, and a long tail. Colors signify and encode this equinine noesis. The various flags and colors of modern racing stables worn by jockeys, those borne by cavalry, and all the flags and banners connected with horse equipage seem almost to be a language of feeling.

There is the sweat and smell of it. In view of these sensuous and erotic qualities, it is not surprising that there are numerous copulatory ceremonies, dances, and mythological matings of horses and people. Fergus was said to have enormous sexual capacity, huge genitals, and needed seven women to satisfy him. Other Celtic legendary heros are linked with horse goddesses in a welter of interconnected foals and human infants, sexual relations and transformations.[12] The passionate association of horses with women, nubile and otherwise, is well known; one thinks of Robinson Jeffers' poem "The Red Stallion." But men too are enraptured, as in Peter Shaffer's play *Equus*. That a horse should also carry one away in sex, out of the mundane world, or to the other world in death are familiar mythic themes, represented in marriage and funereal processions.

It seems at first incongruous that the play *Equus* includes so much horse killing, severed heads, and the maiming and abuse of horses. Yet the great horse cultures of the world venerate the horse in ceremonies including sacrifice. Unlike our own times, in which the secular and psychological importance of horses has replaced its religious significance, such death dealing must be understood in the wider context in which the death of the sacrificed animal is a prelude to a different or renewed life. Such traditional, ritual use of horses extended from Scandinavia to India.

In Celtic cultures the horse shared eminence with the ox, stag, and crane. In bronze age Europe, from Rome to Britain, Epona, the horse goddess, was a major deity, closely associated with dogs and birds. The huge figure of a horse cut into chalk in Berkshire, England, in iron age times is closely related to English-Irish goddesses after Epona. Throughout the Celtic world, in motifs on household objects, bronze figures, and temple sculptures, horse heads suggest decapitation rites. Such a rite was described from twelfth-century Ireland as part of the inauguration of a king who copulated with a killed mare, invoking the power of fertility by bathing in and eating soup made from its body. In iron age India there were burials of horse skulls and limb bones in the megalithic stone circles of Maharashtra. At Pazyryk, in the Altai of western Siberia, horse bones were found in tombs decorated with reindeer masks and elk (American moose) antlers, testifying to the fervor with which the old hunting cultures transferred reverence from their holiest of animals to the newly acquired horse.[13]

The oldest domestic horses are known from northwestern China at about 4,500 years ago, although men may have put bridles on horses two thousand years earlier, based on a single fossil skull showing tooth wear that only a bit could cause. The domestic horse was used at first by grassland peoples of Central Asia and Indo-European nomads from the steppe country between the Black and Caspian seas. To the north, the Scythians, who used a similar trap for catching both reindeer and horse, were reindeer keepers before they became horse people nearly five thousand years ago.

Nomadic pastoralism may have begun with the herding of horses in Central Asia, spreading south with cattle herding after 3000 B.C. Walking, a herdsman can only watch so many animals and cover so much area, but being mounted magnifies his reach. Without a mount a Mongolian shepherd can manage one hundred and fifty sheep; with it, five hundred. Two mounted shepherds can control two thousand sheep or two hundred horses. New horizons appear in every sense. By the same token, the horse was the means of centralized political influence and communication. It was brought to Mesopotamia from the north or northeast in the fourth millennium B.C. In the region bounded by the Black Sea to the lower Volga, the first elements of nomadic pastoralism, pit graves, appear. Until 3500 B.C. the goat-sheep complex, with a few cattle, permeated the Middle East and Egypt. By the turn of the first millennium B.C. the Eurasian steppe had been drying up for over a thousand years. Along the Euphrates, the Sumerians began substituting the onager for the ox a few centuries thereafter. From modest beginnings, on foot with their sheep, the pastoralists became mounted cattle-drivers with iron weapons. By 2000 B.C. the Indo-European Kassites and Elamites had conquered northern Persia, Hittites with horse-drawn war chariots had appeared, and the Hyksos had headed for Egypt out of Asia Minor. Armed mounted migrants from the Asian steppes and the Ukraine transformed the Near East, the Mediterranean world, and established certain bedrock features of Western Civilization.

The old four-wheeled Sumerian chariot drawn by asses gave way to the light, spoked-wheel chariot of Kassite Babylonia in Egypt in 1800 B.C.[14] The Kassites were the major horse people of the Near East, but it was the Assyrians who made the horse the means of organized war. In Assyria the Mitanni produced tablets on horse training in a Sanskrit-like language, the speech of the Aryans. The Mitanni worshiped Varuna, Indra, and the Nasatya twins, who appear with the horse later in India. For them the weather god was the principal deity, opposed to an earth dragon, Illuyankas. The collision between horse and snake had begun. The mythos of the snake would successfully resist the horse in India, but it would collapse in the West.

Indo-European and Aryan nomadic peoples spread through the Middle East, south into Egypt, and east into India. After 1800 B.C., civilizations throughout the ancient world were repeatedly shattered and sometimes completely destroyed by cavalry. Each invasion by the Scythian/Assyrian horsemen produced "widespread abandonment . . . a nadir in settled life" and a "sharp economic and demographic entrenchment." In the second millennium a crucial phase began with the shift from bronze to iron. The Hittites in Mesopotamia, Luwians in Anatolia, Achaeans and Dorians in Greece, Aryans in India and Old Europe—all overran the towns and cities, abruptly ending the traditional crafts, sacking the palaces and going home with their booty or settling to build forts, becoming themselves sedentary, and eventually attacked by other horsemen.[15] For centuries, incursions by the mounted descendents of nomadic pastoralists were "a new living war machine, the armed man on a horse." Diverse as they were, these horseriders shared an ideology of master and maker of animals, economies of stock breeding, and a remythifying patriarchy. They established the model even for nonhorsemen, self-styled "pastoralists," such as Hebrew invaders into Canaan, ruled by a caste of Levite warrior priests, hierarchic, male dominated, violent keepers of the lore of conquest and slavemaking, bent on "utter destruction." [16]

Ishmael and his followers were associated with horses. King Solomon, the archetypal horse dealer, kept forty thousand chariot horses, more than twelve hundred in his own stables, in defiance of Judaic law, which forbade them as idolatry. The sedentary and seafaring peoples of classic Greek civilization were periodically invaded and transformed, beginning as early as 2000 B.C., by "a wave of shamanist horsemen." Consequently, the Greeks acquired horse pastorality with its organizing power and masculine authority, subduing the native power of the goddess in Greece along with her earthly avatar, the snake. Horsemen ruled over healing in all its traditional forms—if not by removing the feminine practices then by controlling them.

Mohammed (d. 632) was captivated by horses as well, which were central to the Jihad, the holy war. He incorporated the old Arab worship of the horse by the idols Ya'uk and Ya'bub. Horse keeping became a tenet of the faith and was involved in many sayings of the Prophet. The horse in the Koran is "the supreme blessing" and was the machine at the heart of the conquest that swept the West out of Persia and drove to the Atlantic Ocean. A horse, "Bukraq," carried Mohammed to heaven, just as a mare will be ridden by Jesus at the end of time.

Among the Old World nomads using horses were the Scythians of the seventh and eighth centuries, horse keepers and milkers whose burials sometimes included hundreds of horses, the Huns of the fourth, fifth, and later centuries, the Avars of the sixth century, the Hungarians of the eighth and ninth centuries, and the Turks of the tenth century.[17] The Turks are the horse lovers who invented the modern noseband that blocks breathing, as well as users of the metal bits with serrated mouthpieces favored by the Assyrian nobility.

The sequence from Neolithic hearth to priestly control to the political state reflects centralization made possible by the horse and given mythic expression in the stories of mounted heroes. The thirteen snake deities of ancient Egypt, the uraeus and caduceus archetypes, the guardian figures of temples, rural households, and sacred springs, all were trampled into dust by equestrians.

WHOLE EPOCHS OF Western history could be organized under the banners of Horse and Snake. The love of horses by women has a special irony in these times of gender sensibility, "women's studies," feminist politics, and its critique of "male" values. Considering how many women like horses, the historical alliance of horses with patriarchal cultures seems paradoxical. It was not only Jews and Christians who promulgated snake-hatred and horse-worship; all pastoral societies hate snakes. But if the horse is not a true feminine symbol and merely a projection of aggressive masculinity, is then the snake woman's best friend? The answer is a qualified yes. But its unlikely acceptance as such by many women marks the degree to which "stud" values in modern society, projecting the fear of snakes and the love of horses, with their assimilation of three thousand years of snake demonology and pastoral propaganda, compounds the insult of being mythically associated with snakes with that of being disqualified as spirit kin of horses.

No other two animals clash with such vivid fallout in art, legend, and mythology. Wherever there are cowboys or other horse pastoralists—in Australia, the U.S.A., Argentina, the Ukraine—there is a public display of loathing of snakes, just as the enchantment with horses expresses a profound consensus. In the long evolutionary epoch and natural history of snakes and horses however there is no sign of conflict between them. It is a struggle in the human mind between symbols—part of the myth of societies who take their cultural style from mounted nomads, even when they themselves have long since become shopkeepers.

Against all this mob of horsemen, astonishingly, in India the snake did not lose its sacred place. *Nagakal*, the stone forms of nagas, half-human/half-serpent, which embody the snake incarnations, are still profoundly revered and ritually attended, as are live snakes. Despite the equestrian hubbub of Aryan invaders, the horse faded: too wet, too hot. Moreover, the indigenous Indian capacity to absorb and accommodate outsiders is seen in the failure of horse culture to sustain itself in the village agriculture of the monsoon subtropics. It is as though India's diversity, like a vast sink, overcame the Vedic/Aryan nightmare, even when it was abetted later by Islamic monotheism. Today, in rural southern India, snakes inhabit the still-active

Siva shrines, composed of "sacrifices" of clay horses, as though here only, in all the torn wounds of the earth, the snake and horse had recovered their prehistoric neutrality.

In the West, however, the association of woman, snake, and horse was given mythic scope in the preclassical Greek myth of the Medusa, the snake-haired Gorgon who was decapitated by Perseus. In him was embodied the great betrayal of the older traditions of the goddess. He was the crucial figure in the arrival of the Olympian gods—Zeus, Poseidon, Athena—as well as the city-state and its culture after about the seventh century B.C. He was "born" in northern Greece, where village traditions emphasized the powers of shamanism in spiritual trance, health and protection through magic, the presence of ancestral ghosts, matrilineal traditions, and the dominant sacred image of the goddess with her temple snakes. Perseus was part of that ethos, garbed as a shaman with wallet, cap, sandals, and shield. He was helped by Hermes, who had a parallel heritage. Hermes was reared by *thriae*, or bee-maidens, who taught him soothsaying. Healing was one of the arts of such diviners, and shamanism was not dominated by men but a calling open to anyone. In trance, the healer's spirit was said to leave the body and travel to the Other World in flight, seeking divine help in defense against famine, disease, or other evils.[18] That village tradition was ultimately denied and betrayed by Perseus and Hermes.

The reality was that Persian, Old European, and Mediterranean societies, small-scale, isolated and sedentary, were overrun by pastoral invaders. In time, keeping horses, horse sacrifices, gifts, exchanges, and wagers came to be part of life. Horses turned farmers into mounted herdsmen, tax collectors, and warriors. The idolatry of the horse and the repudiation of the snake was played out in this story of Perseus. Perseus and Hermes became the founder heroes of the new Mediterranean patriarchal society. Perseus refused to offer a horse gift at the wedding of Hippodameia and rescued Andromeda from the snake-haired "monster," Medusa, the "woman ruler." He then gave Medusa's head to Athena, the androgenized goddess devoted to celestial divinity and war, to put on her shield. He founded Mycenae and became, says the classicist E. A. S. Butterworth, "the very type of the skillful, and fundamentally cynical, politician."[19]

Medusa was a form of Kali, the devouring, feminine goddess of India, where a parallel history occurs, brought on by invading Indo-European horse people riding out of the Danubian/Caspian outlands. In the *Rigveda* "the ecstatic is the horse of the wind, the friend of the god of the tempest, winged by the gods."[20] When her head was lopped off, there sprung from

Medusa's neck two horses, Pegasus with wings and Chrysaor the centaur. Chrysaor represents the cult of the patrilineal worshipers of Zeus, a tradition focused on horses and mounted men. Pegasus is another matter, for it was winged, therefore a composite figure joining birds and horses, a reference to the older, deeper, and more dangerous strain of shamanic practice than the ideal of rationality and military consciousness. The flying horse appeared elsewhere in the shift of religious focus from feminine earth to male sky. The Asian Yakuts considered any white-shouldered horse to be a "winged" divine steed of the spirit.

Centaurs are also part of the defeat of village-centered culture and the rise of centralized states, representing the conjoined warrior and his mount. Closely related to the centaur are the old Greek images of sileni and satyrs; the latter, goat-men in classical mythology, had earlier been forms of the centaur. In the words of A. David Napier: "That the Greek Satyr and its mythology have specific Indo-European correlates seems most likely not only when we consider Indo-European equine symbolism, but also from the simple fact that the horse was not known in Mediterranean countries much before the second millennium B.C."[21] A man-eating horse, the centaur was derived from the sight of Thessalonian horsemen riding into battle. As a principle of the celestial, eternal power of man and horse, the centaur became part of the zodiac as Sagittarius.

As one, the grafted man and horse combine in the orgiastic passion of the cavalry charge, the dream of flight, ravishment, intoxication, destruction. No doubt the horse literally fostered these feelings and events by its own violent behavior. Nor is this history all military and stud masculinity, for the wrathful mare and loyal gelding are part of the equine complex, as indeed are the doltish ass and mule. Still, the stallion dominates the concept, its wild whinny echoing a million years of prodigious freedom and beauty, breath-catching and heart-wrenching.

Stories of erotic horned horses, or unicorns, go back two thousand years, from Asia, Persia, and Europe. Resembling an erect penis, treasured as a panacea and stimulus for male potency, the "alicorn" was both amulet and trophy, part of a larger lore of phallicism. The unicorn's "existence" was given credence by narwhal teeth from the arctic seas, beautifully spiraled incisors that grow to a length of four feet. As Europe knew little of the narwhal or the rhinoceros, their horns provided "evidence" of the unicorn's existence, along with the occasional skull with a deformed horn from goats and other livestock. In the medieval wedding of erotic and religious literature, Christian unicorns appear in the tale of its hunting by the Benedictine abbess, Hildegard of Bingen, in 1156. An animal so pure that its horn can detoxify water poisoned by a serpent, it is sought by the king. Its capture requires the help of a naked virgin, using a mirror to lure it with its own beautiful, narcissistic prick-pride. In some versions it sucks her breast while she grasps its horn; in others "she openeth her lap" (in Greek, the womb) for its head, and the hunters close in.

In the era of courtly love, the rambunctious eroticism of ungentle males—like the coiled spring of the stallion's personality (in fact, like the stallion's own lust and immense phallus)—could, it was said, be tamed by a chaste woman. In religious terms, God the King, with Gabriel's hunting horn, blows the hedonistic spirit of the unicorn toward its control by the madonna-virgin, where the hunter, the Holy Ghost, captures it. Somewhat more uncouth is the rough similarity in this imagery to the custom of using a monkey to sooth the hunted rhino. Psychiatrically, the story is a family drama involving the castration (killing or capture) of the father (the unicorn) by the sons (the hunters) with the help of a deceitful, perhaps oedipal, mother. The old man is finished off in the midst of intercourse, his sexual powers appropriated by the new king.[22]

To recapitulate: centralized power and growth ideologies, along with a metaphysics of the Word and Wrath of God, emerged with the amalgamation of the theocratic states of the Near East and the mounted warriors from the north. The tribes who galloped south out of Asia, smashing, murdering, and enslaving, a "Kurgan wave" of mounted "Aryan" invader-horse-warriors from the plains, provided what would become civilization's cavalry.

CENTURIES OF FLUID alliances among themselves in endless skirmishes over water and grass had saturated pastoral concepts with the messiah complex: the conviction that help and rescue (like calamity and destruction) would come from the outside, riding over the hill, as kindred, debtors, gods. And disasters came riding in, the very model of which was the mounted army. Samuel says in the Bible: "These are the divine chastisements of the sin of man." God's wrath has its animals. The storm of vengeance would be unpredictable in action, full of allegorical, thunderous galloping, the blind fury of beasts and swords, terrifying to foot soldiers who faced the charging horsemen.

In those early days of the first dense human settlements, humankind was vulnerable to the epidemic diseases which were often silently harbored in their own livestock or invisible in polluted water. Thousands would die as suddenly as if slaughtered by a horde of Mongols. Survivors would no doubt wake up screaming, trampled by their "nightmare." The same inhabitants of the new cities were dependent on huge fields of grain that could be destroyed overnight by weather, fire, disease, and insects. Black famine could fall on people as though it were another wave of cavalry sweeping through the streets. The prototypes of famine, disease, the wrath of God, and war represented the terrifying clamor that announced the arrival of death as a traumatic spectacle of killer horsemen.

Calamity on horseback was summarized in a book of seven parts, according to Zechariah and John in the Bible. In Revelation the great apocalypse involves the opening of a book with its seven seals. These are the judgments of God upon history. In the first, following a lion, that older scourge of the Neolithic hamlet, came a white horse, ridden by the Son himself, conquering and scourging. The second seal reveals a red horse, the blood to be spilled (which is to say, already shed) in the assaults against the archaic state, the horse that removes peace with its power. The third is the black horse of famine, scarcity, and poverty, whose head resembles that of the locust and whose rider carries a balance indicating God's vengeance. Finally comes a pale horse ridden by death: the plagues and epidemics, like the pox, to which the peoples of the first urban civilizations and later the indigenous peoples of the New World would succumb by the thousands. Such phalanxes of warriors were hell itself, killing with war, hunger, disease, and retribution. All perceived as born on horses, these divine chastisements would scar the Western personality forever. Horses both symbolized and energized the chief components of history. As villages were periodically shattered, the inhabitants were slaughtered and enslaved, their fields of grain spoiled and rotted, themselves infected by exotic strains of communicable disease. The sixth seal of the book reveals the debacle of earthquake and flood, the latter visited upon them from the upstream watershed effects of logging and overgrazing.

Other seals reveal other havocs. The ninth is the torment of the damned by the unleashing of locusts, which looked like horses, breathing fire from their mouths—indeed as the heads of locusts do resemble those of horses. Empires, trod under the hoof of distant kings, would put burdens on the soil and its ecosystems, culminating in locust outbreaks. Human destiny was no longer prefigured in myths of eternal return, but always hung in the balance of conflicting tyrants, battles, and the sway of nations, energized by horses.

To the east, the incursions of horsemen would not be different. The bronze age world was swept by solar deities impatient with archaic earth ceremonies. The universe, originally said in India to have come from parts of the cosmic serpent, Vrtra, was dismembered by Indra, much as Tiamat had been cut up by Marduk in the Sumerian myth. "The sacrifice of the primal serpent of the oceans is the measuring out of time, and the divisioning of space," says Jyoti Sahi.[23] Indra, in the pastoral tradition, seized power by killing his father. The chief Aryan god, the "destroyer of cities," he came to India from the Urals, far west and north, and the central Asian steppes. He gave humans intoxicants—the sacred soma. Intervening in the older cosmic order of Varuna, he was a shaper of iron, a violent drinker, dog-keeper, and fighter against drought demons. Whole regiments of horses accompanied the buccaneering Indra. They correspond in history to the Aryans who invaded India from the northwest, bringing myths of celestial chariots and mounted gods and ritually sacrificing a consecrated horse after allowing it to run free for a year, its journey symbolizing the sun, the area it covered equal to the king's domain. This ceremony, the *asva medha*, central to kings as the representatives of solar dynasties, introduced the horse into burial rites.

In Indian tradition the horse is energy—the wind, waves, the sun. In the myth of Sagara, which gives the funerary ritual its forms, the horse disappears into the underworld where it is pursued by the king's sons, who are burned for their sins by an ascetic, Kapila, the sun. Vayu, the wind, rides a chariot drawn by red horses. Among the sacred presences in things, the two Asvins, the dawn "sons of submarine fire," benevolent healers, are also day and night, heaven and earth. *"Asva"* is Sanskrit for horse. The Asvins are sons of the mating of a mare and a nymph, harbingers of dawn. Soma, too, was horse-mounted, the spirit of a liquid intoxicant, who performs the horse sacrifice and is allied with Surya and his spiritual tribe, the Aryan Asuras whose demonic leader rode the horse, Uchchaisravas. Savitri, the morning and evening sun, is shown in a golden car drawn by horses. In the final age before destruction, Vishnu will ride a white horse named Kalki. Like their Middle Eastern cousins, most of these Vedic Aryans hated serpents and are associated with celestial rather than earth phenomena. They are hard-drinking, light-skinned, and obviously very horsey.

Later episodes in the inflated self-importance of mounted cultures and their death-dealing were played out in the New World, first in the tyranny of pre-Columbian llama-mounted Incan Indians in Peru, whose history is closely analogous to the horse-riding Aryans. Then the Spanish and Portuguese swept through with European diseases and their war horses and dogs. According to Garcilaso de la Vega in his *Commentarios Reales*, the Spaniards could not have conquered the Americas without the horse.[24]

In North America the horse was not only the bearer of invading white authority but the insidious (and celebrated) instrument of intertribal domination. According to Elizabeth Lawrence, the Crow Indians were "preeminent among Indians affected by the horse." As a result, the horse "made the Plains people much less submissive" and elevated the notion of the "daring and ferocious equestrian raiders" whose need for pasturage reinforced old rivalries among the Sioux, Arapaho, and Cheyenne. As it had for centuries, the horse became "the chief occasion for strife," partly because of the normalization of horse stealing. Of course these "qualities of the shepherd—courage, love of fighting, contempt of hunger and hardship"—had "always" been celebrated. The need for trainers and guards increased with class distinctions and the "reassignment of kinship groups" and the "location of families."[25] With acquisition of the horse in the eighteenth century, differences in wealth among many indigenous peoples were related to the number of horses owned. Tribes became visibly more authoritarian, and "dependent families would be expected to follow orders." There was a cultural shift from one centered in hunting/gathering toward nomadic pastoralism. Like their most ancient horsefellows in Asia, the Comanches ate horses but did not milk them (unlike later Turkish and Mongol pastoralists).

The horse greatly increased the killing of the buffalo by Indians as well as whites, as it had many species of animals in the Old World. Hunting was easier and less time was spent at it and transporting meat. At the same time it exalted "glory, praise, and prestige" in a sense very unlike the older hunters. Weapons were modified, and war among the Plains tribes increased, taking on an "epic grandeur." While the Crows revered their animals, some tribes abused them. For invading Europeans the horse was the seat of power, as it had been in the Old World since bronze-age invaders rode into the Middle East, Europe, and India, and among Native Americans horseriding spread aggression and the centralization of power. No ideologies of peace and brotherhood withstood it or the warrior mentality it nurtured.

Nor was Africa immune. Typical of the effect of the horse as property and symbol throughout the world, among the Mosi of Ghana today horses are kept entirely for aristocracy, war, and sacred rituals. Legend says the horse invented money, taught men commerce, and invented beer. An observer says: "All power symbols are crystallized in the role held by the Mosi horse." On parade, it represents authority, exists for war, is never harnessed, has superhuman foresight, and is an omen in its behavior. The rheum from its eyes is rubbed on human eyelids to cause visions. The horse is closely linked to the family head, to ancestors. Only the chiefs and the rich have horses.[26]

Horses continue in our time to be partners in hounding, as cavalry, mounted police, racing with its overwrought, half-demented "thorough-breds" and their bone-snapping "breakdowns," the rodeo bronco, goaded by a painful strap and gouged with spurs, the performers in circus and parade, dressage, the hunter and jumper. Such exercises in domination by means of the slavery of the horse, the means of control of cattle and human pedestrians, costly care that associates them with wealth and aristocracy— all are mixed with the exuberant, vestigial love of horses, reminiscent of a murderous, unrequited love gone crazy. It is as though we are born with a sublime and tender adoration which by some derangement or horrible curse finds its expression in destruction. Its technological manifestation is "horse-power," the result of combining the horse with the wheel, first on chariots and then on waterwheels, grinding stones, the whole array of wagons and buggies, and the rest of the industrial machine.

Horse racing has thrilled people for thousands of years—as though escalating the intrinsic tempestuous beauty of the individual horse, magnified by the spectacle. In Siena, Italy, the race is raised to the level of primary festival of the urban community ethos. An annual race, the Palio, in the central square, distills the competition among intracity clans. It is said that the race so expends the emotions of the city's youth that the crime rate is much reduced for the whole year. The event seems to sublimate the adolescent crisis of identity and male rite of passage in a wild scramble of horses and their jockeys around the city's plaza. Pageantry, parades, singing, and all the other adjuncts to the festival may themselves be more important to the citizens' need for expressions of loyalty and vicarious initiatory test than the race itself. Important aspects of their lives are confirmed by the Palio—the order and structure of society, the role of chance (or God's grace) in human affairs, and perhaps the ambiguity of nature and culture in the figure of the horse which is both tool and temperamental mystery. Like all horse races the Palio epitomizes a violent, competitive, and essentially masculine preoccupation, dressed out in its most vibrant and alluring form.

THE INNOCENCE OF the solitary equestrian, the nobility of the cavalier, the beauty of the postilion, the handsome bearing of the mounted policeman, the nostalgia for the horse-drawn cab, the dash of the jockey, the elegance of the show rider, the romance of the caballero and cowboy—all seem distant from that violent past of burning cities and hounded people, as though the horse had emerged without a shadow into the sunshine of frolicking foals and pure equid beauty. Such mellifluous images warm our Paleolithic hearts and convince our technophilic egos that all is right with the world. If the horse was at or near the center of human imagination in the Pleistocene it is no less so in the idiom of Progress, the genesis, vehicle, and symbol of the civilized world, far too deeply embedded in the human psyche to be denied. In its spurious or domesticated form, the equine furor remains at the heart of the whole thrust of the modern world.

Its bucolic history as friendly drudge is still with us also. Should it be part of a new, agrarian, human ecology? In an article called "The Horse—Is It the Symbol We Want?" Robert Rodale asks whether the farm horse should be integral to a sustainable, small-scale agriculture. He notes that its inferiority to the tractor would be a conscious way of making the farm more labor-intensive, as "we have already labor-saved too many farmers into oblivion." Is the horse, then, agriculture's ideal direction? No, he says, agriculture's long-standing problem is soil erosion. Twenty pounds of topsoil washes away for every pound of Washington wheat grown, five for every pound of Iowa corn—losses masked by the use of chemical fertilizer that works between now and the total loss of the topsoil, at which point such fertilizers would no longer make any difference. "Can the horse be blamed for what is happening to the land? Yes, at least in part." In England, Jethro Tull, who conceived of planting annual grains in rows and invented the grain drill at the dawn of the eighteenth century, wrote the first book about farming with horses to show that production could be doubled by using the horse to plant and cultivate, replacing the old hand-sown and hand-hoed methods. Living where rains fall gently, "Jethro Tull didn't foresee the way his methods would give the people of America the power to figuratively eat soil . . . that intensive tillage and the clean culture of hoeing would expose the soil of another continent to much greater risk of erosion . . . the invention and use of chemical fertilizers and no-rotation agriculture, which would harm soil structure and create even more potential for erosion . . . or maize, whose rows in America, running up and down the hills, would become channels for torrents with the power literally to tear off the skin of a continent. . . . The ghost of the horse walks up and down almost every one of our fields. . . . It's important that we don't accept the horse as a symbol of alternative agriculture. For the horse is not really part of an alternative. It's at the root of the system we're now using."[27]

The destructive aspect of horses is a subdued theme in the modern world. When the U.S. government decided to reduce the number of feral horses on public lands by killing them, a great outcry arose for their protection that resulted in relocation plans which had only limited success. The inability of the defenders of those horses to feel as deeply for the larger community of life, which the horse destroys, demonstrated the depth of its power and appeal to our Pleistocene minds, as well as the difficulty of weighing that abstract "community" against the most vivid of all animal figures.

If pastorality were merely a mode with domesticated hoofed animals as the means and objects of an economy, it would indeed be little more than a type of agriculture. But it is a proto-philosophy. Despite the urban revolution and the mechanized world, it is strange, says Gordon Brotherston, "how thoroughly pastoralism has been inscribed in the twin ideological supports of western culture: the Greco-Roman classics and the Bible."[28] The opposing values represented by Cain and Abel set in place the style of pastoral life as superior to tilling the earth by which we all became not only grazers in our skimming of the earth's resources but masters and advocates of the hoof, the great lacerator of sod and floodmaker, in all its forms.

IN THESE IMAGES of the horse as the instrument and symbol of the human pursuit of a yielding earth, we are returned to the figure of the hunt. The disastrous compression of nature in the guise of the hunt can be misleading, for the hunt in its primal form, men and women on foot without iron weapons, went on for millennia without this catastrophic turn. The problem was not the hunt as practiced by Paleolithic peoples in Pleistocene times, nor was it the killing of beautiful deer. As all hunters know, the deer is the sweetest of game, the companion of Artemis, the white deer of the reverent Winnebago and dozens of other tribal peoples to whom it brings all good things, the deer in the heart of the Sierra Madre woodland in which the fantastic and harmonious mythology of the Huichol Indians is played out, the honored Celtic stag, the mediator between humans and the animal persons as depicted in the figure of the antlered shaman of the ice-age cave at Trois Frères in southern France.

The saga of the dark side may be said to begin with the employment of dogs and horses in the hunt. The patron saint of hunting, Eustace, was a mounted hunter who cornered a stag, in the spread of whose antlers appeared the crucified Christ. The ripples flowing out from this startling medieval image are with us still. Deer, run by dogs, eventually turn and stand before the hunter rides up. This final pause, even if brought on by exhaustion, is a little like a self-sacrifice, so that Saint Eustace's vision of the deer facing death seems a parable of Christ's sacrifice for humanity. The cornered animal is a plain enough analogy to the Christ hanging in its rack, but what of the horse and hounds? Are they merely the contextual associates of this medieval mélange?

The mounted saint confronting the stag is strangely connected to calamity. Man, horse, and dogs, all against the single deer, prefigure an overkill which will in time cut down the wild landscape and even the hunter himself. It is as if the history of all the destructive machinery and pollution of the modern world were concentrated in this one image. It is the centerpiece of Russell Hoban's novel, *Riddley Walker*, in which the world has been devastated by nuclear war as a result of the hunt by physicists for atomic structure. Hounds and horse represent the first tools of that dubious achievement. What began with the hunt then shifted to the "hunt" of the soil and the political use of equestrian power. We realize that without human control of the hound and horse, the final confrontation would never have occurred, as that combination destroyed an equilibrium that had existed for hundreds of millennia between people and nature, the human hunter and his prey.

At the beginning of Hoban's story, humans have gone feral and the transition from scavenging and foraging to agriculture is being made swiftly this time. Language itself has regressed and reemerged. In his wandering, Riddley "tuk 2 grayt dogs with him thear nayms were Folleree and Folleroo." Most dogs in this twilight world of the year 2330 live in wild packs dangerous to people, though others have befriended humans. Protoscientists are already sifting the residue of the ruined world for the secrets of the control of nature, probing the symbolic meaning of the figure of Saint Eustace in the painting which they find in the wrecked Winchester Cathedral and believe to be a religious relic. "On the stags hed stud the littl shynin Man the Addom in be twean thay horns with arms owt stretcht & each han holdin tu a horn." The tiny Christ is both Adam and atom. The beleaguered stag says: "I am the Hart of the Wud. Nuthing wil run from yu enne mor but tym to cum & yu wil run from evere thing."[29]

The defiant voice of hounded nature, the stag, the atom in its embrace, is the voice of nuclear power in the atomic bomb. As more of nature is domesticated and controlled, it becomes more compressed and dangerous. In the bomb, all the former wildness is concentrated. Humans, in the form of Eustace, have come from being surrounded by nature to surrounding nature. The wild has not disappeared, but at first it is squeezed into forest relicts, its power not diminished but contracted, isolated, pushed to an explosive end in the very heart of the minerals themselves.

It is ironic that nature's revenge is not embodied as a monster with huge teeth and slavering jaws. We think traditionally of great storms and tides and other dangers as predatory demons, but these images are traceable to the bronze-age tyrant emperors and their evocations of enemy power as lions and dragons. We now see that it is not the competitors or hunters of "man" but his victims who will kill him for his failure to keep within the limits and scale of his own natural being. And so the deer speaks for the atom. (As Konrad Lorenz shows us, the herbivores are the cruelest of all when pressed, for only among the carnivores with lethal teeth and claws are there inborn inhibitions against killing, tied to signals of submission. Rabbits are murderous in combat if the lesser of two combatants cannot run away, as it will be relentlessly kicked to death.[30] When Albert Schweitzer playfully pushed on the head of one of his captive antelopes, it turned its pointed horns toward his stomach and the good doctor, his back to the fence, sweating and grunting, barely escaped being impaled after a struggle of some minutes.)

The moral of Hoban's story and the Eustace myth is not about the passion for the hunt but about its perversion into a megalomaniacal quest which began with domestic animals and plants as its objects and instruments. When the cultivation of wheat and barley encroaches too much on the native plants, compressing them upon themselves, their contracted wildness explodes with pests, diseases, locust swarms, and the ergot fungi that drive whole human communities crazy. The use of chemical pesticides and fertilizers escalates something begun with those first grains, finally releasing the fury of chemical poisons in the soils and the living body.

The history of ecological catastrophe begins with the "hound," which comes from the same root word for "dog," meaning "to seize" or "tenacious pursuit." The first domesticated wolves were not pets, guards, companions, or meals but fellow hunters. With dogs, the first domesticated animals, the "conquering" of nature started toward its final calamity. As David Noel says: "What is truly being hounded, harried and crucified in this ambiguous little masterpiece is not Christ but nature itself . . . a dawning sense that hounds and hunting might have something to teach imaginations locked in the global nuclear stalemate."[31]

The allegory does not end with the middle ground, but with the pursuit of nature to its core. With chemistry, the seizure of the earth shifted to compounds and molecules. The physicists, the high priests of the subatomic hunt, have taken the final step by hounding the elements, which, like the stag in the painting, turn finally to defend their wildness. At first surrounding pockets of domestication, nature itself is at last surrounded, its irrational forces contracted in the Bomb, to that point "where it finally must turn and confront us with the very unimaginable wildness we thought we had hunted down and eliminated."

THE BENEFIT TO animals of being domestic is fictitious, for they are slaves, however coddled, becoming more demented and attenuated as the years pass. Among them, the native canine qualities of foxes, jackals, wolves, and coyotes slip away. The dog is the perverse and dysfunctional wolf, the main instrument and symbol of the destruction of the wild wolf. Wolves are not evil dogs, but the other way around: the dog is the corrupted wolf. Having first trained and then unleashed Folleree and Folleroo, our aggression comes to its apotheosis in the picture of Saint Eustace, the moment of the Fall. Before the hound, men hunted with their minds and were on holy ground. With the dog, an equilibrium was lost. The canine itself was the first victim. Mounted the hunters became slavemakers, and everything was relentlessly hunted, as the two slaves became the weapons against the earth.

Argument as to whether the dog or the horse is "man's best friend" is beset with the irony of a choice in which the animals are perceived as competing toys, as though Folleree and Folleroo were merely playmates of Trigger, Silver, and Black Beauty. Indeed, the two compose a somewhat graver alliance: the means in human hands that hounded and horsepowered the earth into polluted, destabilized, and homogenized environments. As wild forms they will always be locked in our hearts. But neither is man's true friend, nor woman's either, as they energized and symbolized the destruction of the Gaian sensibility—that humility and nurturing ethos which resists the pastoral exhortation to overtake, control, and contain.

PART VI

Counterplayers

No longer cohering to a single vision of the world, we are, for better or worse, on our own in defining our relationship to the Others. We can rummage through the old myths, but we must make from the debris of history and nature our own story and our own ecology. Except for an obsolete notion of supremacy and domination, there is no thread between us and them in modern life. And so we must create afresh our idea of who we are in the context of the circle of animals.

19 | The Miss Muffet Syndrome: Fearing Animals

I can bask in the warm feeling toward nature inspired by a growing conquest of man's oldest biological enemies.

LLOYD J. AVERILL

IF THE STATEMENT by Lloyd Averill about our "biological enemies" is not surprising, it is because it so concisely bespeaks our time. We are surrounded and invaded by living foes: they attack our gardens and our crops, kill our livestock and pets, threaten us with unexpected bites and stings, transmit and cause diseases, raid our stores and pantries, creep through our homes, and endanger us in the oceans and wildlands. Our counterattacks are violent and sweeping. Take no prisoners; life is a war.

Even so, bad animals have their uses. I do not mean only ecological uses, with which the technophilic attitude has so much difficulty coming to terms. I refer to animals big enough to be seen in the landscape, toward whom our antipathy is not justified by their occasional threat to us. They are clearly our unconscious proxies for something else. Spiders are a good example—as though they were invented to remind us of something we want to forget, but cannot remember either. Snakes are even more so. Most such animals do not inspire love or fear directly. Acutely scrutinizing and attending raptly, children seeing their first snakes do not loathe or love but simply wonder. The mood and words of grown-ups direct and exploit the child's aroused awe, shaping and suffusing passion with terror or reverence, approach or avoidance.

It has been argued that the human fear of snakes is instinctive. Some animals—the mouse, for example—do not need to learn to flee from snakes. Chimpanzees are alarmed by spiders and snakes, perhaps instinctively as a result of their long coexistence and the danger of poisonous bites. But chimpanzees (and mice) do not go about seeking these "enemies" in order to stamp them out. In our modern bias we may suppose their passive resistance to be merely helplessness, but I believe their behavior models for us a special way of perceiving. It is as though they know what we have forgotten: one *needs* one's opponents. Such animal antagonists are necessary counterplayers, as Alan Watts observed, with elegant insight, a generation ago.[1]

Balaji Mundkur, writing on the cult of the serpent, speaks of the human reaction to snakes as "morbid revulsion or phobia." He sees "the primacy of fear in animal-oriented aspects of religious life" as a motive for much of our religious activity. If fear is the basis of religiosity we can see, he says, why "serpents," however dangerous, are widely worshiped. He notes that we experience a spontaneous physical response to snakes which shapes our belief in the snake's metaphysical or demonic powers.[2] But the words "cult" and "serpent" suggest (to most people) superstition masquerading as a true religion: religious/literary cant for an evil, snakelike thing. Oddly enough, however, snake hysteria is not typical in some countries in the tropical world where in total snakes cause more than thirty thousand deaths per year. In South America the Aztecs raised attention to rattlesnakes to a phobia as part of their morbid fascination with violent death. Yet elsewhere, as in rural southwestern India, cobras are venerated and their presence ceremonially celebrated every year.

Apparently there is a widespread feeling of the *significance* of snakes that outweighs their possible danger. Research on "what children fear" indicates that 80 percent of the children acutely feared snakes, which seems to support the contention that the snake is spontaneously disliked, an old enemy of humankind. "Aversive behavior" toward snakes is shown by adults: 40 percent of college students identify snakes as their most feared animal, although they experience snakes not so much in the field as in their imagination, stories, dreams, and hallucinations. Even the youngest group of children in the study, five years old, had already absorbed their caretakers' emotions, body language, and other responses.

People who are regularly harassed by stinging insects or threatened by real bears at the door do not seem to hate them with a rage, wish to punish or exterminate them, or see them as evil. Such an attitude seems to require more than an inherent caution. (Nor does one see among the animals themselves, such as the large, intelligent antelopes, evidence of fear in the presence of their predators, when they are not actually being chased, or hatred that results in vengeance.) The statement "I hate snakes!" or "I hate spiders!" suggests something beyond the prospect of being actually threatened or bitten. One might be expected to say, "I hate black widow spiders that get into your shoes." But even that statement qualifies the circumstances and goes beyond the "flight or fight" response to real danger. Sweeping statements of generalized hatred have a tenacious, floating hostility that seems to attach to an appropriate target, so that the objects of scorn represent something other than themselves.

The matter seems to turn on a cultural will to illusion. "It is the folk-figure that has been popular with philosophers," not the real wolf, says Mary Midgley. The imaginary cruelty of savage animals has been used "to illuminate the nature of man," mapping him, she says, "by reference to a landmark that is largely mythical."[3] This "nature of man" includes our bodies, our behavior, our family or friends who are too close for us to accept their faults, our social, religious, or national groups, the uncertainty of the human condition, or some diffuse and unmanageable quality of our circumstances such as death, illness, and evil. "Man has always been unwilling to admit his own ferocity, and has tried to deflect attention from it by making animals out to be more ferocious than they are."[4] What happens, she says, is that we confuse the symbol (the wolf for human ferocity) with the thing symbolized. She blames this habit especially on moralists and philosophers, for whom the "beast Within is a lawless monster to whom nothing is forbidden," and quotes this linguistic abuse of animals among such civilized luminaries as Nietzsche, Dr. Johnson, Tolstoy, Plato, Aristotle, and Kant.

Barry Lopez traces the hatred of the wolf in European-American societies in the context of the vulnerability of domestic animals coupled with a demon-hunting ideology.[5] Calvin Martin observes that something like it in times of great social stress occurred among North American Indians, allowing them to join the white hunters in overkilling beaver and buffalo.[6] Such profound psychological stress, due to catastrophic epidemic diseases, resulted in a relaxing of customary circumspection and a collapse of respect for the sacredness of the animals. Plato was among the first to write of the gods as being only good. To us this notion does not seem strange. But when there are bad as well as good gods, it is not so easy for evil to be identified with the physical or animal world as opposed to the spiritual. Once the deity had become good and envisioned anthropomorphically, the antithesis could be perceived as animality. In this way the hatred of wolves arose in a world perceived as already divided in cosmic conflict, an attitude which served evangelical religions well as they advanced into "heathen" lands. Psychologically this process has been described as a form of male chauvinism in which the *anima*, the feminine aspect of the unconscious (in the form of Pandora according to the Greek version), releases a horde of malignant beings in a world rife with the struggle between good and evil. The residue of this philosophical dualism can be seen throughout history.

Our society centers its aim of controlling organic nature on the "natural" aspects of ourselves, sex, death, birth, defecation, growth, hedging them about with rites of cleansing and avoidance and projecting their dangerous qualities onto animals, which become a kind of code for them. This need "to distance our social life from animal origins" is what anthropologist Mary Douglas calls "the purity rule."[7] Some scholars "glorify speech so much . . . as to locate the source of our intellectual freedom in the power of abstract verbal symbols. . . . Consciousness of the knowledge we owe to our animal being is veiled by the purity rule." The tainted world of the flesh, above which our morality inspires us to rise, is more clearly seen in the animals than in our ambiguous selves.

Food rules, especially those having to do with meat, are one of the means by which this concern with natural pollution and corruption is brought into everyday life. Foods are classified as unfit to eat or acceptable in certain circumstances, as normally edible, as requiring special treatment such as the separation of meat from blood, as dependent on differences in the edibility of parts of the body, on the separation of milk and meat, on the type of preparation of food, on the customs regulating the use and storage of utensils. Some animals may be simply abominable, some fit for ordinary meals, while still others have religious significance. All of these degrees of impurity in the world are not only represented by the animals but are seen as intrinsic in them.

Familiarity does not heighten the fear of "beasts," which requires a vague blur without definition in order to attach a free-floating anxiety to it. The beast may not represent a projection of one's own faults so much as it does the desire to fix social and psychic stress on an enemy against which one can take revenge. The anger over one's situation which has no concrete form seeks to find or invent one. Mary Douglas describes this heritage in terms of the distinction between purity and pollution. For centuries the difference between "human" and "animal" lent itself to the recognition of impurity in ourselves. As the legacy of the old moral bestiaries was replaced with scientific study, the divisions between us and the beasts "have turned out, one by one, to be optical illusions."

L. BRYCE BOYER, a psychologist who has studied Apache cruelty to dogs, concludes that we are all vulnerable to a kind of "archaic sadism" that results mainly from certain aspects of childrearing. Among the Apaches this happens especially when there is trouble with a brother or sister. When they quarrel with their siblings, children are threatened by parents' stories of the *jajadeh*—the whippoorwill who is represented as a masked dancer and who, they are told, will carry them away and eat them. Likewise there is a punishing ghost inhabiting the owl, *awshee*. Apache hostility is redirected to the dogs, who are more available.

Apache stoicism and "psychiatric haplessness" are expressions of a widespread syndrome of redirected violence induced by frustration. Athabascan personality traits represent "adaptation" to environmental conditions, expressed in aggression, stoicism, independence, and suspicion. Boyer links these traits with "schizophrenic childrearing." Punishment by animals for social misdeeds is balanced by the reverse: the punishment of animals for having themselves been a victim of society. Such psychotic logic is not conscious. It involves a "mobility of cathexis" that "produces instantaneous displacement of one idea for another. . . . There is no negation, no doubt and no degrees of uncertainty."[8] Such behavior is no respecter of culture or race. In a George Booth cartoon an old man sits on a park bench, his dog nearby. A beautiful young woman walks by. A moment passes, and suddenly the old man gives the dog a swift kick.

"Hating" animals may not only refer to the projection of one's own faults but extend to animosity for certain human groups, the animal serving as a scapegoat for venting these feelings. The differing perception of animals as an expression of social classes may be a reason for the emphasis on "kindness to animals." Beginning in the sixteenth century, the bourgeois saw themselves as distinct from both peasantry and the elite, the latter groups both being eaters of songbirds. For the middle class, living neither as cultivators nor country gentlemen, "animals were something you read about or looked at, rather than things you handled in everyday life."[9]

Cats have been favorite victims of class distinction, perhaps because the cat's own aloofness seems arrogant to anyone already enraged. In Paris, in the late 1730s, printers' apprentices suffered because large printing houses were swallowing up the small shops, reducing the number of employees and increasing absentee ownership. Labor was changing from a partnership to a commodity, and print shop managers were abusive and intolerant. The indigent journeymen had little security and were often fired after a job was completed, regardless of their skill. The workers could not negotiate their plight with the absentee owners. Finally the apprentices found a way to "get their goat," (a saying that came from a sound, like bleating, the French made when razzing someone). Many of the bourgeois printers' wives kept cats, some as many as twenty-five. These were coddled, fed roast fowl, had their portraits painted, and generally lived better than the apprentices. The hated cats were a thorn that pricked the workers' resentment by their nightly howling in the alleys.

Torturing cats was not unknown as a pastime. In the charivaris of the French, as in the German *Katzenmusik*, could be heard a parody of their cries of pain. During the summer solstice—the cycle of Saint John the Baptist—the Courimands (cat chasers) of St. Chamond ignited and then pursued flaming cats through the streets or danced around burning cats tied to a pole. In the city of Metz baskets of cats were ceremonially burned in a *feu de joie* (bonfire) until 1765, culminating a procession of town dignitaries and riflemen. In Saumur children roasted cats. In Aix-en-Provence they were tossed high into the air or hanged on miniature gallows. The impetuous person was said to be "patient as a cat whose claws are being pulled out" or "whose paws are being grilled." In short, cruelty to cats was popular, related in part to an older witchcraft in which one could neutralize the power of the witch's cat by maiming it—and killing it would bring misfortune to its owner. Even without the witch the cat could prevent bread from rising, smother babies, and spoil fish. Cats were enclosed alive in the walls of new houses—a bit of evil to protect from evil.

With all this behind them, and the insolent cats at every hand, the Parisian printers broke into an orgy of felicide. In impromptu festivals they massacred the captured animals in symbolic ravishment of the master's wife, for French references to "pussy" have the same sexual meaning as in familiar English slang. All this came to be known as "The Great French Cat Massacre."[10]

OUR SPECIES, LIKE others, has a native sense of precaution and distancing from other animals. Our learned "terror" of spiders, bats, and snakes exaggerates regard. Psychiatry is full of examples of projection and surrogate targets, substitutions that relieve the individual caught in a double bind, proxies when the real object is unavailable, and metonymic abuse of innocent victims of fury and frustration.

A typical target of direct abuse is the snake, a part of an earth and underworld complex. Anathema to the champions of heavenly cosmologies, snakes thereby represent the whole of the complex. Likewise, vampire bats represent the metaphors of social parasitism, the ambiguities of good and evil analogous to the boundaries between light and dark, bird and mammal, cave and sky, and in addition may embody the whole human/primate history of fear of the dark.

Spokesmen for scientific natural history and education mistakenly suppose that you can allay such abuses by information. While some informed people may have decided that sharks are really not so bad, they yet may find reasons for not swimming in the ocean. Traditional island peoples widely regard the shark as a spiritual being whose analogous aspect in the human self is an embryo. That sharks rip open the body of their victims has a double horror of an unborn monster ravaging the self. Sharks are neither intrinsically terrible nor sacred, but they are utterly fascinating and therefore a perfect candidate for encoding extreme feelings and concepts.

The fearful aspect of animals is valuable culturally, but it is not a function of their natural history nor a motive for "war" against them. It may be more dangerous spiritually to try to destroy such beings by "education" than to live with them, for the vacuum they leave is in our capacity to image, to find metaphors in the natural world, and to affirm the world as a good place. It is as though the mythology of horrible animals were a necessary way of objectifying something that is not their own but ours. Such images of opponents are necessary to the wholeness of things.

So go ahead and hate a few animals and you will feel better afterwards. There is nothing wrong with those horror stories and films that exaggerate spidery legs, bloody fangs, slimy underparts, cold bodies, monstrous size, maggoty manners. But after the giddy terror, in the solace of your protected world, consider their gift, like the necessary role of the evil snake in the play of redemption. Don't make your fear and hatred part of your direct response or public policy toward the animals themselves. The value of hating and fearing animals, like that of revering them, is essential not to our ecology but to our thought.

20 | Cuckoo Clocks and Bluebirds of Happiness: Animals as Machines

The more that society is vested with power, the more it despises the organic processes on which it rests.

MARY DOUGLAS

SERIOUS THOUGHT ABOUT the nature of animals in the seventeenth century attempted to define them and to differentiate them from ourselves. It was asked: Do animals have souls? Feelings? Minds? Language? Self-consciousness? If not, how do they differ from machines? These questions seem intended to indict or to justify attitudes about their souls and their fate. As science, theology, and humanism peeled away the animals' capacity to be human, another segment of society was busy humanizing them as surrogate people, substitute children, and living toys. The differences between the reduction of animals to mere mechanical objects, justifying experimentation and industrial husbandry, and the sentimentality directed toward the animated dolls typical of the cartoons of Walt Disney seem to be opposed: the experimenter at one extreme, the lover of helpless baby animals at the other. When examined more closely, however, they are seen to be cut from the same cloth.

Thinking about birds is a good example. They are at once the models for the little lawn ornaments whose propellerlike wings are driven by the breeze and for the robin on the windowsill who has come to wake us up. Birds are perhaps the most esthetic and spiritual of animals (unjustly perhaps, given the unexamined possibilities among beetles and mollusks) and yet were the first to be emulated by modern machines, when Da Vinci drew plans for aircraft and René D'Honnocourt designed his mechanical birds. Reducing living beings to mere objects and diffusing them into the arts as baby-faced toys—both impulses follow the Cartesian idea that an animal is more like an automaton than it is like a different species. Industrializing emotion in the form of the stuffed toy and the children's stories of Peter Rabbit corresponds to veiling the laboratory animal as a willing benefactor, on the one hand, and obscuring the real life of farm animals behind a fictional facade on the other. Animal stories for children are so highly conventionalized (and mechanistically iterated) that altered phrases in their recitation are met with strong protests, as though they were made of some rigid material.

Anthropomorphic stories and images are valid means for objectifying elusive qualities for children, but they are not adequate substitutes for myth, the kind of stories that adults need. The infantile atmosphere of modern society—its gluttony, instant gratification, and simplistic values, its unwillingness to risk or endure, its lack of a cosmology that recognizes limits and otherness—continues to demand that living things be cartoons, warm and lively in their imagery, preferably projected by or seen as electronic and mechanical devices. Our graduation from the anthropomorphic toy animal as a tool or machine of mindmaking should be a step into the lives of real, nonhuman, nondomestic animals. Such is our own biology, the Pleistocene claim on our brains, that we mature into nature.

Although they articulated the concept of animals as machines, the idea was not simply invented by seventeenth-century philosophers and scientists. Even though mechanism was most succinctly defined by René Descartes in *Discours de la Méthode* in 1637, his thought was anticipated by religious traditions of the soullessness and demonology of animals. Laying the blame on science is a game played by "humanists." Whatever atrocious things are done to animals in the name of vivisection, literature, and art are made possible by a social climate and acceptance. Descartes found the best model of animal machines in the gardens of St. Germain, where water pressure caused mechanical animals to move. Art has often anticipated what science was about to discover—for example, that animals are merely arrangements of parts.

Philosophy and religion all along insisted on the radical distinction between human and animal. Duality has always been a necessary step in category-making as part of the procedure in all taxonomy, as when children learn to recognize and name things. But once divided, things must be reconceived as organic wholes with shared qualities or else one remains lost in endless dichotomies. As a philosophy, duality divides endlessly, redefining "human" by sets of yes or no, culminating in what David Ehrenfeld calls "the arrogance of humanism."

Conceptualizing animals as toys for children and cuckoo clocks for adults emerges from humanism. One would expect the naturalists to have opposed this calamitous error of seventeenth-century philosophy, but biology itself lost its holistic, sixteenth-century genius and succumbed to the reductionist approach of the new mathematics and chemistry. Medicine would conclude that the human body (our animal nature) was merely a machine. With Count Buffon and later behaviorist biology and psychology, the animal was emptied of secrets. As John Berger says, the behaviorists imprisoned "the very concept of man within conclusions drawn from artificial tests with animals."[1] The distinction between animals and humans resolved itself into a material similarity and an incorporeal difference. This marginalization of animals pushed them toward the edge of our "real" purposes, except as machines that could be disassembled in medical research or assembled for childish amusement.

The final reduction of animals from epiphanies to mere objects has only come about in recent centuries, although they remained "symbolic" in a kind of literary gloss. Degradation of animals as beings began long ago, however, as the first denominations of money were conceived in terms of the market value of animals. "Pecuniary," or money value, is derived from *pecus*, the word for cattle in ancient Rome. In Sumeria, sixty shekels equaled one manu, the amount carried by a donkey. A goat constituted five shekels, an ox twenty. Coins in Egypt and Greece were stamped with the heads of animals to indicate their value. The Indian rupee is from the Hindi word, *rupa*, for the zebu, a form of cow. The great archaic states of South America and Eurasia were based on load bearing, grist-mill turning, and pulling, to which animals were bred as the first machines.

Interpreting all life as mechanistic would await Greek and Arabic mathematicians and would not become widespread until the early Renaissance recovery of classical scholarship and the rise of mechanical power. This leads back us to Descartes' *Discours de la Méthode*.[2] Within the world of science (where the civilized public now lives) the mechanists won out. Laboratory animals were used widely in demonstration and research, and biologists speak a vernacular of bodily processes such as "the mechanism" of digestion or of nervous control and so on. The idea implies that animals are only machines, and this mechanical concept of life inspired hostility to the theory of biological evolution, which assigned man to the animal kingdom.

In an engineered world, things are made; in an organic world, they grow. To makers there is something ominous about life as an unfolding, guided by inner, invisible processes. Yet these self-directed events in the life of the organism have been subordinated to the maker-philosophy. "Inspirational" stories of "The Ugly Duckling," "Rudolf the Red-nosed Reindeer," and "Jonathan Livingston Seagull" all assert that one makes or remakes one's self according to the competitive spirit of domination and personal initiative.[3] The lesson of these stories is that nature—organic, genetic, intrinsic— is defective. As a result, the given physicality of our own being is somewhat embarrassing.

WALT DISNEY'S WORLD, from Mickey Mouse to Bambi to Benji, is an easy target for highbrows. Many intellectuals and naturalists deplore the humanized and infantilized creatures with their exaggerated eyes and cuddly shapes. It is not difficult to discover in Disney's cartoon films a betrayal of the "natural history" or objective view of animals.[4] None of the criticism seems to diminish the appeal of the animals to those watching films or to educate the mainstream English, French, American, or Japanese audiences and sojourners to the various Disneylands. Indeed, I know a number of highly educated people who go again and again. People do not seek there a true reflection of the world, nor even an escape from it, but some kind of encapsulation that assures us of its logic, a meaning of Others in the common ground of a manufactured idiom. The juvenile notion of shared life is simply "being like me." Mother Goose animals, who talk, build houses, plan, and fear storms, are elements of an easily grasped mutual world. They become a kind of code for human society, a first step toward organizing the totally human scene with its confusing subtleties and gradations. Good bluebirds and bad hawks are okay at a stage in our personal development. But why should adults be so addicted to the same cartoon stereotypes?

FOR FIVE THOUSAND human generations animals served our species as delicate signs of the way the world goes, as elaborate metaphors and symbols, as spiritual beings, and as themselves—beautiful and imponderable counterplayers in a mysterious cosmos. In the immensity of time, humans acquired, deep in their hearts, the expectation that animals signify. Insofar as they mediate between our conscious selves and the archetypal figures of our innermost being, childhood is a time of "archetypal reminiscence" dependent on animals.[5]

We now struggle to reconcile our culture's story and its machine paradigms with the "ecological imagination of childhood."[6] Failing to graduate from that first ecology, our childlike perceptions remain childishly inappropriate. Our culture is not ecologically mature; instead of an adequate cosmology, we are in adolescence subjected to the myths of the animal/machine, heroes of progress and domination, and the dualisms of ideology. Animal "powers," in the Hindu, Tlingit, and old preclassical Greek sense, require a strong mythic foundation based on humility in a world of Others. If we are unable to accept other races, nations, and creeds, consider how far we are from recognizing ourselves as one of many species. Strengthening the ego to endure a radical, plural otherness is what the old tribal calendars of instruction, endurance, and ceremonial initiation are all about. The task is to move from happy endings to mature serenity in a world without happy endings.

Anyone who has dipped unto the verbiage on animal rights and animal protection has learned that the blame falls on the seventeenth-century scientist-philosophers and writers like Samuel Butler and Richard Leigh who convinced their readers and their technologist heirs that animals are nothing but machines. The agonizing reversal of this brutal presumption has inundated us with new essays on "friendship" in which we strive not to give up our power but to include the automata under an umbrella of charity. The solution, we are told, is not to deny that animals are machines but to learn to love machines. This is the real end of the cuckoo-clock-land of the animated cartoon. Those bent on convincing us that computers can think and feel are the Disneyphiles in their workaday world. What a dazzling thought for sophomores, who need not be asked to throw off civilization's five thousand years of biological jingoism, but only to love with a single force the tweety-bird peeking out of the Victorian clock, the tweety-bird in the animated cartoon, the tweets from their computer, and the bird out there tweeting in the tree, all as engineering triumphs.

Of course, visitors to Disneyland do not think Mickey Mouse is a "real" animal but simply enjoy him as a kind of delightful burlesque. One might even argue that he represents a figure like that of animal-costumed dancers in tribal ceremonies. If so, there is cause for concern. Ceremonial "animals" in traditional small-scale life do not present a safe, disinfected, and friendly illusion, although there may be humor and irony in what they do. There is nothing unknown or withheld from us by Mickey and his fellows except the vast subterranean works that keep Disneyland and its subordinate "lands" operating.

The only difference between the Disney world and the Cartesian view is that Descartes had an old-fashioned suspicion that machines were things, not beings, while the futurist premise is not only that animals (and we ourselves) are machines but that machines are beings. It is not a conflict between living tissue and springs or circuits but a world composed entirely of cyborg or bionic entities needing compassion. That supposition is not really new. Beneath it lurks the same old assumption that the world and life are made, not grown, that technique is the key. The futurists think of the extremely improbable performance of two and a half billion years of our genesis as if it were a fabrication, a *making*.

To this, Romain Gary replies: "In an entirely man-made world, there can be no room for man either. All that will be left of us are robots. We are not and could never be our own creation. We are forever condemned to be part of a mystery that neither logic nor imagination can fathom."[7]

The Great Interspecies Confusion

If we accept the philosophy of respect for life with its view that organisms exist in their own right as fellow members with us in the world community of living things, we must be guided constantly by the discipline of ecological observation, otherwise we are in danger of being rather silly.

SIR FRANK DARLING

IN THE SEVENTY years of its existence, the pages of the *New Yorker* magazine have hummed, whistled, and barked with cartoon animals. In a sense I suppose it should not be surprising that the most urbane of magazines in America would accommodate its readers with images of the non-human life from which they are so distanced. But the cartoons are not about animals. They are comments on what might be called "Knickerbocker Ecology."

To begin at the beginning, what goes on with people and animals in Manhattan started with bringing them home in Jericho and Mohenjo-Daro. As long as animals were understood as prodigiously Other, were watched, studied, and eaten but never brought in to live with us, they could be respected for the differences that made them the most powerful metaphoric source of our self-consciousness. The boundary between nature and human culture seemed firmly fixed; there was no confusion about the two systems. But as soon as animals entered familial life, were tended, protected, bred, herded, fed, milked, castrated, sacrificed, bought, sold, and acknowledged as part of the homestead, we became less sure about the distinction. As companions, slaves, and dependents they did not seem so different from other humans as their wild relatives; indeed, they seemed rather like human idiots, who were also found at home before the era of institutions for the feebleminded. The more they were domesticated, humanized, and incorporated, the less that distinction served analogy and the more it gave way to a kind of crippled sociality.

T. H. White managed to sustain the old rewarding inquiry that allusive difference creates. In *Merlyn and the Animals*, King Arthur, grown old and dispirited, returns to Merlin's cave under a great oak, where all the animal teachers of his childhood are gathered. With the wizard's help, the old king, discouraged with humankind and disappointed in himself, returns to the healing genius of nature, keyed to the primal intellectual activity: watching animals. Merlin transforms him into a goose and then into an ant. The ants were not themselves to be judged as an authoritarian society, but served only as concrete analogues to one kind of political organization, just as the geese modeled a more egalitarian, social world.

Merlin is a Celtic form of elder who precedes the shaman, that boreal combination of the sorcerer, priest, healer, guide, and magician. In modern culture, which urges us to expect gurus, we all seem to want pilots who will make dangerous spiritual excursions on our behalf, rather than wizards who send us directly to other realms on our own, as Merlin did Arthur. Wizards offer no universal panaceas. They are autochthons—part of the earth. As a measure of our discontent we long for messiahs rather than for wise elders or shamans who come out of sleep from under a hill or from the village outskirts—but in our case from other worlds, to help us out of this one. Jesus Christ, Mohammed, and the Buddha, with their doctrines of salvation, escape, or access to the absolute, had no need for those old mediating animals. White, like W. H. Hudson and John Muir, never gave up the hope of recovering the enormous mystery of the animals and their capacity to teach us.

Some gurus offer to end our existential malaise by identifying the spirit with the smallest quarks and or the quasars at the distant edge of the universe. According to the Maharishi Yogi and his scientific friends, for example, the abstractions of physics are about to marry the abstractions of mystical theology, to close the circle connecting the least and greatest limits of nature with the Great Spirit, and give us a Final Answer. The new physics and its "history of time" curiously shadow forth not White's Camelot but the medieval abstractions that inclined Saint Benedict to overlook sparrows. This splayed fixation, one eye on the infinitely great and the other eye on the infinitely small, omits the middle ground with all its inhabitants.

Perhaps the best motive for clinging to the elder or the shaman, however overloaded with modern misconceptions, is that he or she continues to work with plants and the voice of the drum in concert with animal guides who keep us in middle earth where we actually live. The supposition that we can solve the dilemmas of philosophical and personal maturity with epiphanies of final celestial and atomic entities denies what we are—beings linked to the living Others for our health and understanding. Human societies for hundreds of generations believed that souls enjoy sojourning in bodies—seeing, tasting, smelling, touching, loving, and even suffering—and that the dead are evoked by speaking their names and are present in drumbeat, gourd and turtle-shell rattle, bone flute, horn bugle, and conch trumpet. The oldest evidence of culture may be the wearing of feathers or the making of music which took its mode from the songs of birds, frogs, insects, wind, and water. In such primal philosophical discourse, watching birds displaying must, at some point in protohuman consciousness, have evoked curiosity, shown ways of signifying by means of analogies to human society, and inspired conceptions of the lively organization of cosmic powers.

Such performances suggest rites. Ritual is not solely a human possession. It is widespread among animals—that is, stylized communicative acts derived from ordinary self-care or mundane activities. The peculiarly human expression of ritual is that it performs a narration, an enacted metaphor, incorporating other species' ritual movements by mimesis. The various beings in myths are understood to relate to one another in social roles.

Perhaps enactments of important traditions, in which "things are said and done," arise directly from observations of animal movements and calls, translated into silent signals and back again into sound, leading to such anthropological notions as "Hunting and the Development of Sign Language,"[1] those spontaneous movements that might indicate "Primate Communication and the Gestural Origin of Speech,"[2] and the transfer of gestures that enhance speech to hand movements that produced the plastic and visual arts, as seen in the animal imagers of the Paleolithic caves. Dance, past and present, says André Leroi-Gourhan, is "an expression of concepts about the natural and supernatural organization of the living world." People "undoubtedly knew of the division of the animals and human world into confronted halves and conceived that the union of these two halves ruled the economy of living beings."³ When, as in the view of tribal societies, other species are "peoples" and other human groups are like other species, the friction between cultures is reconciled, usually in ceremonial acts of accommodation and watchful attention. In the modern view this practice degrades other people by turning them into animals.

The separation of wild natural systems from human social systems, in which the relationships between different wild species were seen as poetic interpretations of the terms relating different human social groups, was one of the great intellectual achievements, rivaling, say, the "invention" of tools. This analogical poesis collapsed when animals were brought into the household, where they became metonymic and social rather than metaphoric and ecological. Among the ridiculous extremes of this incorporation, for example, animals would be subject to human customs and expected to obey laws. There are records of more than 150 trials of animals in the courts, like that of a fly in Savoy in A.D. 1580.⁴ Today such judgments are no longer the business of the courts but of land managers—the farmers, loggers, engineers, public health officers, and cattlemen who judge "pests," decide on the eradication of mosquitoes, the reduction of nuisance prairie dogs, the tracking down and killing of sheep-eating coyotes, or the sacrifice of whole communities of life for shopping malls.

The captive and domestic forms are generally seen as suborders of human society. To be socially related is to be kin—as parent, aunt, offspring, or cousin—or to be benefactor, friend, protector, caretaker, patron, protégé, sexual mate, well-wisher, business transactor, student, teacher, slave, working partner, boss, subordinate, savior, entertainer, and many other things. Animals have served all these ends. Legally, they are inept persons whose rights need formulation and protection by policy and law because they are widely abused, as are many women, children, the poor, and the elderly. The people of democracies abhor slavery and yet they surround themselves with slaves—defined as "one bound in servitude . . . submissive or subject to a specified person." If you ask the owners of dogs, cats, and birds whether their pets are slaves they deny it. These people share the intuition that we learn from animals, yet they have no way of recognizing the deformity of this intuition in the presence of deformed animals. Because of the loss of otherness in the animals we keep, there is a terrible emptiness in which they mirror our lives instead of informing them.

IN RECENT YEARS animals in cartoons in the newspapers and other media have become so popular that roughly two in five cartoons involve animals.

At the same time, the boundary between them and ourselves has never been so uncertain. The closer they get, the more puzzled we are. We cope with this tension, as with other forms of ambiguity, in humor. Perhaps we are the only laughing animal because we alone have structured self-consciousness by reference to other species. Maybe laughter began with the irresolvable ambiguity inherent in a sense of difference embedded in likeness.

"Krazy Kat," "Pogo," and other long-running cartoon strips address this tension in a curious way. By invoking analogy between the "forest society" and human society they borrow energy from the archetypal interspecies metaphor. But the situations are not denotative; they are simply parallel, so that what emerges is burlesque or satirical commentary in which the animal kingdom can no longer counsel. This shortcoming of the animals is not lost upon the artist or audience. In Gary Larson's cartoons, their failure to live up to our deepest expectations is embodied in the animal nerd: the bulls urging their friend in the ring to "go for the cape," the "stupid birds" leaving the nest by climbing down knotted sheets, the dogs unable to puzzle out the use of a gun to shoot a cat in a tree. Most of Larson's cartoons translate biology into society literally, making human circumstances grotesque. Instead of the wolf/deer predation system evoking an analogical social contract that refers to marriage, the wolves are seen studying a natural-selection diagram before attacking the deer, a parody of think-tank engineers at the Ford Motor Company. We civilized humans suffer a chronic loss of the old metaphoric distance and laugh at our own frustration. Larson does not solve the dilemma of the literalizing of relationships but reveals its irony.

MY INTEREST IN the *New Yorker* magazine has always been its cartoons. I confess as an outcity lowbrow I habitually flip its pages for them. But the distancing of New York from the wild world is the snake swallowing its tail, for these cartoons are often of animals. Urbanity produces eccentric and absurd human behavior that can best be mirrored by baroque projections onto the zoological realm, especially as people and their animals in or near Manhattan apartments—genres that might be called "people and dogs" and "man and parrot."

Confusion about our own identity is nowhere better represented than in the spectacle of a bizarre duet with animals on the twelfth floor of an apartment building on Manhattan Island—if not London, Paris, Rome, and Tokyo. At first it would seem that the aliens in the cities are the dogs and birds, but is that so? Dogs especially have been modified by breeding that is the equivalent of scores of thousands of years of "natural" evolution. Oddly enough, people have never been domesticated; that is, bred under conditions of absolute control according to an objective. Biologically, then, the people may be wilder than the animals: still their old Pleistocene selves, keen on animals, but blind to the post-Pleistocene breakdown of the boundaries that once kept live animals physically out of human households.

Because they are poorly adapted to modern life, the people in these situations are crazier than their pets. The latter are makeready therapy for masters who are confused because their genetic constitution calls for roving through savannas and small-group life under the stars. The pit bull in George Booth's blue-collar settings knows this, his eyes rolled up in an "Oh, my God, what will these idiots do next?" Dogs, given human names, rather than Rover or Spot, take on their owners' idiosyncrasies in exaggerated ways. They do not work. Their snobbery with each other is tuned to upscale sensibilities. They are preoccupied with the quality of their food and with routines of highly structured "going out," at which time they take their owners on a leash. Children are not involved.

Some final step has been taken with respect to the ten thousand years of the human/canine treaty. All of the dog's services—hunting, herding, pulling, companionship, guarding—have vanished. But he stays as a companion, mirroring the irony of success without labor. In New York he is the last major nonhuman remaining—a treasure for those for whom all essential otherness is in jeopardy, not only as external refreshment, but as access to their own wildness and that part of their personality otherwise invisible, the context by which they attempt to mark the boundaries of their own being. Exclusively human company is too lonely to endure, and these apartment dogs are the last defense against it. Without them and a few other animals, we are everything from the vanishingly small, catatonic, subatomic selves, crouched in the closet, to the megalomaniacs whose earth-circling noosphere expands outward to the rims of black holes and the far edges of time.

AS FOR THE parrots in the *New Yorker* cartoons, their perspective on their owners comes from a greater evolutionary distance. Their reptilian eye looks on the world of the hearth with no such vested interest as the dog. They are not companions, in whom there is confusion about likeness and difference, but speakers who seem to be making sense and whose parlance suddenly goes awry. New Yorkers are fascinated with parrots because of the frailty of speech—its inevitable misunderstanding, its limits as communication among us who invented it, only to find that, soap-opera-like, everything spoken adds to the glop of confusion, immobilizing the mind as much as it moves it, in words and phrases by which we conceal, deceive, and fumble meaning away. In the world of contracts, newspapers, literature, leases, and advertising, language is the emperor's new garb. In the backroom of their minds, New Yorkers fear that all the talk, which perpetually fills the air, amounts to the parrot's gibberish in the guise of thought.

The wild parrot is of course a mimic. The bird may indeed learn what "cracker" means, but parrot "talk" is iterated sounds. This imitative ability probably plays some biological role in the lives of free-living parrots, perhaps as the passwords of local groups. In New York, parrots seem at once oracular and ridiculous. The urban person cannot quite believe that what is said does not—should not—make sense. Educated to think that all confession, all Freudian slips, all accounts to psychoanalysts and other counselors, all testimony, all "tongues," mean something, he is transfixed by the parrot's stereotyped words in human language as well as the parrot's characteristic "talk" of its own. He thinks that perhaps it is like cocktail party chatter: trivial on the surface and yet revealing if you listen carefully.

As the great cacophony of New York, London, Moscow, Hong Kong, and Berlin swirls up around its inhabitants, they *must* believe that the babel they hear is the hum of the universe. They are buried not only in communication but, hopefully, meaningful noise. They go home from work and listen to music, the ultimate integrator of a dissonant world. But healing needs also a certain capricious witness who can remind the urbanite, the latent victim of neurosis and derangement, what folly his verbosity has led him to, and that the sounds of the city are not music. He could keep an iguana or a turtle, who silently lose their minds out of despair, but their mute gloom is hardly an antidote to the erosion of the heart and other organs by vibrations in the air and to the urban angst that speaks of the uproar of Broadway as a "lullaby." Deprived of the arboreal splendor of tropical forests, of other parrots, and of flight itself, the parrot is the paragon of survivors who will live thirty years in a room with occasional human company, and go on chatting as if life were intelligible. Among the humans around the parrot, those who see the irony in this little parody may outlast the others.

22 | Final Animals and Economic Imperatives

Drive out the natural and back it comes straightaway.

GASTON BACHELARD

EVERY CULTURE HAS its ultimate animal, a single figure that brings together the unique and disparate elements that distinguish one culture from all others. Represented as a living form, it implies a unity—the most powerful effect of the animal on the mind. The internal logic of the choice of an ultimate animal can be understood fully only in terms of the values and perspectives of particular cultures, but the overriding shadow of economy so shapes them that all hunter/gatherers share a basic outlook, as do all planter/villagers and all pastoralists.[1] These shared factors are strongly influenced by the quality of the natural environment. As the planet deteriorates under its burden of humanity toward a convergent end point, the whole of our species may finally enshrine the goat as its supreme animal.

Of the three basic economies—foraging, herding, and village agriculture—the oldest is undeniably foraging, although all three continue to be present in the modern world. A fourth category—urban peoples (locked to surplus agriculture)—is a composite focused on certain images of provender and the political paradigms entailed by chains of subsistence, dependence, and commerce in crops of plants and animals. This dependence is obscured from them, however, and city people tend to chose more abstract or wild animals such as lions or eagles as their banner.

In a compelling account of the symbolic animals in three African tribes, Roy Willis' *Man and Beast* compares the meaning of animals in the cultures of the Lele, who hunt and gather, the Nuer, who tend cattle, and the Fipa, who farm and trade. The pangolin, ox, and python are their respective supreme animals. Willis says: "I argue that these three beasts symbolize, for these three societies, the ultimate value—what we might call the 'meaning of life.' If my analysis is correct, what these animals symbolize is, respectively: transcendence of individual personality in pure, inner selfhood; transcendence of individual differentiation in pure communalism; and pure becoming, or developmental change, both social and personal."[2]

Willis reviews the general characteristics of each group: their notion of the means by which the world's dualities are resolved, the role of nature as a source or model for power in society, the place of diet as a code for belief, attitudes toward outsiders, the concepts of personal identity, their commitment to change versus stasis, their village layouts and personal naming systems. Only in the context of these aspects of their different ways of life does the choice of sacred animals of hunters, pastoralists, and farmer-tradesmen become clear. Willis then proceeds to analyze the meanings of the pangolin, ox, and python (summarized in accompanying table).

Willis' insight into the historical contribution of different cultural patterns shows the economic twist that defines the meaning of the Others. It demonstrates that the perception of animals is embedded in a milieu—the world, power, diet, identity, outsiders, feelings, and so on. Every culture makes its own fabric or scene; and yet, surveyed at a distance, they fall into a small number of types, based on livelihood, that take their features from the economy in its largest sense. Animals differ as metaphors "in different production systems."[3]

AS WE HAVE seen, foraging shaped the human interest in animals. That humans in all ways of life read animals as a special body of signs is the result of a human ecology that emerged with cognition itself. Hunting/gathering—or foraging—is the prototypical form in which the sacred significance of animals became part of the human condition.[4] Some anthropologists use "foraging" to mean a kind of gleaning and scavenging prior to hunting/gathering, a roving gleaning that was supposedly typical of the prehumans. But foraging here is a less clumsy term for "hunting/gathering," human

The Meanings of Pangolin, Ox, and Python

	Lele Tribe	*Nuer Tribe*	*Fipa Tribe*
GENERAL CHARACTERISTICS			
Economy	hunting/gathering	pastoral	farming/trade
The world	duality resolved	balanced duality	oppositions
Power	totemic	patrilineal	elected male
Diet	taboos	few taboos	few taboos
Outsiders	avoided	truculent	accommodative
Individual	identity = village	intense	private, competitive
Change	static clans	repressed	embraced
Feelings	rhetorical	confessional	hidden
Self	in larger reality	spear and ox names	persona via talk
Identity	kin and clan	vs. outsiders	interact with outsiders
Village	placed huts	very ordered	random, corporate
Culture/nature	integrated	twins	opposed
Name	matrilineal	spear/ox	object-related
SYMBOLIC ANIMALS			
Animal	*pangolin*	*ox*	*python*
Its role	unified	symbolic duality	plurality
The wild	important	distant mirror	unimportant
Society	analogy to wild	bovine idiom	male/female household
Humans	one species of many	cattle link to family	central among all species
Taxonomy	elaborate	based on kraal	provisional
Cattle	anomalous	intimate, symbolic	useful
Ideology	forest-centered	territorial	pragmatic
Philosophy	wholeness/parts	cattle-mediated	human-centered
Animals	psychosocial	detached	domestic only

groups without agriculture, archaic and modern, whose foods range from tiny seeds to large mammals.

Foragers look to the natural world as a vast treasury of forms and processes, ideas and analogical material, for thinking through their own lives. Some foragers are "totemic," thinking that a balanced opposition of nature and culture has systems of parallel differences. All animals in the forager's world are wild; domestic animals are neither kept nor eaten. Though known and studied with lifelong attention, animals remain a sacred gift, mediators, punishers, incarnations, and the subjects of endless discourse and speculation. Tim Ingold, an authority on pastoralism, claims that the distinction between hunters and herders is that the former are interested only in dead animals, the latter in the living. But he ignores the challenge to thought by animals in the lives of hunters, and the time spent watching, speculating, and discussing animal ways.[5] In the forager's cosmos, human concepts find metaphors or expressions in the zoological realm. Humans feel themselves to be watched, always observed, as the eyes and ears of animals are everywhere.

A modern example of such a people is the Lele, the Central African hunting/gathering people cited earlier. Although dependent on what they can find and catch to eat, their intense scrutiny of the natural world is described as an ambitious effort to conceive a comprehensive totality, including humankind. Anatomical and species systems serve as symbols for social groups. The idea of ingestion is for them at the heart of the integrative project of humankind. One might suppose such a "savage" people, "at the mercy of the elements" and without food storage, to be gobbling up anything edible, but the contrary is true. Not only are they discriminating eaters, but they are stern keepers of dietary prohibitions. Eating rules are symbolic links of the people to the larger world. Foragers have a paradoxical attitude about distance between themselves and animals, even though certain species—or all species—are seen as kin. They are circumspect about approaching and touching, a kind of interspecies hesitation. "Luck" in gathering and hunting is seen as a sign of social health, since spirits control both the game's disposition and the fertility of women.

Hunting and generating, male and female, village and forest—the Lele conceive of life as double imprint. Animals and human social entities are grouped in pairs. The body and mind are nourished by foods that are both symbolic and nutritional. However disparate their daily experience may seem, or how incongruous the events of their lives, the Lele believe that the forest is a perfectly integrated complex, the ultimate ensemble, the place of the resolution of contradictions, the center of the cosmos. Duality does not become an opposition.

For them the pangolin embodies this harmonious unity, resolves all ambiguity, and is therefore sacred. Having the scales of a fish or reptile yet giving milk and warm-bodied, living both on the earth and underground, a fearless yet benign being, the pangolin is the forest's cohesive force, the defeat of all contradiction, just as the forest itself comprises the whole of life. If the forest is the ground of Lele being, the pangolin is an enactment of it, translating its principal lesson into flesh and blood, which it transcends as the mediator connecting nature to human society. It presents equilibrium between disjunction and unity, multiplicity and integration, opposition and alliance, individuality and collectivity. The pangolin for the Lele gives life to a "pervasive ethos of ultimate oneness."

Other foragers elsewhere make other choices of a unifying animal—the bear, elephant, whale—but these too personify the most important principle in the universe: its wholeness.

VILLAGE FARMERS LIVE a very different life from that of foragers. Linked to the outside world through middlemen, many of which live elsewhere, it is not a noble world "of the soil" as in the imagination of the distant, urban, industrial world. Although farmers are still close to "nature," they perceive their relationship to it in terms of growth and control. The peripheral world of foraging and the power of animals to signify continues— but signifies differently.

For the Fipa, the Central African traders noted earlier, personal identity is a relative matter of initiative, acumen, and chance. Speech creates shields and disguises personal feelings, so that each individual creates his or her own persona which represents them. All distinctions and categories are provisional. Strangers come into the village bringing money, goods, and news. Meaning is not inherent but emerges out of interaction. Opportunities descend from the outside or from planning. The Fipa have their equivalents around the world.

For them the python most succinctly represents the wild animal and bodily world. It is spectacular and irrelevant. It seems to embody the residual unknown powers of the natural world, brought into town as if subjected to the intellect. The Fipa worry about sorcery and its control, perhaps seeing their own control of the snake as signifying the domination of rationality over the deeply instinctive forces, the nonintellectual self, and the darkness of wild nature made concrete in the python.

PASTORALISTS ARE SHARPLY different in outlook and personality from the foragers and farmers. Clarence Morris has observed: "A shepherd's rapport with nature is different from a farmer's. A farmer cannot rescue growing grain from threats of natural visitations; he is helpless when floods cover his field with mud, when drouths suck juice out of his stalks, and when locusts devour his every stem and leaf. A drover, however, can herd beasts up from high waters and out of blighted pastures. An old Chinese joke jibes at a dull-witted farmer who killed young sprouts by tugging at them to help them grow. A herdsman often, however, compels cattle to be productive; he manages matings, he takes milk, he forces fattening, and so on. The shepherd with so much control does not think of his husbandry as a concomitant part of a larger natural process; he is likely to believe he plays the role of an outside user of nature."[6]

Just as there are gradations from foragers to planters and distinctions among planter cultures, there are wide differences among pastoralists, ranging from farmers who take some grazing animals into high country during the summer, to nomadic herdsmen, to modern ranching.[7] So they live from one extreme in relatively humid environments in sedentary communities to desert and mountain ranges in which frequent movement is normal. Still, as a group, pastoralists have a profoundly different notion of the meaning of animals than either hunter/gatherers or farmer/villagers. They breed, protect, and drive groups of hoofed forms which are objective wealth, analogical nonhumans, and at the same time members of their own sociality. In extreme pastoralism a single species is so intensely used and observed that a kind of condensation of the universe emerges, a "bovine idiom."

This idiom has connections to the personality. Herding peoples tend to be aggressive. Herd animals themselves are species with strong dominance hierarchies, ruled by males who fight among themselves and resist intruders. The nomadic Fulani, a milk-drinking people of sub-Saharan Africa, boast of themselves as fighters and are engaged in "relentless wars of conquest . . . facilitated by the aggressive personality which they share with other pastoral cultures."[8] The young Fulani boys are trained to beat the cattle and praised for their courage. Among themselves a social code of challenge or insult both limits and requires physical attack with sticks. Boys fifteen to twenty-five engage in a ritual *sharo* contest, an exchange of thrashing with sticks that leaves permanent scars.

The bullish behavior of the cattle is part of the picture, requiring the herders to control the tough bulls by voice, threat postures, and sticks. The first requirement among the young cowherds is that they be able to drive, separate, and manage large temperamental animals among which fighting and copulation are major drives. Observers have noted: "If the herdsman has the sort of personality needed to display sufficient aggression to maintain his position as dominant over all the cattle in his herd, we might expect that his interactions with people would also involve assertive and aggressive behavior." This work, "quickening a kind of masculinity," is somewhat like sled-dog cultures in the far north.

Patriarchy, pastoralism, and the perception of animals are not related simply in the subordination of women and animals in a single mix of severe and impassioned authority, but in the variable situation of young men in the same net. Once the herding system has extended itself beyond the immediate reaches of the village, the power of elderhood is diminished by the independence of sons, who are chafed by poverty until they can steal enough cattle to be on their own. A system of having animals rather than marketing them is the engine that drives the social machinery, and its oversize herds degrade the ecosystem, diminishing the resources over which the cattlemen contend.

Distancing from the village, with its traditional restraints, often mediated by women, is made possible by mounts. In Incan and then Spanish South and Central America, in European North America and Australia, in Scythian/Mongol Eastern Asia, the Semitic Middle East, and Somali and Taureg North Africa, men riding cameloids, mules, and horses, driving goats, sheep, and cattle, accompanied by dogs, have galvanized a "human" ecology based on competition with the wild animals and among themselves.

Cattle keeping is not a stable enterprise. It is subject to "the constantly present menace of mass loss of cattle through jute (severe frost following a thaw), drought, epidemics, enemy forays, etc.," with losses of up to three-fourths of the cattle. Nomadism has limits of "internal development" when based on cattle breeding. Its overextended demands result in economic impasses and class struggles, the necessity of decentralized dispersion, conquering agricultural lands, and alliances which disintegrate, reform, and decay again. The convertibility of livestock lends itself to quick wealth, to social differentiation, and to regular wars over cattle and pastures.

As members of a species who are bilaterally symmetric and bisexual, the Nuer, Fipa, and Lele people perceive the world as binary (as do we). The Fipa farmer/tradesmen ignore this reality whereas the Lele foragers reconcile it as the transcendent unity of the wild forest. But the Nuer pastoralists find the dualistic world a source of perennial tension. Opposition between heaven and earth, body and spirit, men and women, wild and tame, like paired combatants in a field of battle, are balanced in their cattle. "Within the inclusive category 'cattle,'" says Roy Willis, "there is a further bi-polar opposition which reproduces, in little, the form of the basic Nuer distinction between domestic and wild animals."[9]

The cosmos in all of its oppositions and twinness is represented for the Nuer in their cattle. Moreover, the cattle mediate between human and cosmos. All pairings, all mirrorings, all symbioses—the very idea of unity and division—may be espied in the distinction between bulls and cows, bulls and oxen, not as equally valued, but as dyadic. Impervious to outsiders, isolationist with "a deep sense of superiority," the Nuer see themselves and their cattle as forever projected against a background of wild nature and other peoples. The cattle herd "is" human society in this imagery: the ox or cow is the person and descent, traced in bulls, implies the agnate genealogy of the patriarchal Nuer. All social processes are defined in terms of cattle. As their cattle scatter to graze and coalesce to travel, the Nuer disperse and concentrate in season.

PANGOLIN, PYTHON, AND OX subsume the worlds of the Lele, Fipa, and Nuer tribes. Elsewhere, the choices are different but continue to be defined by the economy. Pigs represent the human vision of life in many small-farm, horticultural, subsistence societies. As forest omnivores, the pig and bear have a special place in cosmological systems. The bear receded with civilization; but the pig, because it lives in social groups, can be incorporated into human society.

As late as the Middle Ages, wild boars were among the most holy creatures—intelligent, omnivorous, powerful, beautiful, and respected in association with the goddesses of the Old World. The pig was sacred to the bear goddess, Artemis, and, in the north, to Freya, the goddess of the underworld. Attis, the lover of Cybele, and Adonis, the lover of Aphrodite, were killed by boars. Anaeus was castrated by the Caledonian boar, just as, in Egypt, Osiris was killed by his pigheaded brother, Set. How are we to understand all this death and wounding by pigs? Is it the goddess, in a rising tide of patriarchy, striking back? It was not only the Jews who refused to eat pork, but the followers of Adonis, the Egyptians, and the Muslims, perhaps in order not to outrage the goddess—or to internalize her qualities.

Among subsistence farmers the pig remains central to the calendar of life, lean and hardy when not penned, little different from its wild form and yet keyed to human ceremonial and social life.[10] In northern Europe, more than any other animal, it enabled usage, access, and modification of the post-Pleistocene primeval forest. In the wet protein-less deserts of the tropics it is a remarkable source of amino acids.

For farmers in the modern sense, from medieval peasants to Iowa agribusinessmen, the pig has become pork, the first garbage disposall, from the lowly scavenger of the streets to the great blobs who spend their life indoors, from Albert Schweitzer's pet, "Josephine," executed by the great doctor in masterly peasant fashion, to numbered individuals in a variety of breeds, none of whom could last a week in a wilderness, north or south.

For the farmer's city counterpart, "pig" and "swine" have the worst possible connotation. From the corpse- and shit-eating brute, sunk in mud, to the menacing boar and the slob of a sow, through all the images of greed and intractable pigheadedness, pigs have come to represent the degraded status connoting "animal" or "flesh" or "body."

For pastoral people pigs are useless, slow, and difficult to drive, as well as being the hated animal belonging to planters. Of the many possible explanations for the disdain of pork by the Jews (who perceive themselves as traditionally "pastoral"), there are various theories—from empirical protection against trichinosis, to the pig's ambiguous taxonomy as a non-cud-chewing form of "cattle," to its guilt by association and earlier status as an animal sacred to the great goddess, the "harlot" in Hebrew eyes.

Although the animal is celebrated by some poets, clearly the degradation of the pig is related to its obesity and sloth in captivity. Kept forest animals are anomalous; the basic human settlement is a clearing—which is not pig habitat, so they have to be fenced. The deterioration of the lean, powerful, wild boar is bound to the conditions of human settlement. Its omnivorousness, so adaptive in the wild, becomes a flaw in captivity, where it eats what even goats would reject, and it lays on fat in its deplorable gluttony. Due to its diet, its feces (like those of humans), unlike the sweet manure of horses and neat pellets of sheep, pile up stinkingly in its confined spaces. Its efficient rooting, clean body, and neat mudbaths in the wild become the fetid wallow in the barnlot. In tropical settlements, where it is prized meat, it is so vexatious a plunderer of gardens and invader of dwellings and killer of puppies and chicks that it is periodically slaughtered in celebrations of human catharsis. Yet its keen intelligence and individuality are such that, being butchered, its frantic struggle and mortal screams are like murder. In the industrial farm this pig dilemma is hidden from most of us, but it is there: the animal who most poignantly and profoundly represents, in its terror, psychopathology, and brutish voracity, a culture dominated by bouts of greed and anxiety, relieved by intervals of torpor or frenetic activity called "recreation."

Other animals could be identified as the bellwether species in their particular places and times.[11] As time runs out for the fertility and spaciousness of our planet, as wild species vanish as though in funereal procession, as pets themselves become too expensive, let us leap forward to the end of the next century to discover its supreme representative.

THE GOAT ASCENDS as the planet loses its soils and forests and becomes increasingly homogeneous, and we can foresee a uniform, worldwide, more or less impoverished landscape and culture that elevates a survivalist to the status of its metaphysical symbol. The landscapes of the world desert will increasingly resemble those fringe habitats of the Near and Middle East, where the wild goat lived and became the first domestic animal. At the end of affluence, as the world becomes desertified, the goat will return, and with it the great drama of ecological and historical domination, initiated with the dog and horse, will reach its last act.

Wild members of the genus *Capra* once occupied terrain too rough for antelopes and gazelles. As a specialist in steep, rocky places in subequatorial mountains, it developed not only superior footwork but unsurpassed gleaning ability and a good brain for the strategies of survival in difficult environments. Goats stand by us as the ultimate, practical, therapeutic companions who will lick our faces when all else fails and (unlike dogs) convert nothing into something. Although loved for its companionable qualities, it remains close in appearance and behavior to its original ancestor, a creature of stony places which became the animal of peoples on the periphery of the human habitat.

More than most domestic animals, the goat has retained most of these adaptive traits in habitats so meager or so despoiled that the vegetation is composed only of the tough plants that can manage thin soils, undependable water, and wide latitudes in physical conditions. The teeth, hooves, agility, and helpful digestive microbes of goats all were factors in adapting to such places. The same teeth and hooves are the instruments of keeping such places barren, since goats, when constrained, eat all the plants down, the bark off the saplings, and the organic debris, while their hooves cut the soil loose from the slopes.

Livestock keepers prefer cattle or sheep, but the latter need more level ground, more water, and richer grazing. All three flourish in rich pasture, and there is much overlap in herding practice, especially with mixed flocks of goats and sheep. Away from such productive places, the cattle drop off first. Some breeds of sheep do all right on poor range but lack the necessary intelligence. At the outer limits of habitat in the desert or its mountain equivalent, where the diet may include the toughest, most minute plants, bark, leaves, twigs, and almost any organic debris, where it takes a good brain to shelter from blizzards and predators, the goat thrives.

In such environments human groups are smaller. As meat on the hoof the goat is packaged in more numerous and smaller units than cattle. The staggering of milk production throughout a herd ensures a more sustained supply than that of cows. If the young need human help they are easier to manage than calves. Individual losses to predators or disease have less impact on the whole. The smaller hide is easier to work or carry than a cowhide. When feeding, goats remain in a more compact group than cattle and can be more easily overseen, defended, or rounded up. Being more intelligent than sheep or cows, they respond more readily to people and easily learn to take directions. They develop distinct personalities, too, which makes them interesting and attractive to humans. Their playfulness, independence, and cantankerousness give them individual qualities. Goat milk can be used by people in many forms and is lower in fat and cholesterol than that of cows. The meat is excellent. Bones, horns, sinews, soft organs, skin, wool, urine—all have uses. By gathering roots or berries, killing small animals, or growing a few crops people can live quite well on goats.[12]

Great claims are made for the goat's "ecological" efficacy.[13] That they do not patrol the streets instead of cattle in Third World countries such as India is both practical and historical. Goats' droppings are small pellets, harder to collect than a "cow pie." Manure is a major cottage industry in India, primarily as fuel, but also to fertilize gardens. Thousands of women gather, shape, dry, and sell patties of cow manure. Cows, despite their size, being slower, stupider, and more benign, nuisance that they are, accommodate better to traffic and street stalls than the nimble goats. The big heads and large soft eyes of cows are endearing and reassuring. Their calm demeanor suggests sanctity or repose, and their benign behavior contrasts to that of the horny goat whose climbing and butting are amusing in the countryside but not in an open marketplace. Sly and quicker, the goats are a problem in town, easier to steal, less predictable in front of traffic, not protected from slaughter by religious custom, and more vulnerable to attacks by dogs.

The cow's religious role is long and complicated, growing from pastoral obsessions with the symbolism of several species of South Asian bovids. The cult practices of Indo-Europeans, coming from the eastern fringes of the

Caspian Sea, no doubt influenced indigenous Indian respect for the cow as opposed to the lean and smelly goat, with its suggestion of poverty in its frugal ways and the dearth of plant life typical of its habitat. Goats may not be so rich in their associations, but they are more fecund. They have more copulation and more babies more often than cattle. The sexuality of billygoats is equal to that of bulls, more conspicuous, and less easy to control. The liberated personality of pastoralists may relish this lasciviousness more openly than do the repressed townsfolk.

Goats live so well and supply human needs so completely in marginal places that enthusiasts invite us to go "goatwalking" in the desert.[14] Goatwalking in North America is advocated by Jim Corbett, who extols the virtues of seasonal, nomadic, pedestrian pastorality, attracting the disenchanted. His success rests on the superiority of goats to sustain the carefree life and on his political activity—herding illegal immigrants, primarily refugees from Central America, into the U.S. by way of the Southwest deserts, as the "Sanctuary Movement." The exodus of the Hebrews from Egypt and their wandering as a pseudopastoral people is an inescapable analogy. Thus does ideology link itself with goats, as it no doubt did when the first sons of Abel, surely goat-keepers, assumed the preferred place in the eyes of God and drew upon themselves the scorn of effete city folk and the envy of Cain's sons, the farmers. So too does the twentieth-century goat lend itself to civil disobedience, as though its own irascible personality were an extension of a protest movement. It is no wonder that it was the darling animal among the dropouts and communal hippy farmers of the 1960s.

Meantime, the world's deserts expand. The greenhouse effect and the diminished ozone shield enhance this process. Even if, by a miracle of desalinization, we manage to irrigate vast regions we will create ecological deserts because of the loss of wild species and reduced diversity. As the world's forests shrink, they are replaced by scrub rather than grass. Net soil erosion on planet Earth far exceeds soil formation, so that the underlying rock comes peeking out, except for the scattered and leathery tissues of a sparse herbage or the thickets of the woody epilogue of a vanished forest. The immense pressure of the exponential increase of human protoplasm, though it progresses at different rates here and there, will eventually require gleaning every last straw.

Wild animals are being replaced on terrestrial earth by human artifacts and towns. Like wild animals the "useful" animals require resources. As time passes even they will drop away according to a hierarchy of needs, just as horses, for example, have already disappeared in impoverished nations and cattle in the more desperate countries. The resources for keeping horses and cows diminish along with the room for wild things. All the dreams of politicians and economists notwithstanding, the growing mass of humanity is as inexorable as a glacier, competing with and replacing nonhuman animals. By the mid-twenty-first century, there may be nothing but "Third World" human communities. The North may be covered with a veil of acidic detritus, the South with dust. The goat may be everywhere, along with such insects as consume its dung. Perhaps the oceans will have swallowed most of the topsoil and will be mined by millions of rafting small-scale fishermen and their families, whose goats, dancing on the decks, will provide milk and munch the parts of fish which people cannot eat.

Dogs, being more nearly omnivorous, scavenging, and increasingly needed as prosthetic social and psychological devices, will last long. But the goat is a good companion, too, and a better gleaner. In a scarified earth, efficiency is everything in converting an ever larger share of the sun's energy into human protoplasm. In the barren ground, the piles of waste, the raw margins of the productive enterprise, the skinned earth, the tiny yards— who else could be our nonhuman companion and protein-maker except the avatar of poverty?

Apologists for the goat, as in most cases of environmental abuse, cite only the (human) good that flows from it. Indeed, the United Nations has not hesitated to applaud the grazing of goats on marginal land. According to John Klars, the goat is "the channel or concentrating agent by which vast submarginal areas contributed to the betterment of a few agricultural foci." Because it eats anything, "in such a landscape, where the potential for sedentary agriculture is at best minimal, the goat assumes truly magnificent usefulness." And so, we are told, "the white knight of survival may . . . still come riding on a goat."[15]

23 | Rights and Kindness: A Can of Worms

They too are persons . . . entitled to respect because they exist as ends in themselves, not just as objects for us to look at or as means to any of our ends.

MARY MIDGLEY

ANIMAL RIGHTS IS one of the most visible and dramatic campaigns in the modern world. Although abuses of animals have a long history, urban industrial society has added so many ways to ill-treat them that public outcries have become loud and organized against the whole range of horrors. Advocates of "animal liberation," "animal ethics," "the humane movement," "animal rights," or "animal protection" have had striking success in creating public outrage and reaction. Even so, as a major philosophical umbrella for articulating the relationship of our species to others, as given voice by Mary Midgley above, it is disastrous.

Evidence for cruelty to animals is easy to find and to trace. The Jewish Bible warns against mistreating one's animals—clearly a prohibition against known problems and therefore evidence of abuse. But its modern expression arises with modernity in the eighteenth century. Experimental science, widespread pet keeping, the increasing application of technology to animal husbandry, and the marketing of livestock all produced their own forms of abuse and reaction against them. Enlightenment science with its new objectivity set a ruthless mood in the laboratory that to many seemed little different from the long-standing customs of cat-burning and the beating of dogs and horses. Much has been made of René Descartes' denial of a soul and sentience to animals, but it is clear that science was no different than popular culture when it came to cutting up animals. New thinking about the mind and consciousness played an important role. Deliberations on pain, and the discussion of the moral status of animals by Montaigne and others, contributed to Victorian confrontations, particularly in English street demonstrations against medical training and experiment.

Concern for animal welfare came about in part as a trickle-down from bourgeois pet-love, urban eyes on cruel drovers, the spectacle of bear and dog fights. Animal protection emerged in the twentieth century as a footnote to political and civil liberation movements, as a complex, educated wrath over the general conditions of animals in research, on the farm, in slaughterhouses, in the household, in the streets, on fur farms, in traps, the "harvest" of seals and hunting of whales. In modern congested cities, millions of animals became victims of negligence and vicarious punishment, growing out of abandonment in which unwanted litters become feral populations living on rats and garbage.

While animal keeping on small farms may often have been benign, there was never an ethical issue among peasants as to the physical control, marketing, and slaughter of the birds, fish, and mammals that constitute the peasant's world. With twentieth-century "industrial" agriculture, however, the practices were organized on a scale previously rare and without the spontaneous caring that was usual among farmers. The breeding cycle became controlled to a degree and with measures that went far beyond anything known to old-style farmers. Domestic animals are typically squeezed into pens, cages, and stalls, castrated, debeaked, dehorned, injected, altered with hormones, force-fed, and exercised only when it is consistent with the desired product. Restricted throughout their lives, slaughtered in terror, even their own social lives are destroyed as the animals spend little time in groups. The industrialized farm crowds chickens for slaughter into small containers in the dark or puts egg-layers into artificial light/dark pens. Sows are harnessed and in-stalled, piglets quickly taken from them, and finally enclosed in darkness as they are fattened. Calves for veal are tightly confined and even sheep and cows are increasingly constrained. Their foods are increasingly unlike their natural food. They are shipped, warehoused, and dealt a mechanical, impersonal death. Deformity and cancer are widespread. The "pig" is little more than insentient pork, without access to sunlight and open fields, dazedly enduring monotony and isolation, with an attenuated or narcotized nervous system that reduces it to a numbed body and a vegetative brain.

Suffering and the broken heart, being largely subjective, are difficult to measure. For farm animals, free life in an outdoor setting is nearly gone; even on the range, herd animals must be fed, vaccinated, branded or tagged, and protected from the leg-breaking holes of burrowing rodents, from odd predation on their young, and from their own immobility and stupidity in bad weather. Their brief lives end in impersonal handling and degradation.

The "farming" of animals includes the fur farm and the wild game farm for private hunting, target practice and dog-training practice, restocking, and meat. The recent tide of caring and kindness extends to zoos and private displays, circuses, trapping, the keeping of pets, and hunting. From attention to simpleminded apathy, ordinary horse-whipping and careless brutality in the preindustrial world, through Hogarth's drawings of the stages of cruelty, to neurotic hunters' misplaced zeal, scientific and industrial exploitation, and the insensate side of pet keeping, we see the vast landscape of oppression as the motive to protect "our animal friends." Added to all this is the trade in wild animals and their parts for folk medicine and aphrodisiacs, scaled up in the past decades from regional to international, subject to magnified abuses by well-heeled gangs with automatic weapons and distributional networks.

Among the nations, Sweden points the way toward which civilized countries move. It has the most stringent animal welfare laws, guaranteeing the freedom of cattle, pigs, and chickens from the worst horrors of factory farming. Cows have grazing rights, pigs a right to a clean bed, and chickens a right to get out of their cages periodically. Except for the treatment of disease, no drugs or hormones may be used on them. Slaughtering must be done humanely. Curbs are imposed on scientific experiments using animals. By 1990 the Swedish legislature was considering prohibiting the use of hormones to increase milk output in dairy cattle; by the mid-1990s the target was genetic engineering. The Swedish leader in this movement in the 1980s was Astrid Lindgren, the author of *Pippi Longstocking* and other children's books and grown-up versions of children's stories for the widest-circulating newspaper in Sweden.[1]

Elsewhere, much of the animal rights movement addresses the use of animals in research. In America, Tom Reagan's book, *Animal Sacrifice*, for example, is directed to this issue. With respect to domestic animals, no sane person could quarrel with his concern, and the moral validity of his perspective is obvious.

BUT WITH TOM Reagan and many others, the message of animal protection does not end with captives. In the projection of this hearth rug altruism onto nature the proponents of "kindness" fall into the traps created by their own, often infantile, imagination. No doubt animal welfare, sweeping up derelicts from the street, has an important moral weight on its side. All the traditional motives related to mercy, compassion, and kindness apply, but only when the reform is limited to the management of enslaved animals. To project this logic onto wild animals is to envelop the natural world in the fantasy. One of Astrid Lindgren's newspaper stories is written from the viewpoint of God, "who made man to take care of the animals," and who is disappointed in our performance. We are cautioned to do better. After all, "taking care" of the animals does not mean allowing them to be at risk by running free. It means fixing the conditions of their lives and protecting each of them from human cruelty and natural hazards.

This idealism is defective because it leaps from the wretched, abandoned cats and dogs and their "suffering" as experimental objects, and from the blighted lives of cattle and poultry in factory farms, to hunted victims—to "lost" whales supposed to be helplessly blundering about in harbors or beached at low tide or tardy in their southward migration and trapped north of the arctic ice. It exaggerates into a ghastly caricaturing of the medical oath to save all lives and end all pain everywhere at any cost. Attempts at such animal welfare address a truly colossal debacle, but the reaction is muddled in piety and its dream of the survival of *all* the buffaloes and *all* the stray cats.

That a moral outrage against suffering cannot be directed against the "suffering" of populations and ecosystems is ironic beyond laughter and tears, because the destruction of those natural communities is immeasurably more important than all the pain in all the cats everywhere.[2] But the extension of "kindness" to nature exhibits a deracinated clinging, a neurotic zeal that omits all of "nature" except the individual—the projected self as it gropes for terms of responsibility, justice, and affiliation in animal welfare: protection, rights, kindness, ethics, liberation, reverence for life, sanctity of being, confounded with the semantics of "friendship," "stewardship," "kinship," "humane," "fellow creatures" and mixed in analogous politics of human political and social liberation.

"Protection" is among the least complicated of the terms. It refers to the freedom of animals from misuse, ranging from the threat to an endangered species in some far area of the earth, to the trade in captives and skins, to a local population of butterflies threatened by development, to saving rhinos from the hired guns who would cut off their horns for the apothecary market. Beyond that, a fierce obsession takes hold—to protect wild antelopes from the weather or starvation in a harsh winter, to save baby rhinos from marauding lions, rare mollusks from human sewage, wild moose from cattle diseases. Is there any end to our benign outreach that imagines you can "save" all individual animals? In this fairy-tale world of "saving," death is banished forever.

"Liberation" invokes the end of oppression, especially in the political and military sense, where nations, races, and classes are subjugated. But in order to be liberated one must be captive, which is well enough for sheep and turkeys, but the captivity of lynx and peccaries is the "captivity" of a dangerous wild nature, a world of risks and early death. Of course, "liberation" can mean "to take from its previous owner," as when soldiers "liberate" property. Into what alternative status other than the barnyard do we liberate animals from their oppressor reality of food chains, parasites, disease, and storms?

"Rights" implies some kind of cosmic rule prior to any contracts among users, legislation for protection, or decisions to liberate. It refers to something intrinsic or given by God or Nature.[3] Animals may be said to have rights to continue life and freedom because that is what we say they have. A chicken may be given rights to walk about freely for two hours each day by the Swedish legislature. If we wish, rights apply as well to a goldfish in an aquarium, a bacterium in a biological laboratory, a zoo animal, milk cow, or pet cat.

Wild animals do not have rights; they have a natural history. To specify the right of baby lions not to be killed by hyenas or the right of spider monkeys not to be eaten by Peruvian hunters gives the lions and monkeys a legalistic or pseudopolitical reality. One may legislate on behalf of them, dictate a denial of rights, create rights, cancel them, or never think of them at all, but they do not exist outside of those human considerations. Talking of wild animals' rights is like running in sand. Rights of one animal infringe upon the rights of another. If the issue is truly their welfare (and not some act of guilty compensation on our part), the legal and moral rights of forests or mountains would be more to the point, but only as a rather tenuous metaphor. Otherwise, the whole matter of rights becomes as conspicuously silly as it is obscure in reference to a cockroach. The only way such rights make sense is in terms of restrictions on my actions toward cockroaches or mountains—which I am prepared to accept—but they do not have to be described in terms of "rights" at all nor assumed to preexist in the order of things. Somehow these indigenous "rights" appear by magic when stated.

Philosophers argue the matter of animal rights in legal abstractions as well. Joel Feinberg says that duties toward animals follow a right, which in turn results from a claim. A claim becomes a right when it is acknowledged to exist from some prior circumstance or when its moral basis is accepted due to one's conscience or God's direction. Such a moral basis, he says, is seen in the statutes against cruelty. Since animals lack duties, moral standards, and understanding and are not "persons" with "interests," they cannot be participants in their own case. (To consider them thinking beings, he notes, is a primitive overregard for beasts.) Being so deficient, they belong to that category of helpless beings whom the law recognizes as deserving protection, like infants or imbeciles. Some rights, he says, precede legislation, being "natural," such as being free of unnecessary pain, free of having their movements restrained, free of lack of privacy, or free of being wantonly slaughtered. The idea of justice centers on biblical/legal retribution. The recipients of justice must *deserve* what they get. Deserving assumes accountability, and a claim to be made. Thus the idea of rights correlates with justice. A modification of the relation of justice to rights is the concept of the common good. On this basis it can be applied to animals.

But are rights really prior to our announcement of them? Where were they in the Pleistocene? The Paleozoic? Rights do not make sense without reference to a source. The dictionary says that rights are "that which is due to anyone by law, tradition, or nature." But law and tradition can just as easily be murderous and tyrannical. If there are "natural" rights of an animal, they are the right to participate in the ecosystems in which they evolved, a "right to exist" (only as long as they live, which for many newborn elk and gnus is just a few minutes). Is the right to be free from cruel treatment expressed only in terms of human restraints, or is it a right to be free from being eaten slowly by a parasite or rapidly by a predator or postmortem by a vulture? If so, what about the predator's right to a meal? What of the rights of the grass which is eaten by the deer?

Theology answers that we recognize all these values, but some are more important than others and we are therefore obliged to judge them. So said Albert Schweitzer when he poured kerosene on the army ants while espousing "reverence for life." We are back to the biblical and humanistic hierarchies; so much for "natural" rights. Since the idea of liberation grows from Western ethics, the power of rights seems to be determined by how much a species of animals resembles humans or contributes to human life. But evolution is not a ladderlike structure leading to our species; it is tree-like, with the humans at the pinnacle of nothing but their own family branch, on which they are the only living species. The perverse theme of modern rights is that locusts are less important than dogs, snails less than badgers, and all of them together less than one human life. When rights ethics tries to abandon such a hierarchy, it creates purists like the Puritans or fanatics like the Jains, whose real concern, as Schweitzer put it, is not the animals but our immortal souls.

"Ethics" as applied to animals is a more formal philosophical statement or code of human obligations to them, including their proper use, their rights to be, and even, perhaps, the necessity of their liberation. Ethics is formally a part of philosophy, as invented for the Western world by ancient Greeks, just as the Hebrews invented our morality. Ethics, conceived originally to deal with relationships among "men," or between them and their gods, may include the logic of human relationships to animals. Ethics lends itself to the concept of justice and to the justification of rights. It explains the logic of a morality. The "ethic" includes opposition to giving AIDS to experimental chimpanzees, to the wearing of furs and skins, to laws against dogs riding in open trucks, to fox hunting, to the overworking of horses in "western" movies, to the reduction of feral burros in national parks—all revealing the deeply felt absence of the natural in our lives, expressed as high-sounding transactions with nonhuman life.

THE PROBLEM OF infringing on one species while protecting another has been "solved" by arguing that animals have both intrinsic and instrumental value. The first is of a higher sort. Intrinsic value implies something above utility. All creatures have intrinsic value, but some, being more sentient (like us), have greater intrinsic value than others. Thus an elephant has more intrinsic value than a fish. But what is the value of an elephant in the eyes of the fish who breed in its footprints in a stream: is it intrinsic or instrumental? How are we to act on the intrinsic value of my dog's fleas? Intrinsic values are ideal abstractions, like goodness, beauty, and truth. They are useful in order to understand the limits of idealism in an ambiguous world.

Instrumental value has caused even more trouble, since that is what cows and dogs have for us. As an ideal it may have some integrity on the family farm, but it produces untold injury when the instrumental values of animals are incorporated into the world as commodities and the instruments of the factory. Ecologically, all things have instrumental value. These are not necessarily gross and materialistic and selfish—that is what we have been making them, turning them into objects, units of meat or hides or energy or experimental victims. Only in captivity and museums do animals exist as objects. Outside, all live in natural systems and are instrumental to one another. The question is not to act on their "intrinsic" value in isolation but on their place (and our place—our limits) in such systems.

And then there are the "friends" of animals. "Friendship" can have a certain bizarre reality between master and slave, but it is a one-way window otherwise. Few individual wild animals recognize members of another species as "friend," except perhaps to approach in shy curiosity or becoming habituated to tolerance or benefiting from its warning calls. Self-styled "animal lovers" argue that wild things learn to fear man, that otherwise they would all be cozy friends in a kind of garden world. They point to foxes on St. Nicolas Island, or iguanas in the Galápagos, or boobies on Guam. But the examples all come from islands with no large terrestrial predators, populations in which the genetic disposition of normal vigilance which keeps such vulnerable animals at a distance from anything big has been gradually dimmed, just as the capacity for flight wanes among some species of birds on islands. The little fawn that does not run is not evidence that the deer has not learned fear; it is evidence that flight is a poor choice in its circumstances and that its nervous system is not yet developed. Most such animals on larger land areas have inborn fear responses, especially by the time they are able to run or fly. They can learn not to fear humans or other large forms by habituation, but they are not "born that way." Young mammals also follow cues from their parents and their companions in this regard.

Like humans, many animals develop inherent avoidance when they are big enough to discriminate among species and have become mobile—responses that can be escalated into fear or diminished by habituation. The wild African chimpanzees and gorillas who accepted Jane Goodall's and Diane Fossey's proximity, and that of all the cameramen who have come after, were not their friends any more than they were friends of the bushes around them. Thinking of ourselves as a "friend" of distant whales, oceangoing sea turtles, or "vanishing" rhinos is a playlet in the head in which "I" am being "friendly" by my charity, giving money or appreciating them. A wild animal's "friendship" toward humans usually consists in going on as it was, or being lured in with food or captured. The relationships between species are not enmity, friendship, corporation, or family; they are simply not social in the sense of amicable attachment and generosity that the word generally means within our own species.

Of all the terms of the humane movement, "sanctity of being" has the best ring to it. It seems to say: "Let be; respect all creatures as they are; exercise responsibility or even compassion for life by not interfering," as Mary Midgley says, "as they exist, as ends in themselves." But is "being" a noun or a verb, or both? Are only individuals the object of this letting be, or does it include populations, species, natural communities, the biosphere? Does it include *our* being—just being, or being something? In these semantic tangles (and quicksands) philosophers live and play; their rational ploys and ambiguities pin us to our skulls. Their streetside counterparts, moved by small tragedies, carry banners, barricade fur salons, shout slogans, "Let them be!"

As models of people with such an "animal ethic," the Jains are often cited because of their doctrine of *ahimsa*, or "harmlessness." Wearing masks to avoid accidentally inhaling and killing insects, stooping to remove tiny creatures from the footpath, abstaining from all meat and even eggs and those parts of plants that cause the death of an embryo, the Jains appear to be the very acme of the love of life. But when the Jain philosophy is examined more closely, it is seen to be motivated by an avoidance of defilement. Jains are among the world's great dichotomists, whose abhorrence of all things organic drives them to despise the living stream. For them the "pure crystal" of nonmaterial essence and the "dark pollution of the material world" are in profound conflict. Organic life, unfortunately, mixes the spiritual and physical, which the Jains would keep separate. Their attempts to save worms from being trampled signifies their desire to avoid all possible connection with the natural world, even on the bottoms of their sandals. Their "harmlessness" is estrangement. As Frank Darling says of them: "The Jain, bent double peering at the ground before each step lest he should kill anything, would need eyes of an order not granted to us to see the still smaller creatures in his way. There is no room in our philosophy of responsibility for preciousness."[4] One does not have to dig far into Christian roots to find a parallel with Jainism's world rejection—and to suspect that the sweet charity of animal protectionism is energized not by religious insight and commitment to life, but by a free-floating compassion, more cathartic than ecological, soured by the contamination of birth and the fear of death.

"DEFENDERS" OF ANIMALS, animal rights and protection advocates, antivivisectionists, vegetarians, opponents of cruelty and torture, authors of a new ethics, antagonists of hunting, and promoters of "kindness" to our animal "friends"—my skepticism with this vast moral outrage, with its smarmy cloak of affection and fellowship, may seem strange in a book attempting to witness the essential place of animals in our humanity. Altruism is undeniably important; cruelty, callous keeping, rationalized butchery, overmechanized husbandry, social and legal disregard, corporate, commercial, and scientific exploitation, and the ordinary street torture of animals are all reprehensible. Given the reality of such cruelty and the advance of the modern paradigm of "kindness," why do these expressions of concern and caring make one uneasy? Why do polemics on behalf of the dumb beasts seem so inadequate as ethics or morality, and why does this philosophical equivalent of the veterinary romance of *All Creatures Great and Small* stick in the craw? As social phenomena and selfless intentions, such sentiments are clearly an expression of widely shared feelings. But this gentle and generous concern, the extension of civilized regard for a relatively helpless, kept assembly, however appropriate in the ethos of captives, is mostly bad ecology that can drive out the good.

With respect to nature as a whole, if kindness, friendship, and caregiving are not the Way, then what is? Is it "bonding" with animals? Mary Midgley advocates a new sense of bonds with animals.[5] She calls the idea of human superiority "species solipsism," resulting in the "species barrier" or "species gap"—that is, "the powerful tendency to resent and fear all close comparisons between our own species and any other." What we need is a "species bond," she argues, like that of "Elephant Bill," who worked with Burmese elephants for thirty years, one of those "who spend their lives dealing successfully with extremely demanding animals." In view of the subtlety of the elephant personality, she characterizes the thesis that animals have no consciousness as "ignoring elephants." Unlike some animal protectionists, she has the good sense not to blame our harmfulness on evolutionary theory ("nature red in tooth and claw" or "survival of the fittest") and to understand that Charles Darwin was himself "a broad and generous spirit" whose work made possible not only modern ecology but the modern naturalist's sensibility and recognition of mutual dependency and cooperation among animals, as well as the overall positive function of predation, parasitism, and competition.[6]

Part of the problem with Midgley's perspective, and that of most animal ethicists, is its simplistic repudiation of killing animals. (Elephant Bill actually loved elephants so deeply he had no doubt that some had to be shot.) Life sets traps for philosophers. Is "bonding" just a perspective? Midgley uses "bonding" to mean the recognition of a responsible kinship, but the term has long been used in the scientific study of animal life. Baby animals imprint on their parents, so that they direct their social behavior as adults toward the right species, becoming capable of an appropriate bond. If reared by a different species, a surrogate keeper, they misbond and develop social abnormalities that render them incapable of associating with their own kind. Unfortunately for "ethics," the natural community of life is not a bonding between species.

Another writer has argued that keeping animals was the tribal peoples' means of "bonding." Tame raccoons, moose, bison, bears, coatis, parrots, eagles, and so on demonstrate, he says, that pet keeping is not simply a feature of "Western wealth, decadence, and bourgeois sentimentality."[7] But those captive, tamed wild animals were simply prisoners, objects of beauty or curiosities, but not welded to their keepers by anything more than clipped wings and deformed social bonds. These captives were never confused with the vast realm of wild animal life that surrounded such villages. To propose the tame animal and its owner as a model "bond" for human/wild animal relationships is nonsense.

"Bonds" evoke images of babies and their mothers, of fraternal loyalty, of priestesses in secret rites with their goddess, of blood brothers, old war buddies, or cellmates. "Interspecies bonds" implies something felt by both parties, though it could mean commitment to animals without expecting any such commitment in return. Moreover, bonding suggests powers stronger than commitment, something forged in unconscious ties that transcend or precede loyalty, emotional as well as intellectual, yet having formal obligations. The term is used by psychologists to describe mother/infant dependency or sibling loyalty. But a bond is not only a connection. It is the deposit of valuables risked by those charged with a crime to guarantee their presence in court. Bonds are legal instruments defining rights in terms of restraints on the freedom of the bonded to be themselves—that is, for animals, to be wild. The primary symbol of a bond is a tether. It is difficult not to see in this reference to bonding by animal protectors a kind of barnyard hypocrisy. Am I to feel interspecific bonds to all other species? Am I bonded to bacteria as well as to horses? I am connected to bacteria by descent (not to horses) and I am host to their depredations and benefits in my own tissues. I suspect that life on earth would end without them. But I make no deposit against the risks in their personal lives. Their individual tragedies are not my tragedies.

It is as though all the virtuous and well-intentioned sense of obligation toward a few species of domesticated animals were seen as a guidepost for attitudes toward all animals everywhere. Organizations like Kids in Nature's Defense (KIND), sponsored by the National Association for the Advancement of Humane Education, pledge "Creatures great and small, we must defend them all." It sounds good, but domestic pets are isolated from their ancestral ecology, like bits of tissue in a petri dish in a laboratory. The dream of bonding has, in a twinkling, taken "bonding" into the wilderness as though all life were kittens abandoned on the street. The clinical regimen for captives—therapy, care, and consoling—becomes the key concept. The world is to be a protected and prophylactic zoo, at once medical and moral, the object of charity and the anodyne for guilt.

How are we to understand this misplaced zeal? I think it has to do with death—the final, unacceptable reality of the natural world. This is where the anti-organic stance of our culture and our civilization leads in the end. "Protection" is a repository of the protector's own dread of the cycle of birth and death, which close our dreams of perfection in a visceral and protoplasmic envelope.

One writer cites Plato's Athenian speech on the preservation of the whole in contrast to John Calvin's statement that "All other things can properly be put in the service of man."[8] Putting them in our service does not mean only their wool and milk but their vicarious existence, their tendency to die, as an arena in which we can fight death. But for such thinkers, recognition of the whole is itself the true test of human eminence: "The dignity and distinctive value of the human species that an anthropocentric vision sustains cannot be ignored."[9] Somehow the other members of that whole seem to remain invisible before its august image.

A large part of the logic of animal rights involves the similarity of animals and humans—how they feel pain, communicate, think, respond emotionally, and so on. But when it comes to killing and eating animals, which all other carnivorous and omnivorous animals do, we suddenly become very different from them, superior in morality, intellectually estranged, nature's stewards, custodial keepers and architects, or we become vegetarian, a state of beatific mastication which we left six million years ago in our evolution from herbivorous primate ancestors. The double bind sets our membership in the natural community and its ecology against our desire for a warm bath of "animal friends."

Socially, there is some plain foolishness in the logic. In "Beasts for Pleasure," Maureen Duffy says: "The animals have sunk from being members of the family or at least dependent servants to being automata and for this the popular interpretation of Darwin's conclusions is largely responsible" and "after thousands of years of living with them we know almost nothing about them."[10] In other words, evolution and ecology cannot be our guides, and the wisdom and knowledge of indigenous peoples do not exist for her.

Hunting animals is anathema to most "animal lovers." In his book *In the Company of Animals,* James Serpell remarks, "With the exception of the so-called 'whaling industry' and the sporting activities of the learned rich, hunting as a way of life has more or less vanished."[11] Hunting/gathering cultures are indeed at the brink of oblivion, thanks to ten millennia of genocide at the hands of farmers and pastoralists, but hunting has not ceased to be important in modern life. In North America alone 50 million people hunt and 100 million fish every year, apart from the indigenous cultures, from Point Barrow and Ellsmere Island to Arizona and New Mexico, who have never quit believing in the sacred hunt.

Hunting is an easy target. The harassment of English fox-hunters, like the picketing of the fur stores, is easily connected with resentment against the rich and provides opportunities for demonstrations and the media. The commercialization and perversions of the hunt—the game hogs, the drunks, the shooters of cows, the facades of camaraderie—make the war against hunting both easy and facile. Despite recent studies of the religious and ethical sensibilities of hunting/gathering cultures, opening new paths to the modern ecological conscience as exemplified by modern hunters like Aldo Leopold and Theodore Roosevelt, there remains a vociferous opposition to hunting as degenerate and atavistic. The myth common to many ethnic hunters about the marriage, once upon a time, of humans and animals, says one author, is merely a recollection of sexual relations with animals.[12]

Maureen Duffy furiously attacks hunting, but she reveals more about antihunters than the ethos of hunting. In her are married the outraged feminist and the animal lover. Hunting, she says, is thinly disguised rape. The social functions of hare coursing begin "after the kill when comparisons are swapped, sizes compared as bull sessions compare sexual prowess." The body shape of fox and otter "isn't hard to see as a phallus and both are traditionally sexy beasts and dwell in holes." Killing the hare by breaking its back is "the usual erotic movement of courtship being transferred to death." Hunting is like "masturbation fantasies . . . the build up and the orgasm," and so on. For her, death and sex constitute crimes against nature and women. She is gratified that bullbaiting, cockfighting and the public feeding of snakes at the zoo have declined, but she is puzzled that bird hunting and fishing have increased.

Duffy does not seem to see that those things which have declined were vicarious expressions of peasant and town neuroses, while those which have increased hark back to a positive generic layer of human being. The traditional myths associating hunting and death speak metaphorically of consummation and renewal rather than trivial sensuality. Her description has no relationship to the actual practice of the hunt. Indeed, her view seems to me to reflect a culturewide fear of being a participant in a world where life lives on death. She says that "we need other animals as part of our background as we need unpolluted skies and rivers, trees in architecturally beautiful cities, oil-free beaches." These are the words of a spectator, for whom art and nature are valuable amenities, who imagines herself as in a movie, in a pictorial world, framed, static, preserved from the realities of organic process. As "background" nature is acceptable, removing us from the danger of "barbaric regression."

We are space-needing, wild-country, Pleistocene beings, trapped in overdense numbers in devastated, simplified ecosystems. We project our problems onto mythic forms of barbarism. Whereas the sanctity of nonhuman life was a normal part of small-scale societies for thousands of years, the "world religions," with their messianic, human-centered, and otherworldly emphasis, trampled those traditions and now are beginning to recognize what they lost: sensitivity to human membership in natural communities and affirmation of and compliance with the biological framework of life. Greek ethics and biblical morality, organically alienist to begin with, cannot cope with our circumstances. Our ethics and morality deny the sacredness of the human connection with other life as it is played out in metabolic chains and the numinous presence of animals at the heart of religious experience. When we try to extend our ethics to that with which it is incompatible, we get pictorial and esthetic images of nature, the Renaissance spectator, museum patronage, the culture of abstract appearances and dissociation.

The recent history of kindness to animals is a vast self-reproach. It says that nature is full of innocent persecuted animals who would rather be friends with people, that we must adopt a "reverence for life" or an "ethic" of animal rights as we might put on a clean shirt. It patronizes life, poses the important question of our true relationship to nature as condescension, and confuses a sentimental fiction with civilized enlightenment.

Formal philosophy cannot contain the question because its original assumptions are estranged from nature: the view that a metaphysics of the earth and its life is a primitive error. The neoclassical, rational discourse, the watery stuff of empathy, and the "feeling" for the individual move us into metaphysics without polytheism. We are trapped in reason on the one hand and kindness on the other. Our environmental deprivations deny to much of modern experience and understanding the physical and allusive connections to the Others which are the visible expressions of an inner spiritual community and a larger cosmology.[13]

In the end, the abuse of animals will not be solved by ethics any more than by rebuke or exhortation. Neither logic nor charity can deal with what is, beyond pets and chicken factories, a mystery and an ecology: the ambiguity of life living on death, the spiritual nature of nonhuman life, traditions of human membership in natural communities embedded in place and ancestry. Earth history places us among the animals, as one of them, in food chains and other symbioses which we do not invent, but inherit, and which set our limitations among the Others.

The humane movement is an appropriate response to the abuses of domesticated animals, who need physical and legal protection. The pet is itself a singular redress to urban life and human crowding, a balm and true helper in the miseries of the multitude, but a monster in nature. The therapeutic beast and the urban society are made for each other. The escalation of pets to institutions such as hospitals, hospices, and prisons, like the formalization of humane concern for animals in statutes and laws, corresponds to the general drift of technological civilization in which the countermeasures against our loss of nature are found in industrialized therapy.

But the heart of animal life is gone from animals created in dependency and so emotionally appealing. We lose sight of their exclusion from the larger "nature." Wild animals have so receded from human experience that they seem as peripheral as sparrows and cockroaches or appear only in television, in art, or as captives—either in calendar art or in sensational, intimate views of the lives of eagles and dolphins as remote from our own experience as the landscapes of the moon.

James Rachel argues that what we need is "a philosophy that does not discriminate between different species, one that addresses each being on an individual basis."[14] The individualizing of our anguish over animals is reminiscent of the "person/planet" mesopia, with its lack of middle ground of species and communities. After all, how can a population or an ecosystem feel pain? Species are abstracts which you cannot touch or love. There is no place in "rights" for normal death, disease, or deprivation, only happy faces. The fundamentalists of interspecies sensibility hew to a literal text, to the physical problem of the helpless animal, seeing it like the heathen person, to be saved. "Human *kind?*" asks Cleveland Amory, and replies: there is no kindness in killing things.[15] His thesis cannot be reconciled with a world where all must die and all living is at the expense of others. It is a surreal view of "animals" which exist only as ideas and not in the wilderness; hence his feeling is directed to images that live "forever" like those on a Grecian urn.

AT THE HEART of the ideal of animal protection is their "right to be," or their "right to be let be," to serve no human end. Its best expressions are magnificent pieces of rhetoric which perfectly express the detached ethos of the educated, urban mind. It seems to say: why shouldn't we all just leave animals alone (except perhaps for filming them or otherwise appreciating or studying them at a distance), just as the activist animal protectors in their homes, libraries, cafés, and theaters do? Why struggle with the problem of how to relate to animals, especially when it is complicated with the protoplasmic pitfalls of disease, predation, all that ecological/evolutionary quagmire, and all those disturbing primal and ethnic human precursors? Why work out relationships to animals in terms of that morass of prehistory and the demented (or fallen) creation so interwoven with death? In Cleveland Amory's world our kindest act is avoidance, our deepest obligation protection at a distance, our best satisfaction a friendship like that of Petrarch for Laura, without response, a comfortable, ecstatic remoteness, its recompense of the heart rather than the stomach. We can stand back from it all and enjoy "nature" in art and literature and science, a subject matter in a great museum refuge and art gallery. What a truly civilized idea! With the finality of disconnection.

In this way the ethics of "let be" deals with the enigmas and perennial inquiry, finalizing the game by freezing nature in place and removing ourselves. But the true vocation of humankind, to puzzle out reciprocity, requires that we know, as the elders of a million years past knew, that there is no "solution," but instead an ongoing participation. Bystanding is an illusion. Willy-nilly, everybody plays. This play contains that most intimate aspect of the mystery—our own identity—signified in finding ourselves in relationship to the Others.

A hideous overabundance of humans and our demands on energy and space diminish the place for other species. The loss of wildness, extirpation of species, reductions of natural populations, extinctions, compression of habitat, and poisoning of life by air and water are the tragic circumstances in which we see animal protectionists as indulging in a kind of sentimental morality that is more important to them than the world of animals. As Paul and Ann Ehrlich and Garrett Hardin have been telling us for thirty years, the ri-

diculous code of medicine that prolongs human life at any cost and advocates death control without birth control has damaged life on earth far more
than all the fox hunters and cosmetic laboratories could ever do—perhaps
beyond recovery—and leads us toward disasters that loom like monsters
from hell.

Human political rights are meaningless as interspecies relationships.
"Liberation" means nothing to a calf elk about to be eaten by a wolf or a
salmon about to be eaten by a man. "Bonding" to animals is a willful, Dis-
neyish dream. Most of the advocates of these ideas have never watched wild
animals closely and patiently, have little notion of their intelligence,
otherness, or the complexity of their lives, cannot imagine combining holiness as killing them or celebrating them by wearing their skins, do not recognize the flesh of animals as a food sanctity, or perceive animals as a means
of speculative thought, referential analogy, or immanent divinity.

The Many and the Fuzzy:
Plurality and Ambiguity

The Greeks, being sane, were pantheists and pluralists,
and so am I.

D. H. LAWRENCE

THERE IS MUCH talk of "biological diversity," but the phrase is a sleeping tiger, a coiled philosophical snake not unlike the one in the Garden of Eden. Just as the snake is both renewing and lethal, it represents the paradoxical aspect of animals: their uniqueness and, inversely, their resemblance to ourselves. Therefore identity—theirs and ours—is, finally, the underlying "problem" of the relationship of humankind to the rest of the animals.

This theme of identity is further complicated by the ever changing bodily aspect of all living things. This disconcerting characteristic is put at bay only if the true self is a fixed, inner essence, a notion that is crucial to the metaphysics of eternal life. In the historically rooted cosmos of the Western world, the schizoid declaration "I *have* a body" (like "I *have* a past") reveals that the "I" stands apart from the rest of it. The ailments of the spirit in such a world are acedia—the boredom of monotonality—or hubris, experienced as the imperative of dominating the Other. The world of one deity and one history is a uni-verse only because it was an initial whole, broken, betrayed, and squandered. Yet such a universe is fundamentally dual, part of which is "in God's image" and the remainder a flotsam of inferior qualities and beings.

Plato is widely remembered for the thesis that our senses mislead us. Indo-European tradition had already concluded that impressions are but a swirl of illusion. Both rejected an earlier, primal cosmology in which what was heard and seen was accepted as veridical. The West has argued that neither is as reliable as script. The primacy of The Word is the insistence of literate cultures that the ambiguity caused by our innocent acceptance of perceptual reality is intolerable. As a result, that culture makes us feel we must choose between the terms of a contradiction, between appearance and truth.[1] Dissociated essence is more important to us. We believe that what endures is the true reality, as opposed to what changes, that not only is appearance misleading but age disfigures, decay corrupts, birth contaminates, and therefore nouns are more reliable than events or the relationships among them. This superiority of quintessence over guise was implicit in Indo-European languages and explicit in Greek philosophy as well as Near Eastern theology.

It is odd that writing, so seemingly neutral, should tacitly contain such a biased view of the world. Socrates, the enemy of writing, understood. Later, Bertrand Russell and Alfred North Whitehead also recognized the invidious subjectivity of writing in which things got in the way of events, words replaced what they represented. Aldous Huxley says: "A man must do more than indulge in introspection. If I would know myself I must know my environment." The separation of immutable idea and inconstant flesh was contrary to an ancient notion that "the physical universe is somehow a model for conscious intellectual activity."[2] Why did this shift to a rigid monolithic principle occur? Why did personal identity lose its grounding in place and in reference founded on enactive metaphors of the natural world?

In the cultural ethos of an urban rhetoric of change and historical time, confidence in the natural world as the essential clue to self-understanding diminished. The natural world failed. In the Near Eastern root-time of Western history, dereliction of the land, eroding fields and damaged watersheds, precarious crops, deforestation, and epidemics that seemed to rise up from the soil itself, all created an ideological malaise among the displaced and deracinated slaves, wandering merchants, missionaries, bureaucrats, and soldiers. Nature itself seemed to mirror the cacophony of voices, creeds, and heresies and the broken cries of dislocation. Only the sky, like royal power and its massive architecture, remained immutable, a world one could trust. The sky's authority would persist, the weather as unpredictable as a temperamental monarch, and the stone monuments with their inscriptions would endure where reed and mud villages with their communities of life, antelopes and songbirds, were gone with the wind.

Monuments such as pyramids were the symbol of progressive subordination, immolation and sacrifice, religious escapism and messianic dreams which transcended the ambiguity of nature while pointing to the heavens.[3] Recorded in writing, this hostile mindset of religion and politics kept alive the anguish of dislocation, redirected to struggles for centralized power in which social and personal dilemmas have no ecological reference. Except for the clay tablets, in desert cities where the only landscape features were signposts put up by bureaucrats, destroyed or altered every generation, nothing "holds."

Our alphabet began with pictograms, but in time it became unrecognizably stylized and distanced from its organic origins. Words fashioned from ABC's ceased to resemble what they symbolized. The deserts on the march, the decline of the earth's fertility, the disintegration of the riches of subsistence culture—all encouraged the suspicion that appearance is illusory and drove symbolic form away from the concord of multiple meaning. The alphabet confirms separateness and fixity rather than movement. A word in ABC's "remains strictly itself," as Aldous Huxley noted, while a pictogram of the same subject is two things. A person, or another organism, can "naturally" be two things if we understand reality as a kind of embryological miracle. Diversity assents to larval ambiguity, while the written word forces choice. Opposition to "nature" is based on the assumption that substance or appearance are misleading if not evil.[4] These roots of scholarship now find nature peripheral. Their secular expression is an economy of consumption, control, and growth: "the foolish metapsychology of fat men in a declining culture" characterized by aggrandizement, overweight, overpopulation, overkill, cancer, and technological escalation and proliferation.[5]

The mythologies of empires, East and West, share the story of a ruined paradise, of humanity betrayed by "naturalness" in the form of weather, disease, and demographic calamities. History is the ideology of expected catastrophe and hope for something beyond. The result is an irresolvable ambiguity in which mortal life, instead of a great gift, is characterized by deprivation. Christians and Muslims look to the end of the world, Buddhists to a denial of life itself. All "world religions" divorce themselves from chthonic centering and, in doing so, reject the earth as a home and the significance of its organic domains. Since wild animals are themselves bound to niche and place, they are subject to the judgment of a divided consciousness, relegated to the inferior world, corresponding to the instinctive strata of the human psyche, repudiated as lesser phenomena.[6]

Wild animals had been the archetypes, the medium of communication of the hidden aspects of experience. Like dream-bearers, the animal "message-carriers" protected the mind from a final schism and sustained threads of connection. When the rational mind keeps such travelers through the "gates of the underworld" locked out, "destructive unconscious material" erupts into our lives. Our "advanced consciousness" blocks the immemorial mentors, who appear not only in nature but in dreams and trance. Their multiplicity is "the worst that can happen" in a "cosmos of unity" that demands fixed subordination and the exile of the uncontrolled.[7]

In cultures in which animals incarnate the diversity of spirit and the flow of forces, there is no problem with changing appearance, with metamorphosis and transformation, because the animals show the way. Tragically, the most dramatic loss of animal life is now taking place in traditionally polytheistic, small-scale societies, invaded from the outside, where respect for the animals has crumbled under the pressures of human population growth, the illusions of progress, and subversion by multinational economic development and the contamination of greed.

Our acceptance of multiplicity can do for the world's animals what all the laws, humane societies, zoos, and films cannot. The model is prehistoric. The music, paintings, and sculpted forms of primal cultures are two things at once: an image or sound and what it represents—not a duality but a synthesis. Dance and pictographs embody conjunction. The task which this paradox sets for the imagination is the mystery of double identity. It includes substance and essence in such a way that each is illuminated by its polar complement. The resolution of all such oppositions—the relationship of nature and culture, body and spirit, god and nature, human and animal—is, in the human mind, incarnate in animals.

Suspicion of appearance is the reason why Western religion is hostile to drama unless it is clearly secularized and subordinated to virtuosity and personal style. Theatre is a rascally deception, its players a suspect company. Buddhism and the Vedic Upanishads are similarly suspicious of the visible manifestations. As Europe replaced pagan idolatry with Judeo-Christian idealism, with its Judaic, Islamic, and Puritan Christian prohibition of "graven images," an image vacuum arose, a lack of natural forms in ecclesiastical art, relieved only where indigenization crept into the Church. In these Western tradition, souls at death, shuffling off the mortal coil, do not reincarnate; the true spirit does not have successive, multiple expression. One cannot be both this and that. Birth and death cannot be integrated with other traditions in which masked ghost dancers treat the corpse as sentient, speak to dismembered heads, or celebrate death as a birth with dances that mimic fish. In the logic that keeps the deity distant, as well as the Christian idea of a single triumphant sacrifice for all believers for all time, there is a feeling of emptiness, for there is little participatory mimesis of bees, enacted metaphors of frogs resolving the antinomies, sensitivity to the birds as the enspirited dead, or recognition of the soul-significance of the hibernating bear. People are asked to rely on faith in the invisible and intangible, repudiating the beasts on which primal peoples depend as intermediaries, embodying spirits, affirming death, giving form to the mystery of the multiple truth of mortal existence, and acting as vehicles to other realms.

Without animals in these epiphanies, there is "chronic, endemic exaggeration of the difference between our species and others" in which animals represent "a great dark outside nonhuman area"—and hence the entire nature of Others as well as the unacceptable parts of our own nature. Descartes rationalized this "species solipsism" (only humans are self-conscious), a "wildly perverse view" that established the so-called species gap—a "powerful tendency to resent and fear all close comparison between our own species and any other."[8]

Around us are the disasters of three thousand years of the god-king as religious ideal and the politics of androcentric or "male" values. The archetypal feminine does not necessarily evoke snakes or elephants, but "she becomes the representative of multiplicity, who splits us up by sowing divisions, thereby reminding us how complex is totality."[9] The feminine in us, insisting on plurality, embroils us in doctrinal conflict today because we fear plurality. Such quarrels are about diversity and identity. These schisms are necessary, however, "to end the illusion of unity" and to recover "the distinctive multiplicity of the archetypal powers affecting our lives." All true wholeness, examined closely, is plurality.

The feminist emphasis on the goddess should not, however, become an end in itself. Feminism which seeks power rather than a renewed perception of all life as disparate, or the substitution of the Great Mother for the Great Father, is merely another anthropomorphic obsession. The solution is not a shift in gender dominance, in which we remain at the center of a narcissistic cosmos in which centralized power plays out the metaphor of plant and animal domestication. Matriarchy versus patriarchy is one more dualism, opposed to the incarnate heterogeneity of the universe modeled by animals, and independent of the armed might of the calculating father or the nurturant mysteries of the brooding mother. In polytheism the deities are individually limited. Even Zeus, being neither omniscient nor omnipotent, could be fooled. Such a network of power was like an ecosystem: it was patterned but its interrelations constantly shifting in detail—the stuff of the perennial scuffle of the Greek deities.

In a multiple world I too am multiple. Otherness is inside and out, a part of myself and a part of the outside world. David Miller observes: "Multiplicity is absolutely basic in Jung's description of the psyche. . . . Psychic structure is polycentric. It is a field of many lights, sparks, eyes." Like stars in the astrological universe, glowworms in New Zealand sea caves, fireflies in a summer night along the Mississippi, this "polycentricity of the psyche, its many constellations with their many faces, were represented once by a polytheistic pantheon." Borders are distinct but permeable. There is "mutual interpenetration and interfusion. . . . The gods blend."[10]

Thus do the sacred figures transform, overlapping in power and role, just as birds are reptilian in their feet and mouths, the fetuses of humans have tails, and the sacred beetle is first a white worm. Sacred multiplicity disperses power and thereby increases accessibility. Short on omni-things, such as omnipotency, and on mono-things, like monopoly, it affirms disjunction as necessary to transformation; it allows for the gray areas of ambiguity and changing places; it reflects a cosmos in which life itself is more sacred than any one kind. In such a world the self is plural, contrary to the observation by the nineteenth-century philosopher Soren Kierkegaard, who, in his book *Purity of Heart Is to Will One Thing* renounces all doublemindedness as the "best available judgement of civilized humanity," because "the Good is one thing."[11]

How can *the* good not be one thing? Because the good is plurality. In the myths of multiple powers there are two modes, ecological and social. Sacred beings may be plants, animals, the wind, mountains; or they may be men, women, and children. They may also combine the modes, or change from one to another. In their stories, most pagan cultures mix the two motifs. Historical societies eliminate the first except as parable, subordinated to humanism, and finally to a single figure. The polycentric psyche corresponds to a world that is a whole dynamic unity. Like an organism, it is complex within and without, not monolithic like a piece of basalt. Plurality allows for change and kinship without losing kinds, a world at once coherent but diverse, even when, within itself, the plants and animals use, destroy, and sustain one another just as members of a human family sometimes battle and disagree. It is a "biological diversity" not easy for modern culture to accept, but it is the world in which we live and one that resonates with our inner experience.

A FEW YEARS ago a book called *Man's Presumptuous Brain* traced the rational control of human bowels in the nineteenth and twentieth centuries. The smooth muscle lining of stomach and intestine is normally regulated by the autonomic nervous system without our conscious control (although its spasms certainly respond to our emotions). The alimentary tube is our principal encounter with the world, and its health reflects the quality of our lives with at least as much sensitivity as a good artist. Because of sedentary habits and bad food, modern people have ailing alimentary systems—from desiccated salivary secretions and heartburn at one end to constipation and hemorrhoids at the other. In a world in which it seems necessary to control nature, it is logical to take command of such a system. From a daily dose of laxatives to yoga breathing to biofeedback, we are eager to set right our undependable organs. How appropriate that the digestive dilemmas herald the uncertainties of our ecological sensibility, since food is the physiological link to the larger thermodynamics of the universe.

It is as though we have concluded that we must take arms against that which we cannot control. No biological determinism for us! Just as we were urged a century ago to become captains of our souls, we must now, as Buckminster Fuller tells us, become captains of spaceship earth, choose between "utopia or oblivion." But our souls are not ships; nor is the earth a spaceship. The earth is more like an embryo, or perhaps it is the outer aspect of the collective and consummate wisdom of 500 million years of smooth muscle. Thus do our meditations return to the first subject of the book: "swallowing the world."

A TREMOR RUNS through the modern world, contrary to the historical passion to place everything within human service, expressed in the rescue of endangered species. It has astonishing energy and momentum in a public most of whom cannot expect ever to see the threatened species: the prosimians of Madagascar, the eagles of Scotland, the snow leopards of Tibet. Peregrine falcons at risk were the key in halting the use of DDT compound as an insecticide. The whooping crane was furiously protected in North America by many who wouldn't know it from a duck. When two whales were trapped by ice in the arctic, an international effort was made to "save" them at great expense. Despite the deep fears about wolves and bears, the grizzly and timber wolf are the objects of campaigns to perpetuate and extend their range.

Something in us is working for elephants and rhinoceroses. In an essay following the publication of his fine novel on wild animals as "the roots of heaven," Romain Gary writes: "Dear Elephant, Sir: Our destinies are linked. . . . You represent to perfection everything that is threatened today with extinction in the name of progress, efficiency, ideology, materialism, or even reason. . . . It seems clear today that we have been merely doing to other species, and to yours in the first place, what we are on the verge of doing to ourselves. . . . The echo of your irrepressible thundering march through the open spaces of Africa keeps reaching me. . . . It sounds triumphantly like the end of acceptance and servitude, an echo of limitless freedom that has haunted our soul since the beginning of time. . . . There is no doubt that in the name of total rationalism you should be destroyed, leaving all the room to us on this overpopulated planet. But let me tell you this, old friend: in an entirely man-made world there can be no room for man either."[12]

In Rudyard Kipling's "Just So" story, the elephant got its long trunk when a crocodile tried to pull it into the river. But it is not really a tale of tooth and claw. In the long view, things may be just so—not a carnage but animals in a harmony of limited conflicts and assimilation, dependency, compliance, and death, a collaboration of coming and going. In the play of such joined groups there is no final victory or loss. In this "conflict of tempered brotherhood" Alan Watts calls for "a frank recognition of your dependence upon beloved enemies, underlings, outgroups and, indeed, upon all other forms of life whatever." Otherwise, he concludes, we are doomed to a schizoid, uncompromising, unscrupulous, bitter idealism and gangsterism, without humor or humanness; or to a gooey philosophy of undifferentiated belonging.[13]

The loss of numen or spirit is in animals the great modern defeat. Now they emerge as remedial or therapeutic, treatments in a world in which we accept punishing bouts of monotony and entertainment, in the expectation of a healing.[14] The recovery of that spiritual aspect is necessary because animals are elements of a deep and forsaken self, the "abandoned child." Numerous myths tell the story of the foundling who becomes a leader, separated from parents in order to move forward into the "hands of the cosmos," when the individual is placed in the care of others. It is, says James Hillman, "the basic cry a person addresses to his environment."[15]

The economic and political backlash against saving eagles and ferrets and rare frogs has come about because such archetypal mnemonics intimidate the rational mind, as if it were a collapse into infantility. Our rejection of "myth" lets us abandon the world of imagination to childhood, as though we could give over the flowery landscape to children at play while reserving for ourselves the grass as grain and the trees as logs. It is the old, false, binary choice: surrendering the cherished symbol of the tree with its mandala-like cross section of the trunk, the imagery of the wooden cradle and coffin, being born as fruit, having ancestral roots, branching. If plants are the context of our life, animals are, as Jung says, "not planted and rooted but roaming, not cyclic only in our rhythms but multiphasic, not only a seed blown but a new volition and sensitivity, a new level of trust in the organism we are." Our growth beyond the rooted fetus is the shift from a plant mode to an animal mobility, and with it all our movement in every sense. If the secret of the plant is growth and place, then the secret of wild animals is embodiment and passage, as though the creature were itself a gesture or a speaking.

It is no surprise that the animals are guides, gatekeepers, couriers, and exemplars, as they are epiphanies of aspects of ourselves. "We can never perform the ego's tasks, coping, struggling, advancing into light and knowledge," James Hillman says, without also regressing to the child."[16] Neglecting this "going back," we truly orphan our past and all past. If we do not go back, cults arise of "renegade psychopathology"—immediacy, irrelevance, transcendence, revolution—and the isolated self becomes beginning and end, rocking between "omnipotent hope and catastrophic dread," parental inflation and burdened worldwide responsibility.

Explicit recollection of our years from the age of two to five is lost from conscious awareness, and yet the primordial images remain. In reminiscence we recover the deep structure of mythologized existence in a community of Others. We must not abandon the child archetype to childhood; rather, being in part child, we must go back in memory beyond our personal childhood to the eternal child. Freedom from childishness requires that we rescue the vision of the world from the "amnesia of childhood." The past of our species yields authenticity, limits the possible, helps us avoid the arrogance of salvation apart from the Others.

As John Rodman observes: "The question 'What has posterity ever done for me?' arises only in people who have ceased to feel gratitude to their ancestors."[17] Recollection changes the brain and stimulates visual centers. Consciousness arises in awareness that what has vanished, seen in the mind's eye, will come back, induced by speech. And so a triad—absence, the spoken invocation, and presence—becomes central to thought itself. According to primal mythology, and therefore deep in human minding, the mystery of the hunted animal is its absence and presence. The forager's deity comes and goes (rather than rising up from the earth, like the gods and goddesses of the planters, or looking down from the sky like the deity of the pastoralists).

Large dangerous animals remind us that we are small in the order of things. It is still possible, Aldous Huxley once rejoiced, to get yourself eaten by a tiger. Without tigers, we become the big animal, subject only to larger, mindless forces—storms, floods, volcanoes, and the titanic insanity of atomic events, random and unselective. The celestial and mineral processes that contain us, unlike elephants, bison, elk, and tigers, are without living purpose. Without animals more powerful than ourselves, there are no intermediaries. No wonder we witness the rise of the absurd in modern art and culture; mindlessness has become our cohort. The missing beasts are those with whom we share consciousness, while their otherness can be as remote as the purposes of a tornado. Somehow, in human thought, all of the elemental monstrosity of the storm is focused through them, taking on shape and life, coherence and order, when we perceive animals as embodiments of meterological, mineral, and celestial power. Midway between ourselves and the colossal events in the sky, the great beasts become interlocutors, whose lives sift the forces of wind and water and fire, seeming to say that all such phenomena ultimately are purposeful and ongoing expressions of a meaningful world. The big animals are momentary embodiments of the atomic vitality that energizes nature itself.

In the presence of the small things, too, we are reminded that we live in the marginal fragments of habitat, roadsides, field borders, vacant lots, "yards" of homes, and other open land. Close by are frogs, grasshoppers, beetles, spiders, snakes, butterflies, mice, rats, squirrels, shrews, moles, and small birds, sometimes a raccoon, possum, skunk, or even coyote. These animals are part of the landscapes of childhood, and they are the mnemonics of a small but diverse and structured world. Just as our physical substance comes from scattered molecules brought together as food, we begin as mental selves strewn around. Everything out there, especially living forms, has its resonance in us and our idea of ourselves, volant guides into ourselves.

It has been argued that only affluent people and rich nations can afford to worry about endangered species. If this is true it still does not follow that such concern is only for another amenity; it means that the rich still have enough margin against want and enough leisure to explore the complex aspects of health. The motive for saving nature may be at bottom selfish, and meditation on nature a privilege. Even so, human well-being is linked to the

Others. Third World peoples may not have the resources to save gorillas, gibbons, rhinos, and a thousand other species, but they, as well as the rest of the world, are poorer for the loss.

"The concrete referent," says David Sapir, "is a vital part of a metaphoric construct. . . . The cock of the walk is a dead metaphor . . . but if we, urban dwellers that we are, knew something about cocks and hens (other than that they are good to eat and that a cock is somehow vaguely associated with the masculine organ), would not the metaphor say something infinitely more precise about the man than to say only that 'he is sexually vain'? So let my closing words serve both as a pitch for animal conservation and as an empirical caveat. 'Animals are good to think'; true, but only where there are animals about with which to think."[18] Without them what becomes good to think? In the absence of the Others, at the boundary between the self-obsessed too close and the nonliving too far, they may be the only bridge. One might imagine a letter of reply if the question were put to them:

The Forest, the Sea, the Desert, the Prairie

Dear Primate P. Shepard and Interested Parties:
We nurtured the humans from a time before they were in the present form. When we first drew around them they were, like all animals, inhabitants of a modest niche. Their evident peculiarities were clearly higher primate in their obsession with social status and personal identity. In that respect they had grown smart, subtle, and devious, committed to a syndrome of tumultuous, aseasonal, erotic, hierarchic power. Like their nearest kin, they had elevated a certain kind of attention to a remarkable acuity which made them caring, protective, mean, and nasty in the peculiar combination of squinched facial feature and general pettiness of all monkeys.

In ancient savannas we slowly teased them out of their chauvinism. In our plumage we gave them esthetics. In our courtships we tutored them in dance. In the gestures of antlered heads we showed them ceremony and the power of the mask. In our running hooves we revealed the secret of grain. As meat we courted them from within.

As foragers, their glance shifted a little from corms and rootlets, from the incessant bickering and scuffling of their inherited social introversion. They began looking at the horizon, where some of us were both danger and greater substance.

At first it was just a nudge—food stolen from the residue of lion kills, contended for with jackals and vultures, the search for hidden newborn gazelles, slow turtles, and eggs. We gradually became for them objects of thought, of remembering, telling, planning, and puzzling us out as the mystery of energy itself.

We courted them from the outside. Dancing us, they began to see in us performances of their ideas and feelings. We became the concreteness of their own secret selves. We ate them and were eaten by them and so taught them the first metaphor of their frantic sociality: the outerness of themselves, and ourselves as their inwardness.

As a bequest of protein we broke the incessant round of insect and herbivorous munching, giving them leisure. This made possible the lithe repose of apprentice predation and a new meaning for rumination, freeing them from the drudgery of browsing and the grip of relentless interpersonal strife. Bringing them into omnivorousness, we transformed them forever and they entered the Game as a different player.

Not that they abandoned their appetite for greens and fruits, but enlarged it to seeds and meat, and to the risky landscapes of the mind. The savanna or tundra was essential to this tutorial, as a spaciousness open to infinite strategies of pursuit and escape, stretching the senses to their most distant reference. Their thought was invited to a new kind of executorship incorporating remembrance and planning, to parallels between themselves and the Others and to words—our names—that enabled them to share images and ideas.

Having been committed in this way, first as food and then as the imagery of a great variety of events and processes, from signs in dreams to symbols in metaphysics, we have accompanied humans ever since. Having made them human, we continue to do so individually, and now serve more and more in therapeutic ways, holding their hands, so to speak, as they kill our wildness.

As slaves we stay close. As something to "pet" and to speak to, someone to be there and to need them, to be their first lesson in otherness, we have shared their homes for ten thousand years. They have made that tie a bond. From the private home we have gone out to the wounded and lonely, yearning for unqualified devotion—to hospitals, hospices, homes for the aged, wards of the sick, the enclaves of the handicapped and retarded. We now elicit speech from the autistic and trust from those in prison.

All that is well enough, but it involves only our minimal, domesticated selves, not our wild and perfect forms. It smells of dependency.

They still do not realize that they need us, thinking that we are simply one more comfort or curiosity. We have not regained the central place in their thought or meaning at the heart of their ecology and philosophy. Too often we are merely physical reality, mindless passion and brutality, or abstract tropes and symbols.

Sometimes we have to be underhanded. We slip into their dreams, we hide in the language, disguised in allusion, we mask our philosophical role in "nature esthetics," we cavort to entertain. We wait in children's books, in pretty pictures, as burlesques in cartoons, as toys, designs in the very wallpaper, as rudimentary companions or pets.

We are marginalized, trivialized. We have sunk to being objects, commodities, possessions. We remain meat and hides, but only as a due and not as sacred gifts. They have forgotten how to learn the future from us, to follow our example, to heal themselves with our tissues and organs, forgotten that just watching us can be healing. Once we were the bridges, exemplars of change, mediators with the future and the unseen.

Their own numbers leave little room for us, and in this is their great misunderstanding. They are wrong about our departure, thinking it to be part of their progress instead of their emptying. When we have gone they will not know who they are. Supposing themselves to be the purpose of it all, purpose will elude them. Their world will fade into an endless dusk with no whippoorwill to call the owl in the evening and no thrush to make a dawn.

The Others

Notes

INTRODUCTION

1. W. H. Auden, "Ode to Terminus," *New York Review*, July 11, 1968.
2. The death of Ben Rook at eighteen in World War II deprived us of a bit of that anachronism we need now, for he combined the paradox of the hunter and a tenderness far beyond that of the modern self-appointed animal protectors.
3. Christopher Fry, *Vision and Design* (London: Chatto & Windus, 1920).
4. Ted Hughes, "Poetry in the Schools," *American Poetry Review*, September/October 1977.
5. E. B. White, "Walden 1954," *Yale Review*, Autumn 1954. See also Robert Erwin, "The Village Apollo," in *The Great Language Panic* (Athens: University of Georgia Press, 1990).

1. THE ECOLOGICAL DOORWAY TO SYMBOLIC THOUGHT

1. Harry Jerison, *The Evolution of the Brain and Intelligence* (New York: Academic Press, 1993).
2. Jay Boyd Best, "The Evolution and Organization of Sentient Biological Behavior Systems," in Allen D. Breck, ed., *History and Natural Philosophy* (New York: Plenum, 1972).
3. Polly Schaafsma, "Supper or Symbol: Roadrunner Tracks in Southwestern Art and Ritual," in Howard Morphy, ed., *Animals into Art* (London: Unwin Hyman, 1989).
4. Whitney Davis, "Origins of Image Making," *Current Anthropology* 27 (1986):193–202.
5. Alexander Marshack, "Comments: Whitney Davis, 'Origins of Image Making,'" *Current Anthropology* 27 (1986):205–206.
6. David Guss, *The Language of the Birds* (San Francisco: North Point, 1985).

2. THE SWALLOW

1. Valerius Geist, "Did Large Predators Keep Humans Out of North America?" in Janet Clutton-Brock, ed., *The Walking Larder* (London: Unwin Hyman, 1989).
2. Clifford J. Jolly, "The Seed-Eaters: A New Model of Hominid Differentiation Based on a Baboon Analogy," *Man* 5(1) (1970).

3. Joseph Campbell, *The Masks of God*, vol. 1 (New York: Viking Press, 1964), p. 58.

4. Tim Ingold, *The Appropriation of Nature* (Cedar City: University of Iowa, 1987), p. 246.

5. Roy Willis, "Introduction," in *Man and Beast* (New York: Basic Books, 1974), pp. 7–10.

6. James Dickey, "The Heaven of Animals," *Poems, 1957–1967* (New York: Collier, 1968).

7. "Designed to be seen" is the felicitous phrase of Adolf Portmann in his book *Animal Forms and Patterns* (New York: Schocken, 1952). Put this way, the emphasis is not intended to reaffirm a conventional god but to remind us that many of the characteristics of animals evolve in reciprocity with other species.

8. Morris Eagle, David L. Wolitzky, and George S. Klein, "Imagery: Effect of a Concealed Figure in a Stimulus," *Science* 151 (1966):837.

9. Perhaps the evolutionary sequence might look like this (with the oldest at the bottom):

socially paradigmatic food	*Homo sapiens*	modern humans
omnivorous attention	*Homo erectus*	ancient humans
carnivorous overlay	*Homo habilis*	archaic humans
scavenger hunting	*Australopithecus*	primordial ancestors
granivorous side dish	hominids	prehumans
frugivorous tastes	hominoids	apes and prehumans
herbivorous, insectivorous context	anthropoids	early apes and monkeys
socially significant food	primates	simians and prosimians

10. This is true even in those garden styles where the wall was removed or hidden, the wall signifying the outer edge of human control. There was, in the eighteenth century, the self-induced illusion that the unwalled English "Gentlemen's Park" or landscape garden matched nature; but the "nature" of the English countryside was the result of six thousand years of agriculture and clearing.

11. Luke Taylor, "Seeing the Inside: Kunwinjku Paintings and the Symbol of the Divided Body," in Howard Morphy, ed., *Animals into Art* (London: Unwin Hyman, 1989).

3. THE SKILLS OF COGNITION

1. Eleanor Rosch, "Principles of Categorization," in Eleanor Rosch and Barbara Lloyd, *Cognition and Categorization* (Hillsdale, N.J.: Lawrence Erlbaum, 1978).

2. David Bleich, "New Consideration of the Infantile Acquisition of Language and Symbolic Thought," *Psychological Review*, Spring 1976.

3. Bleich says: "The cognitive component of the ability to recognize mother—conceptualizing her—is, at the onset of representational thought and language, graded into existence by the need to cope with the affective loss." See Bleich, "New Consideration."

4. Marvin W. Daehler and Danuta Bukatko, *Cognitive Development* (New York: McGraw-Hill, 1985), p. 221.

5. Susan Carey, *Conceptual Change in Childhood* (Cambridge: MIT Press, 1985), p. 43.

6. Ted Hughes, "Poetry in the Schools," *American Poetry Review*, September/October 1977.

7. Experts, of course, recognize gradations among the various species, often of an extremely subtle and technical nature. But the prototypical human perceiver is not a biogeographer looking at a range of skins collected from different places. Seldom does more than one "subspecies" or race occur in a region.

8. Virginia Woolf, quoted in Roberta Crawley and Steven P. R. Rose, eds., *The Biological Bases of Behaviour* (London: Harper & Row, 1971), p. 179.

9. Emiko Ohnuki-Tierney, "An Octopus Headache? A Lamprey Boil? Multisensory Perception of 'Habitual Illness' and World View of the Ainu," *Journal of Anthropological Research* 33 (1977):245.

10. Michelle Zimablist Rosaldo, "Metaphors and Folk Classification," *Southwest Journal of Anthropology* 28 (1972):83.

11. Alexander Marschack, "Upper Paleolithic Notation and Symbol," *Science* 178 (4063) (1972):817.

12. Claude Lévi-Strauss, *Anthropology and Myth* (Oxford: Basil Blackwell, 1987).

13. Cecil H. Brown, *Language and Living Things* (New Brunswick: Rutgers University Press, 1984).

14. Brent Ballin, "Ethnobiological Classification," in Eleanor Rosch and Barbara B. Lloyd, eds., *Cognition and Categorization* (Hillsdale, N.J.: Lawrence Erlbaum, 1978).

15. Merrill Moore, "A Note on Conchology," *American Imago* 3 (1942):113.

16. The pack rat or trader rat is a wood rat of the genus *Neotoma*. It piles up huge nests, mostly sticks, but includes anything it can find, such as keys, wristwatches, and small kitchen utensils. The trade occurs when it puts down one object in order to pick up another.

17. Paul Shepard, "The Garden as Objet Trouvé," in Mark Francis and Randolph T. Hester, eds., *The Meaning of Gardens* (Cambridge: MIT Press, 1990), pp. 148–154.

18. Susan Sontag, *On Photography* (New York: Dell, 1973).

19. I desist here from this line of thought as it leads further from the subject of the

chapter, but I should add that geometrized criteria in art find congenial equivalents in physics, mathematics, and classical music. The love affair between the humanities and modern physics is a symbiosis, the Western celebration of logic as a superior mode of experience.

4. SAVANNA DREAMING

1. See especially Edmund Leach, "Anthropological Aspects of Language: Animal Categories and Verbal Abuse," in Eric H. Lenneberg, ed., *New Directions in the Study of Language* (Cambridge: MIT Press, 1964), and Mary Douglas, "The Abominations of Leviticus," in *Purity and Danger* (New York: Praeger, 1966).
2. Judges 15:4–5.
3. Song of Solomon 2:15.
4. David Gordon White, *Myths of the Dog-Man* (Chicago: University of Chicago Press, 1991), p. 7.
5. Ibid. The foregoing paragraphs on the dog-man are based on White's excellent book.
6. Ibid., p. 4.
7. E. H. Gombrich, *The Sense of Order* (Ithaca: Cornell University Press, 1984), p. 256.
8. Edmund Carpenter, "If Wittgenstein Had Been an Eskimo," *Natural History* March (1980).

5. THE SELF AS MENAGERIE

1. J. David Sapir, "Fecal Animals: An Example of Complementary Totemism," *Man* n.s. 12 (1977):1–21.
2. Eligio Stephen Gallegos, *Animals of the Four Windows* (Santa Fe: Moon Bear Press, 1991), p. 31.
3. James Hillman, "Dream Research," in *Loose Ends* (Zurich: Spring, 1975).
4. David Foulkes, *Children's Dreams* (New York: Wiley, 1982).
5. Ibid.
6. C. G. Jung, "The Archetype in Dream Symbolism," in C. G. Jung, *The Symbolic Life: Miscellaneous Writings* (Princeton: Princeton University Press, 1983).
7. Bruno Bettelheim, *The Uses of Enchantment* (New York: Knopf, 1976), p. 76.
8. Ibid.
9. Ibid.
10. James C. Ferris, *Nuba Personal Art* (Toronto: University of Toronto Press, 1972).
11. Ibid.
12. Robert Bly, *Leaping Poetry* (New York: Beacon Press, 1975), p. 63.
13. Loren Eiseley, *The Immense Journey* (New York: Random House, 1957), p. 166.
14. D. H. Lawrence, "Form and Spirit in Art," in *Collected Stories* (New York: Knopf, 1994).

6. APING THE OTHERS

1. Iris Vinton, *The Folkway Omnibus of Children's Games* (Harrisburg, Pa.: Stackpole Books, 1970), p. 59.

2. James Fernandez, "Persuasions and Performances: The Best in Every Body and the Metaphors of Everyman," *Daedalus* 101 (1972):1.

3. James Fernandez, "Paradigms and Other Speculative Instruments of Social Anthropology," *Reviews in Anthropology*, November (1974):603–614.

4. Elizabeth Sewell has written eloquently of "the inevitable and beautiful anthropomorphism" in which mind and nature are not subject and object but interworking, dynamic systems. Her book, *The Orphic Voice* (New Haven: Yale University Press, 1958), is a vision of "postlogic" as the common ground of science and poetry in the comprehension of nature as a language. Thus she advocates anthropomorphism as a final reconciliation. Its ultimate expression is Orpheus dead and nature singing of him—nature as the hieroglyphic of thinking. Her radical thesis is that poetry does not simply articulate the epitome of mind giving meaning to nature, but the postlogic view that poetry is a natural product and therefore has a natural history, which she calls anthropomorphism. By contrast, Ruskin's concept of "pathetic fallacy," the attribution of human moods to nonhuman phenomena, is a romantic literary device or psychological illusion. The attribution of human feelings to animals may have a more homological basis than Ruskin supposed.

5. Erik Erikson, "The Ontogeny of Ritualization," in Julian Huxley, ed., *Philosophical Transactions B* 251 (1954).

6. Bettelheim, *The Uses of Enchantment* (New York: Knopf, 1976), p. 46.

7. The major alternatives are clear from history: an otherworldly kingdom, either biblical or simply extraterrestrial, the more humanoid the more virtuous; or the mechanomorphic world of physics, astronomy, and mathematics of the past four hundred years (or its modern extension in the quark/black hole mentality which leads some of its spokesmen to think they are about to sit down at the right hand of God).

8. Sewell, *The Orphic Voice*, p. 20.

9. Florence Krall, personal communication.

7. THE ECOLOGY OF NARRATION

1. See, for example, Walter Abell, *The Collective Dream in Art* (New York: Shocken, 1966).

2. Typical rhetoric of this kind can be seen in Fernand Braudel, *The Mediterranean*, vol. 1 (London: Collins, 1972), p. 399, who speaks of ancient France as "overrun with wild creatures" and characterized by "forests, brigands and wild beasts." So much for the alliance of History and Progress.

3. Bertram Lewin, *The Image and the Past* (New York: International University Press, 1968).

4. Orvar Lofgren, "Our Friends in Nature: Class and Animal Symbolism," *Ethnos* 50 (1985):184–213.

5. John M. Roberts, Brian Sutton-Smith, and Adrian Kendon, "Strategy in Games and Folk Tales," *Journal of Social Psychology* 61 (1963):185–199.

6. G. K. Chesterton, "Introduction," *Aesop's Fables* (New York: Avenel Books, 1911).

7. Brian Sutton-Smith, "Play, Games and Sports," in H. C. Triandis, ed., *Handbook of Cross-Cultural Psychology*, vol. 4 (Boston: Allyn & Bacon, 1980), pp. 425–471.

8. MEMBERSHIP

1. "Awakening to their disappointment and yet sensing their own being," says Flo Krall, "adolescents devote much of their time creating outer signs and significance" by arranging their hair and fussing with jewelry and clothing (personal communication). G. E. Hutchinson in "The Uses of Beetles," *The Enchanted Voyage* (New Haven: Yale University Press, 1962), p. 97, writes: "Our enthusiastic ornithologists were arranging the collection of bird skins. Suddenly it dawned on me that I had never realized what an extraordinary number of pigeons are bright green. . . . Many of them have in addition minor decoration in a great variety of other colors, often of a rather startling kind. To me, this realization, though it had no apparent value in relation to anything else that I knew, gave me intense pleasure that I can still recall and re-experience."

2. Julian Huxley, "Ritualization of Behaviour in Animals and Man," in Julian Huxley, ed., *Philosophical Transactions B* 251 (1954).

3. Claude Lévi-Strauss, *The Savage Mind* (Chicago: University of Chicago Press, 1968).

4. Ibid., p. 13.

5. Ibid., p. 17.

6. Ibid., p. 76.

7. Ibid., p. 88.

8. Jane Ellen Harrison, *Epilegomena to the Study of Greek Religion and Themis: A Study of the Social Origins of Greek Religion* (New York: University Books, 1962), p. 121.

9. A. R. Radcliffe-Brown, "The Comparative Method in Social Anthropology," *Journal of the Royal Anthropological Institute* 81 (1951):15–22.

10. Hilda Kuper, "Costume and Identity," *Comparative Studies in Society and History* 15(3) (1973):348–367, and "Costume and Cosmology," *Man* 8 (1973):630.

11. S. J. Tambiah, "Animals Are Good to Think and Good to Prohibit," *Ethnology* 8 (1969):423.

12. Like many such schemes, the infinite inside and outside toward which it leads at each end may show a closure. In this case, the ape in the eyes of the Thais is so

like humankind and so distant in the forest that its inclusion in the scale at all is unthinkable.

13. Gary Urton, "Animal Metaphors and the Life Cycle," in *Animal Myths and Metaphors in South America* (Salt Lake City: University of Utah Press, 1985), p. 275.

14. Gerardo Rerchel-Dolmatof, "Tapir Avoidance in the Colombian Northwest Amazon," in Urton, *Animal Myths.*

15. Wendy Doniger O'Flaherty, *Sexual Metaphors and Animal Symbols in Indian Mythology* (Delhi: Motilal Banarsidass, 1980), p. 257.

16. See Anton Blok, "Rams and Billy-Goats, a Key to the Mediterranean Code of Honor," *Man* 16 (1981):427–440; Yutaki Tani, "The Geographical Distribution and Function of Sheep Flock Leaders: A Cultural Aspect of the Man–Domestic Animal Relationship in Southwest Eurasia," in Janet Clutton-Brock, ed., *The Walking Larder* (London: Unwin Hyman, 1990). The goat/sheep dichotomy, a social metaphor in many parts of the world, is based on behavioral differences between them. Since sheep flocks are not always easy to control, even with dogs, a leader is often identified or even reared, named, and trained. Sometimes these leaders are wethers (castrated males), since rams are troublemakers and are not usually kept with the flock all year; at other times the leader is a female. In some parts of southern Eurasia where a professional tends the flocks of several different owners he brings his own goat or guide animal, reared on a bottle, trained on a rope, given a name, used as a pillow by the sheepherder at night, and fed barley or maize by hand. In other places, such as Crete, the leader may be a castrated goat, with which there is no such cosy intimacy.

17. Michael Wigglesworth, in Kenneth B. Murdock, ed., *The Day of Doom* (New York: Russell & Russell, 1966).

18. Lofgren, "Our Friends in Nature."

19. Paul Shepard, "The Ad-Man's Bestiary," in *Thinking Animals* (New York: Viking, 1978).

20. H. de Balzac, "Preface," *La Comédie Humaine*, vol. 1 (Paris: A. Martel, 1946), p. 4.

21. J. Christopher Crocker, "My Brother the Parrot," in Urton, *Animal Myths.*

22. Terence Turner, "Animal Symbolism, Totemism, and the Structure of Myth," in Urton, *Animal Myths.*

9. THE MASTERS OF TRANSFORMATION

1. See Dorothy Lee, "Codifications of Reality: Lineal and Nonlineal," *Psychosomatic Medicine* 12(2) (1950):89–97, and Walter J. Ong, S. J., "World as View and World as Event," *American Anthropologist* 71 (1969):634–647.

2. This clinging to an unchanging identity may be one of the few true marks of Western thought. Jane Ellen Harrison sees it as characteristic of the anthropo-

morphism of the gods in classical Greek times: "All life and that which is life and reality—Change and Movement—the Olympian renounces. Instead he chooses Deathlessness and Immutability—a seeming Immortality which is really the denial of life, for life is change." See Harrison, *Epilegomena and Themis.*

3. Harold Skulsky, *Metamorphosis: The Mind in Exile* (Cambridge: Harvard University Press, 1981).

4. It is similar to the shifted concreteness I call the "Marjorie Nicolson Syndrome" after reading her book, *Mountain Gloom and Mountain Glory* (New York: Norton, 1963), in which the idea of mountains was more real to her than mountains themselves.

5. L. G. Freeman and J. Gonzalez Echegaray, "El Juyo: A 14,000-Year-Old Sanctuary from Northern Spain," *History of Religions* 21 (1981):1–19.

6. James Hillman, "Archetypal Theory," in *Loose Ends* (Zurich: Spring, 1975).

7. Ibid., p. 41.

10. HEADS, FACES, AND MASKS

1. J. Grimm, *Teutonic Mythology*, vol. 1 (New York: Dover, 1966).

2. William K. Gregory, *Our Face, from Fish to Man: A Portrait Gallery of Our Ancient Ancestors and Kinsfolk Together with a Concise History of Our Best Features* (New York: Hafner, 1967).

3. Demorest Davenport, "I Am Not What I Seem!" *Parabola* 6 (3) (1981).

4. Joseph Campbell, "The Lesson of the Mask," in *The Masks of God: Primitive Mythology* (New York: Viking, 1959), pp. 21–29.

5. Fernandez, "Persuasions and Performances."

6. H. Stewart and Wilson Duff, *Images: Stone* (Vancouver, B.C.: Hancock House, 1988).

7. A. David Napier, *Masks, Transformation, and Paradox* (Berkeley: University of California Press, 1986); Aleene B. Nielson, "Masks, Transformation and Paradox: A Review of the Book by A. David Napier" (unpublished manuscript, 1989), p. 3.

8. Roy Willis, *Man and Beast* (New York: Basic Books, 1974), p. 9.

9. Howard M. Jackson, *The Lion Becomes Man* (Atlanta: Scholars Press, 1985), p. 50.

10. Ibid., p. 115.

11. Nielson, "Masks."

12. Ibid., p. 91.

13. Jackson, *The Lion*, p. 196.

11. THE PET WORLD

1. James Serpell, "Pet-Keeping and Animal Domestication: A Reappraisal," in Clutton-Brock, *The Walking Larder.*

2. The obvious implication is that all pets serve their owners in this therapeutic vein. But in the random play of society at large there are certain costs. Forty

thousand people are bitten by dogs each year in New York City alone. The resources needed by 100 million dogs, 100 million cats, 350 million fish, 22 million birds, and 8 million horses in the United States are immense. Thousands of abandoned, feral, or gone-wild dogs and cats kill native lizards, birds, and other animals. The cost of catching, housing, and killing the unwanted or surplus is $100 million. The pet food bill is $25 billion per year (or one-third the amount needed to feed the world's undernourished people). Five thousand species of wild forms are captured and imported—of which 75 percent die before reaching an owner, including annually 110 million aquarium fishes and 100,000 monkeys. Add to this the human energy and time spent on vaccination, neutering, and ordinary medical treatment, the 4 million tons of dog feces produced daily, the cruelty of keeping the 10,000 big cats in captivity, and the grotesque excesses of special treatment (acupuncture, anxiety syndrome therapy, psychosexual medication and treatment for aggression, caesareans, cataract operations, pacemakers, injections, blood transfusions, abortions, and allergies), dog cemeteries, grooming salons, restaurants, psychiatrists, and astrologers.

3. John Berger, "Why Zoos Disappoint," *New Society*, April 21, 1977.

4. John Berger, "Vanishing Animals," *New Society*, March 31, 1977.

5. Emily Hahn, "Getting Through to the Others," *New Yorker*, December 28, 1981.

6. Walker Percy, *Lost in the Cosmos* (New York: Pocket Books, 1983).

7. Mary Allen, *Animals in American Literature* (Urbana: University of Illinois Press, 1983).

8. Neil Evernden entitled his book on human disharmony with nature *The Natural Alien: Humankind and Environment* (Toronto: University of Toronto Press, 1993). If we are deformed as a species, I suspect it is in our capacity to adopt bad cultural traits and destructive ideologies, rather than the inevitability of doing so.

9. Aaron Katcher and A. M. Beck, *New Perspectives in Our Lives with Companion Animals* (Philadelphia: University of Pennsylvania Press, 1983).

10. Leo Bustad, *Animals, Aging, and the Aged* (Minneapolis: University of Minnesota Press, 1980).

11. Yi-Fu Tuan, *Dominance and Affection: The Making of Pets* (New Haven: Yale University Press, 1984).

12. Paul Shepard, *Thinking Animals* (New York: Viking, 1978).

13. James O. Breeden, ed., *Advice Among Masters: The Ideal in Slave Management in the Old South* (Westport: Greenwood Press, 1980).

12. THE GIFT OF MUSIC

1. Harry J. Jerison, *Evolution of the Brain and Intelligence* (New York: Academic Press, 1973).

2. Henry Wassen, "The Frog in Indian Mythology and Imaginative World," *Anthropos* 39 (1934):613–658.

3. Shirpada Bandoypadhyaya, *The Music of India* (Bombay, 1958).

4. Ibid.

5. Michael McClure, *Scratching the Beat Surface* (San Francisco: North Point, 1982); quotation reprinted in David Guss, *The Language of the Birds*, p. 289.

6. Robert Plank, *The Emotional Significance of Imaginary Beings* (Springfield, Ill.: Thomas, 1968).

7. Robin Riddington, "Beaver Dreaming and Singing," in Pat Lotz and Jim Lotz, eds., "Pilot Not Commander: Essays in Memory of Diamond Jenness," *Anthropologica* n.s. 13(1–2) (1971). Also quoted in part in David Guss, *The Language of the Birds*.

8. Pamela B. Vandiver et al., "The Origins of Ceramic Technology at Dolni Vestonice, Czechoslovakia," *Science* 246 (1989):1002–1008.

9. Steven Lonsdale, *Animals and the Origin of Dance* (New York: Thames & Hudson, 1981), p. 12.

10. Ibid., p. 12.

11. John Layard, *The Lady of the Hare* (Boston: Shambhala, 1988).

12. Lonsdale, *Animals.*

13. But this loss of seasonal acknowledgment is probably not the whole story. Hunting/gathering peoples do not hold civic festivals, ritual celebrations, or violent spectacles as often as do peasant agriculturists. Ceremony among the latter tends to represent ranks in the social order; among the former it reflects the cosmic structure. One suspects that the many fete days and carnivals in peasant agricultural societies are a release from the drudgery and monotony of their dreary work schedules and long hours. It may also reflect their preoccupation with annual cycles of crops, whereas the hunters and gatherers deal much more with perennial organisms. The coarseness of pioneers in ecological succession, exemplified in annual plants, is well recognized as contrasted to the distinctiveness of the organisms in "climax" or mature communities, botanically perennials. That moderns are worse off without the "traditional" festivals suggests their own therapeutic needs.

14. Mickey Hart, *Drumming at the Edge of Magic: A Journey into the Spirit of Percussion* (San Francisco: Harper, 1990).

13. ONTOGENY REVISITED

1. Mary C. Wheelwright, "Notes on Corresponding Symbols in Various Parts of the World," in Franc Johnson Newcomb et al., *A Study of Navajo Symbolism*, Papers of the Peabody Museum of Archaeology and Ethnobiology 32(8) (1956).

2. Anyone venturing on one more analysis of Pooh, after Fredrick C. Crews' *The Pooh Perplex: A Freshman Casebook* (New York: Dutton, 1963), will have to take what he gets when it comes to criticism and satire.

14. THE MEANING OF DRAGONS AND
WHY THE GODS RIDE ON ANIMALS

1. Chris Knight, "Lévi-Strauss and the Dragon: *Mythologiques* Reconsidered in the Light of an Australian Aboriginal Myth," *Man*, n.s. 18 (1983):21–50.

2. Daniel 7.

3. Shepard, *Thinking Animals*.

4. Ezekiel 1:4–28 and 10:1–22.

5. When the powers are represented simply as a lion or snake, the animal or its image is but an avatar. But combined with a human body, as many of the Egyptian sacred figures were, it signifies an individuality and human-made representations.

6. Howard M. Jackson, *The Lion Becomes Man*.

7. Ibid., p. 137.

8. Ibid., p. 138.

9. Isaiah 11:6–9 and 62:25; Job 5:3; Hosea 2:18 and 2:20.

10. Jackson, *The Lion*, p. 199.

11. Genesis 1:26–27.

12. Revelation 9:17–18.

13. Jackson, *The Lion*, p. 74.

14. S. Angus, *The Mystery-Religions* (New York: Dover, 1975), pp. 218–219; first published in 1925.

15. Jeanne Addison Roberts, "Animals as Agents of Revelation: The Horizontalizing of the Chain of Being in Shakespeare's Comedies," in Maurice Charney, ed., *Shakespearean Comedy* (New York: New York Literary Forum, 1980). Shakespeare was probably influenced by Montaigne's essay "Of Cruelty," in which he says: "I abate much of our presumption, and am easily removed from that imaginary soveraigntie that some give and describe unto us above all other creatures. . . . There is a kinde of enter-changeable commerce and mutuall bond betweene them and us." See Michel Eyquem de Montaigne, "Of Cruelty," in Ernst Rhys, ed., *Essays by Montaigne*, vol. 2 (London: Dent, 1938), pp. 124–125.

16. Paul B. Courtright, *Ganesha: Lord of Obstacles, Lord of Beginnings* (New York: Oxford, 1985).

17. Marianne Cardale Schrimpf, "The Snake and the Fabulous Beast: Themes from the Pottery of the Llama Culture," in Howard Morphy, ed., *Animals into Art* (London: Unwin Hyman, 1989).

18. William James, quoted in Victor Turner, *The Forest of Symbols* (Ithaca: Cornell University, 1967) p. 105.

19. Bruno Bettelheim, *The Uses of Enchantment* (New York: Knopf, 1976), p. 118.

20. P. Thankappan Nair, *The Peacock* (Calcutta: Irma KLM Private, 1977).

21. Ibid.

22. Eric Neumann, *The Great Mother* (New York: Pantheon, n.d.) p. 276.

23. Quoted in Lawretta Bender, "Animal Drawings of Children," *Journal of Ortho-psychiatry* 14 (1944):521.

15. AUGURY AND HOLOGRAMS

1. This contrast of history to myth as a fundamental praxis is mapped out in Herbert N. Schneidau, *Sacred Discontent* (Berkeley: University of California Press, 1977).
2. Harrison, *Epilegomena and Themis*, p. 113.
3. Ibid., p. 110.
4. Hesiod, *Works and Days*, quoted in Harrison, *Epilegomena and Themis*, p. 95.
5. Edward Deevey, "The Hare and the Haruspex: A Cautionary Tale," *Yale Review*, Winter 1980.
6. Lévi-Strauss, *The Savage Mind*.
7. Daniel Goleman, "Holographic Memory," *Psychology Today*, February 1979.

16. BOVINE EPIPHANIES

1. The importance of the small and rare "Venus figures" has been exaggerated by the desire in modern feminism for a prehistoric goddess. As Paleolithic objects these number fewer than a couple dozen—insignificant compared to the vast number of animal figures. Even if they do represent sacred beings, there is no evidence of their primacy or proof of status as a queen of beasts or "Earth." Although sometimes referred to as "pregnant," in more objective terms they are fat.
2. The discovery that people could parasitize cows unleashed dreams of affinity that are with us still. A modern veterinarian, for example, has eulogized the taking of milk from other species as "the most intense man-animal bond." He regards the drive to suck as a teat-dependent phase of life, its emotive momentum transferred to cows. Cow goddesses are a result of this interspecies sucking as a symbol of all life-giving and nurturing motherhood. Although lost to drinkers from plastic cartons, this "truly intimate, sensual" event remains deeply imprinted.
3. Martin P. Nilsson, *The Neo-Mycenaean Religion and Its Survival in Greek Religion* (Lund: CWK, 1950), pp. 165, 187, and 316.
4. The ideal of the "sacred cow" has become something of a model for animal protectionism in the West by people whose ethical stance has nothing to do with the ensoulment of animals. One anthropologist wrote a long article defending the sacred cow on "ecological" grounds as a consumer of weeds and plant materials that otherwise went to waste. This is a flagrant but familiar abuse of the concept of ecology as maximum use instead of a complex, stable, biocentric community. If the sacred cow in India were not a manure and milk producer its protection might diminish quickly. In any case, the celebration of maximizing of grazing/browsing/scavenging as a kind of vernacular wisdom is a form of

cow-towing to the subequatorial Third World and exhibition of modern blindness to the ecology of the soil, its invertebrate populations and plant associations, as a truly productive environment.

5. Numbers 25.

6. Amelia A. Prior, "Some Contributions to the Development of the Sanctity of the Cow in Indian Tradition" (M.A. thesis, Claremont Graduate School, n.d.).

7. Tim Ingold, *The Appropriation of Nature* (Iowa City: University of Iowa Press, 1987), p. 267. The transition to animal keeping is seen today in the reindeer-herders of the far north, especially the Siberian tribes: Chukchi, Koryak, and Tungus. Traditions that the herds sprang from the innards of the bear, the similarity between domestic and wild reindeer, all living in a half-wild land, and the mixed patterns of hunting traditions and pastoral customs suggest a stage in the emergence of animal keeping comparable to the Mediterranean reorientation from the wild aurochs to cattle herds, of which only archaeological evidence remains.

8. Jung, *Symbols of Transformation*, p. 103.

9. Harrison, *Epilegomena and Themis*.

10. Ibid., pp. 468–469.

11. Walter Umminger, *Supermen Heroes and Gods* (New York: McGraw-Hill, 1963).

12. Jung, *Symbols of Transformation*, p. 103.

13. Angus, *The Mystery-Religions*, p. 94.

14. Ibid.

15. Ibid., p. 103.

16. Victor Turner, *The Forest of Symbols: Aspects of Ndembu Ritual* (Ithaca: Cornell University Press, 1967), pp. 97–111.

17. Dale F. Lott and Ben L. Hart, "Aggressive Domination of Cattle by Fulani Herdsman and Its Relation to Aggression in Fulani Culture and Personality," *Ethnos* 5 (1977):174–186.

18. James Fernandez, "The Dark at the Bottom of the Stairs," in *Persuasions and Performances* (Bloomington: Indiana University Press, 1986).

19. John C. Galaty, "Cattle and Cognition: Aspects of Masai Practical Reasoning," in Clutton-Brock, *The Walking Larder*.

17. LYING DOWN WITH LAMBS AND LIONS IN THE CHRISTIAN ZOO

1. Florence Murdoch, "Trailing the Bestiaries," *American Magazine of Art* 24(1) (January 1932).

2. Anne Clark, *Beasts and Bawdy* (London: Dent, 1925), p. 60.

3. From the Latin apocryphal Gospel of the Pseudo-Matthew 14, 18–19, 35, and from the book of Isaiah 2:1, 3.

4. Clark, *Beasts and Bawdy*, p. 60.

5. Alun Llewellyn, "Celtic Christianity I," *Aryan Path* 39(8) (1968).

6. Ibid.

7. Aldous Huxley, "Adonis and the Alphabet," in *Tomorrow and Tomorrow and Tomorrow* (New York: Harper, 1952).

8. Lynn White, Jr., "St. Francis and the Ecologic Backlash," *Horizon* 9(3) (1967).

9. Paul Shepard, "Reverence for Life at Lambarene," *Landscape* 8(2) (Winter 1958–1959).

10. Meyer Schapiro, "The Religious Meaning of the Ruthwell Cross," *Art Bulletin* 26(4) (1944):232–245; Psalms 91:13.

11. Isaiah 2:8, 20; 9:19–20; 10:5, 33–34; 11:6–8; 13:21–22; 65:25.

12. Isaiah 11:6–9.

13. *Apocalypse of Baruch* 73:6.

14. Alun Llewellyn, "Man and Mind: Time," *Aryan Path* 40(11) (1969).

15. The whole matter is a kind of non sequitur, somewhat like the fish and the rivers of the American West. Cutting forests in great clear-cut chunks (however efficient it may be as forestry) has resulted in the obstructing of river channels with gravel bars, the shallowing of rivers by the filling of deep holes, the destabilizing of their levels by reducing the waterholding capacity of the mountain slopes and ultimately the water tables, and the altering of the chemical composition of the water itself as runoff from "plantations" instead of old-growth forests. The indigenous fish, which used to maintain populations without outside help, have lost out because of all these factors, so that we must now "help" the fish by rearing them in hatcheries and releasing them by the ton. The rivers with their fish cease to be healthy systems and become aquariums where fishermen "must" catch a large proportion of the (released) fish each year because otherwise they die.

16. Harrison, *Epilegomena and Themis*, p. 12.

17. E. P. Evans, *Animal Symbolism in Ecclesiastical Architecture* (Detroit: Gale, 1969), p. 27; first published in 1896.

18. Jack Bulloff, "A World Safe for Rhinos Is Not Best for Men," *University Review* (State University of New York) 3(2) (Summer 1970).

19. Margaret Mead, "The Island Earth," *Natural History* 79(1) (January 1970).

20. Huxley, *Tomorrow and Tomorrow and Tomorrow*.

21. Robert Eisler, "Jesus Among the Animals by Moretto Da Brescia," *Art in America*, 23 (1935):137–141.

22. Ibid.

23. Schapiro, "The Religious Meaning of the Ruthwell Cross."

24. 1 Kings 17:4–6.

25. Jackson, *The Lion*, p. 203.

26. J. Migne, ed., "Sayings of the Desert Fathers," *Patrologia Graeca*, 65:365A, no. 1787.

27. Ibid.

28. Lawrence S. Cunningham, *The Meaning of Saints* (San Francisco: Harper & Row, 1980).

29. The title of Susan Power Bratton's essay, "The Original Desert Solitaire: Early Christian Monasticism and Wilderness," evidently alludes to Edward Abbey's *Desert Solitaire* (New York: Ballantine, 1971). But there is no real similarity. Abbey never expected wild animals to do him favors or submit to him or to a Christian God. His work celebrates wildness on its own terms and the independence and otherness of animals. He did not talk about "animal friends." Abbey saw his own presence in the Utah and Arizona deserts as the antithesis of the exploitation and destruction brought on by Christian culture.

30. Nowhere in her apologia does the author show any interest in animals. They are all diaphanous abstractions meant to verify the saints' holiness and "teach lessons about Christian behavior." A Jewish equivalent is Bernard Malamud's novel, *God's Grace* (New York: Avon, 1982), which cites Jane Goodall's book *In the Shadow of Man* (Boston: Houghton Mifflin, 1971) as one of its inspirations. Yet his novel denies Goodall's concern that chimpanzees be recognized as true beings and worthy as they are. The novel tells of a lone, post-holocaust man who marries a chimpanzee and attempts to civilize all the chimpanzees, gorillas, and baboons. Parodying a new Genesis and failed Paradise, an allegory of human weakness, the author shares with Bratton not a shred of genuine curiosity about animals, who are nothing but markers in an ideological play. Goodall should be horrified to see his dedication, as her life's work leads to an enlarged sense of kinship based on the discovery of the shared experience of life and the true value of primate otherness, so long shut off by Hebrew and Christian dogma. In contrast to John Collier's *His Monkey Wife; or Married to a Chimp* (London: Hart-Davis, 1957), a funny story about confused identities, Bratton's pious exculpation and Malamud's reaffirmation of a male god's authoritarian dispensation perpetuate the same old human arrogance in the guise of humility.

31. Jacques Le Goff, *The Medieval Imagination* (Chicago: University of Chicago Press, 1988).

32. James 3:7.

33. Braudel, *The Mediterranean*, vol. 1.

34. Revelation 5:13.

35. Marvin Monroe Deems, "The Sources of Christian Asceticism," in John Thomas McNeil et al., *Environmental Factors in Christian Thought* (Chicago: University of Chicago Press, 1939).

36. Jung, *Symbols of Transformation*, p. 57.

37. H. Paul Santmire, *Brother Earth* (New York: Nelson, 1970).

18. HOUNDING NATURE

1. John Berger, "Why Zoos Disappoint," *New Society*, April 21, 1977.

2. Colin P. Groves, "Feral Mammals of the Mediterranean Islands: Documents of Early Domestication," and Sytze Bottema, "Some Observations on Modern Domestication Process," both in Clutton-Brock, *The Walking Larder*.

3. *New York Times*, November 21, 1989, p. B7.

4. Ruth Harrison, *Animal Machines: The New Factory Farming Industry* (New York: Ballantine, 1966).

5. Daniel E. Koshland, Jr., editorial, *Science* 244 (16 June 1989):1233.

6. Eli Edward Burriss, "The Place of the Dog in Superstition as Revealed in Latin Literature," *Classical Philology* 30 (1935):32–42.

7. Ibid.

8. Mahasti Ziai Afahar, *The Immortal Hound* (New York: Garland, 1990), p. 54.

9. Ibid., p.8

10. These were the big, dangerous, beautiful, thoughtful objects of the chase in which the humans were the predators. This conception may neglect, however, our own role as prey. We may also owe much of our intelligence to the necessity of escaping lions, leopards, hyenas, wolves, and their Pleistocene antecedents.

11. But an essential part of the human unconscious was made possible by a predecessor animal, the snake. Unlike the great hoofed forms, it never challenged our capacity for strategy the way the mammals did. The snake was an abiding other, a witness and watcher more than a counterplayer, encountered (in the modern sense) by chance.

12. Anne Ross, "The Horse," in *Pagan Celtic Britain: Studies in Iconography and Tradition* (New York: Columbia University Press, 1967).

13. Elwyn Hartley Edwards, *Horses: Their Role in the History of Man* (London: Collins, 1987), p. 52.

14. R. Gurney, *The Hittites* (London: Penguin, 1952).

15. Marija Gimbutas, "The First Wave of Eurasian Steppe Pastoralists into Copper Age Europe," *Journal of Indo-European Studies* 5 (1977):277–338.

16. Numbers 31:32–35; Deuteronomy 3:3–6.

17. Valentia Pavlovich Shilov, "The Origins of Migration and Animal Husbandry in the Steppes of Eastern Europe," in Clutton-Brock, *The Walking Larder*.

18. Shamans, those darlings of a New Age public, went greedily for visionary, ecstatic flight on horseback as soon as their societies obtained horses. Shamanism itself was betrayed in this development, as the mounted shaman gained in a kind of militaristic battle with evil demons rather than a collaboration with the spirits of the wild. Shamanism parallels the history of civilized politics with its gradual capture by a hereditary line of males.

19. E. A. S. Butterworth, *Some Traces of the Pre-Olympian World in Greek Literature and Myth* (Berlin: De Gruyter, 1966), p. 25.

20. *Rigveda* 10:136.

21. A. David Napier, *Masks, Transformation, and Paradox* (Berkeley: University of California Press, 1986), p. 59.

22. Emile Male, *The Gothic Image: Religious Art in France of the Thirteenth Century* (New York: Harper, 1958), Chapter 10.

23. Jyoti Sahi, *The Child and the Serpent* (London: Routledge, 1980).

24. Quoted in Robert M. Denhardt, *The Horse of the Americas* (Norman: University

of Oklahoma Press, 1975), p. 161. But an analogy in pre-Columbian America, based on the llama, precedes the Spanish. The huge Inca empire of six thousand years ago, from Chile to Colombia, was consolidated by military force and based on the mobility of the llama and alpaca and on "a programme of breeding, distinguishing and counting types and ages of beast down to the minutist detail." This "state army unparalleled in America" depended on a mythology very like that of the Middle Eastern pastoralists. The thousands of conquered, tribute-paying people looked to a "supreme shepherd." "Like their flocks the subjects of Tahuantinsuyu could be considered contained and penned, pastured elements of the great *Pax Incaica*, safe as such from the threat of enemies and the barbaric wild beyond its run." There were hymns, liturgies, and a "Fall," the *Situa*, all equating "flock with folk." See Gordon Brotherston, "Andean Pastoralism and Inca Ideology," in Clutton-Brock, *The Walking Larder* (London: Unwin Hyman, 1989).

25. Elizabeth Lawrence, *Hoofbeats and Society* (Bloomington: Indiana University Press, 1985).

26. Jean Bigtarma Zoanga, "The Traditional Power of the Mosi Nanamtse of Upper Volta," in Wolfgang Weissleder, ed., *The Nomadic Alternative* (The Hague: Mouton, 1978).

27. Robert Rodale, "The Horse—Is It the Symbol We Want?" *New Farm*, February 1982.

28. Brotherston, "Andean Pastoralism and Inca Ideology."

29. Russell Hoban, *Riddley Walker* (New York: Washington Square, 1980).

30. Konrad Lorenz, *On Aggression* (New York: Bantam, 1966).

31. David Noel, "The Nuclear Horror and the Hounding of Nature," unpublished manuscript, 1982.

19. THE MISS MUFFET SYNDROME

1. Alan Watts, *The Book on the Taboo of Knowing Who You Are* (New York: Pantheon, 1966).

2. Balaji Mundkur, *The Cult of the Serpent* (Albany: State University of New York Press, 1983).

3. Mary Midgley, *Beast and Man: The Roots of Human Nature* (Ithaca: Cornell University Press, 1978), p. 27.

4. Ibid., p. 31.

5. Barry Lopez, *Of Wolves and Men* (New York: Scribner, 1979).

6. Calvin Martin, "Subarctic Indians and Wildlife," in Carol M. Judd and Arthur J. Ray, eds., *Old Trails and New Directions: Papers of the Third North American Fur Trade Conference* (Toronto: University of Toronto Press, 1980), pp. 73–81.

7. Mary Douglas, *Implicit Meanings* (London: Routledge, 1975), p. 216.

8. L. Bryce Boyer, *Childhood and Folklore: A Psychoanalytic Study of Apache Personality* (New York: Library of Psychological Anthropology, 1979).

9. Orvar Lofgren, "Our Friends in Nature: Class and Animal Symbolism," *Ethnos* 50 (1985):84–213.

10. Robert Darnton, "Workers Revolt: The Great Cat Massacre of the Rue Saint-Severin," in *The Great French Cat Massacre and Other Episodes in French Cultural History* (New York: Basic Books, 1984).

20. CUCKOO CLOCKS AND BLUEBIRDS OF HAPPINESS

1. John Berger, "Vanishing Animals," *New Society*, March 31, 1977.

2. René Descartes, *Discourse on Method* (New York: Macmillan, 1986).

3. Philip Slater, *Earthwalk* (New York: Doubleday, 1974).

4. The production of "nature films" by the Disney studios, made in "nature," seem on the surface more true to life, but the editing that selectively shields the viewer from the mortality that characterizes the wild world is the real deception.

5. James Hillman, *The Dream and the Underworld* (New York: Harper & Row, 1979).

6. Edith Cobb, *The Ecology of Imagination in Childhood* (New York: Columbia University Press, 1977).

7. Romain Gary, *The Roots of Heaven* (New York: Simon & Schuster, 1958).

21. THE GREAT INTERSPECIES CONFUSION

1. William T. Divale and Clifford Zipin, "Hunting and the Development of Sign Language," *Journal of Anthropological Research* 33 (1977):185.

2. Gordon W. Hewes, "Primate Communication and the Gestural Origin of Speech," *Current Anthropology* 14(1–2) (1973).

3. André Leroi-Gourhan, *Prehistoire de l'Art Occidental* (Paris: Edition d'Art Lucian, Mazenod, 1970), p. 20.

4. Countess Evelyn Martinengo-Cesaresco, *The Place of Animals in Human Thought* (London: T. Fisher Unwin, 1909).

22. FINAL ANIMALS AND ECONOMIC IMPERATIVES

1. I have used the term "economics" in this essay to mean something wider and more inclusive than the modern form given it by Adam Smith. Indeed, the idea that resources, production, capital, labor, and money make the world go around is an offense against the root concepts from which the word comes. In a world run by banks it is not surprising that two words from the same source—economics and ecology—represent conflicting views.

2. Roy Willis, *Man and Beast* (New York: Basic Books, 1974), p. 8.

3. Richard Tapper, "Animality, Humanity, Morality, Society," in Tim Ingold, ed., *What Is an Animal?* (Boston: Unwin Hyman, 1988).

4. Robert Ardrey, "Four Dimensional Man: Thoughts of a Contemporary Controversy," *Encounter* (1972):9–21.

5. Clutton-Brock, *The Walking Larder*.

6. Clarence Morris, "The Rights and Duties of Beasts and Trees: Law Teacher's Essay for Landscape Architects," *Journal of Legal Education* 17 (1964):185–192.

7. Clutton-Brock, *The Walking Larder*, p. 117.

8. Dale F. Lott and Ben L. Hart, "Aggressive Domination of Cattle by Fulani Herdsmen and Its Relation to Aggression in Fulani Culture and Personality," *Ethos* 5 (1977):174–186.

9. Willis, *Man and Beast*, p. 16.

10. Roy A. Rappaport, *Pigs for Ancestors* (New Haven: Yale University Press, 1968).

11. To be exact, "bellwether" is appropriate only for sheep, since the word means a castrated ram, or wether, wearing a bell.

12. When ecologists criticized Hindu culture for keeping so many cows, anthropologist Marvin Harris came to their defense by arguing that these lean animals winnow weeds in villages and cities and therefore turn waste into milk. The goatlike adaptability of the little Indian cattle includes the digestion even of wastepaper and leaves. Harris was correct—the Indian cows in village environs do channel a little more energy and nutrients into the human system. But calling this increased efficiency in converting weeds and cardboard into humans "ecological" is a mistake. It is a common misconception that adding stuff to the human gut is "better" ecology. Exploitation at the expense of the rest of the biological community is not ecology. Cattle in the countryside, as well as the village, displace a large array of wild herbivores—insects, birds, small mammals—who contribute to the quality and stability of the whole. The weeds themselves have other values as sustainers of butterflies and other pollinators, as oxygen producers, and as medicinal herbs.

13. See, for example, John Kolars, "Locational Aspects of Cultural Ecology: The Case of the Goat in Non-Western Agriculture," *Geographical Review* 56 (1966):577–584.

14. Jim Corbett, *Goatwalking* (New York: Viking, 1991). This is one of those essays that—however wrongheaded in its vision of pastorality as the solution to human difficulties—gives color and life to the gray mass of modern book publishing and relief from the usual preoccupations of power and affluence. As one who advocates a hunting/gathering (rather than agricultural) model for human society, I can sympathize with how crazy he and his theme must seem to the makers and shakers as well as the mass who trudge along in the ruts of the usual civilized ambitions, agog in the celebrity world and electronic entertainment.

15. Kolars, "Locational Aspects of Cultural Ecology."

23. RIGHTS AND KINDNESS

1. Steve Lohr, "Swedish Farm Animals Get a Bill of Rights," *New York Times*, October 20, 1988.

2. In one meeting animal lovers picked the work of a psychologist, Harry Harlow, as an example of cruelty. His research involved separating monkey infants from

their mothers to study the dynamics of bonding of infants to mothers, especially as a tactile experience, and the psychopathology that results when they are isolated, including the effects on the next generation when the deprived monkeys themselves become mothers. The insights of this work on the infant/mother interaction, on the needs of infants and their consequences for the social-sexual relationships of adults, especially as it stimulated observation on the care of human babies, and on its effect in enhancing the quality of ordinary family life at a time when both parents work, would be difficult to overestimate. Harlow was a nominee for the Nobel Prize. His laboratory was a model of responsible care. The kind of information he gathered could never have been so unambiguously established from the observation of people upon whom such an experiment would have been criminal. The best relief in a worldwide epidemic of schizoid behavior would be radical changes in child care based on the recovery of small-scale society much like that of primal peoples, among whom social and economic pressures do not deprive babies of necessary nurture. Otherwise we are stuck with chronic social sickness which we had better understand.

3. Several heretics have plumbed the superficialities of animal "rights." Myrdene Anderson's reply to the "rhetoric in which we drown" — "What's Wrong with Animal Rights?" — observes that the argument always favors some creatures over others. Anderson's paper was presented to the Twelfth Annual Congress of the Canadian Ethnology Society in 1985. She says the flawed assumption is that rights exist a priori in the natural order rather than being the invention of cultures. Another perennial curmudgeon, Garrett Hardin, further uncovers the rights-and-ethics bias as self-interest in two essays in 1982: "Ethics for Birds (and Vice Versa)" and "Limited World, Limited Rights," both in *Naked Emperors: Essays of a Taboo-Stalker* (Los Altos: Kaufman, 1982).

4. Frank Frazer Darling, *Wilderness and Plenty* (London: BBC Reith Lectures, 1970).

5. Midgley's excellent book *Animals and Why They Matter* (Athens: University of Georgia Press, 1983) is perhaps the most thoughtful essay on the question of an ecological ethic, yet one looks in vain there for the answer to the question: Why do they matter?

6. See Paul Sears, *Charles Darwin: The Naturalist as a Cultural Force* (New York: Scribner, 1950).

7. James Serpell, *In the Company of Animals* (Oxford: Blackwell, 1986), pp. 52–53.

8. John T. McNeill, ed., *Institute of the Christian Religion*, vol. 1 (Philadelphia: Westminster Press, 1955), pp. 81–82.

9. James M. Gustafson, "Ethical Issues in the Human Future," in David Ortner, ed., *The Laws*, vol. 2 (New York: Random House, 1937), p. 645.

10. Maureen Duffy, "Beasts for Pleasure," in Stanley Godlovitch, Rosalind Godlovitch, and John Harris, eds., *Animals, Men, and Morals* (New York: Taplinger, 1972).

11. Serpell, *In the Company of Animals*.
12. Ibid., p. 26.
13. The carping against Greek rationality and philosophy in this chapter should be understood as aimed at formal philosophy on the one hand and the "philosophy" of zealots for animal rights on the other. Despite all those academic philosophers who have projected domestication and "friendship" as a view of nature, there are some who see how shortsighted and artificial that "philosophy" is. Indeed, the discipline of philosophy has responded with more verve and openness to twentieth-century environmental issues than many other elements of the educated community.
14. James Rachel, "Created from Animals," in Bernard E. Rollin, ed., *The Unheeded Cry: Animal Consciousness, Animal Pain and Science* (Oxford: Oxford University Press, 1989).
15. Cleveland Amory, *Man Kind? Our Incredible War on Wildlife* (New York: Harper & Row, 1974).

24. THE MANY AND THE FUZZY

1. Carpenter, "If Wittgenstein Had Been an Eskimo."
2. Walter J. Ong, "World View and World as Event," *American Anthropologist* 71(4)(1969):634–647.
3. Octavio Paz, *The Other Mexico: Critique of the Pyramid* (New York: Grove Press, 1972).
4. Huxley, "Adonis and the Alphabet."
5. James Hillman, "Schism," in *Loose Ends* (Zurich: Spring, 1975).
6. Carl G. Jung, *Man and His Symbols* (New York: Doubleday, 1964), p. 52.
7. Hillman, "Schism."
8. Mary Midgley, "Beasts, Brutes and Monsters," in Tim Ingold, ed., *What Is an Animal?* (London: Unwin Hyman, 1988).
9. David L. Miller, *The New Polytheism* (New York: Harper, 1974).
10. Ibid.
11. Dean Inge on Kierkegaard; quoted in the Introduction to Soren Kierkegaard, *Purity of Heart Is to Will One Thing* (New York: Harper, 1948), pp. 68–69. Even the "one-thing" thinkers cannot bring themselves to abandon the animal figure in their stories, but the myth becomes homily. Kierkegaard tells the story of the horse who suffers mistreatment and goes to the horse meetings in order to become edified and is always disappointed (pp. 155–157). This fatuous tale incorporates horses only because horses seem to be inherently social. We are to understand that the willingness to suffer is all (p. 121). There is no hope for the horse whose abuse is given as a fact of life.
12. Romain Gary, *Life*, December 22, 1967, pp. 126–139.
13. Alan Watts, *The Book on the Taboo Against Knowing Who You Are* (New York: Random House, 1989). My students often ask whether the people of the Orient

have a better philosophy about animals than ourselves. On the face of it, Tao and Hindu thought seem more open to the spirituality of animals, less mechanistic. The early Buddhists disdained the Brahmanic overseers for the goal of a personal access to the absolute through meditation or "austerities" and physical discipline. Despite the countercurrents in Tao and Zen, the Buddhists seem more otherworldly than the Westerners. Their dislocation from their center of origin in fifth-century India was perhaps due to the burden of abstraction imposed by a system of thought which denied the images, incarnations, and other tangible representations of sacredness. Hinduism seems an even more likely religion in terms of harmony with nature. It is much concerned with events on earth. *Samsara*, the eternal round of life, death, and rebirth, places humans, animals, and plants in a continuum which gives spiritual value to all beings. In a life the individual, whether human or bird, is obligated to be itself in the best sense, to follow the appropriate path, or *dharma*, in any particular incarnation. Complicity in this effort results in upward or downward incarnation in the next life. In many lives of correct *dharma* the spirit, moving upward toward *bhakti*, merges with god, and finally, in *moksha*, is released from further physical return. The objective of it all is escape from life and the world. The Indian god Vishnu, who among them all is most renowned for his animal incarnations, does not abide but arrives. He is engaged, as a fish, tortoise, boar, and lion, in tasks of rescue. Like other messiahs he intervenes when evil gets too strong—to save a sage in a deluge, to recover the lost cream of the milk ocean of the primal world, to recover the Vedas from a demon, and to attack the egocentric demon. Help from the outside—a familiar theme in the West.

14. See Elizabeth A. Lawrence, "Wild Birds: Therapeutic Encounters and Human Meanings," *Anthrozoos* 3(2) (1991). Although I have cited Lawrence's paper to make a point, in fairness I should add that her essay proceeds from examples of the therapeutic effect of birds to discuss basic qualities and benefits of "bird-watching."

15. James Hillman, "Abandoning the Child," in *Loose Ends* (Zurich: Spring, 1975).

16. Ibid.

17. John Rodman, personal communication, 1982.

18. Sapir, "Fecal Animals."

Acknowledgments

I have found writing to suffer the paradox that the more you think you know about the subject the more it resists. This book fought me for more than eight years in more ways than I care to recount. During that struggle to harness ideas and achieve lucidity I was supported and advised through interminable revisions by my wife, Florence Krall, whose devotion and keen eye, as much as anything, made the book possible. When ill health limited my ability to make final revisions and bring order from a chaos of footnotes, Florence, with the help of Lisi Krall and Kathryn Morton, made closure possible.

My thanks go also to my friend and editor, Barbara Dean, whose care and thoroughness would be any author's dream, to my friend and agent, Lizzie Grossman, for suggestions and mastery in finding the right home for this and other manuscripts, and to Don Yoder, whose meticulous copyediting rescued my wayward syntax and saved me from a thousand embarrassments.

Index